Center for the Study of National Reconnaissance Classics

HEXAGON (KH-9)
MAPPING CAMERA PROGRAM AND EVOLUTION

CENTER FOR THE STUDY OF
NATIONAL RECONNAISSANCE
CHANTILLY, VA

APRIL 2012

Foreword

This volume re-publishes *Hexagon Mapping Camera Program and Evolution* as part of the *Center for the Study of National Reconnaissance's (CSNR) Classics* series. The introductory information explains that while the focus of this history is on the Hexagon KH-9 mapping camera subsystem, it actually provides an overview of the entire Hexagon program and a full discussion of the evolution of satellite photoreconnaissance up to 1982. This volume's author, Maurice G. Burnett, drew heavily on existing secondary source histories for his broader discussion of satellite photoreconnaissance. In particular, he relied on Robert Perry's National Reconnaissance Office (NRO) histories that were available at the time. Burnett also turned to the then Defense Mapping Agency and Aerospace Corporation for additional data in support of his research.

Much of the redaction in the history involves the removal of Hexagon's KH-9 panchromatic intelligence images. Although the National Imagery and Mapping Agency (renamed National Geospatial-Intelligence Agency on 24 November 2003) declassified and transferred the KH-9 mapping system primary film record to the National Archives and Records Administration on 20 September 2002, the KH-9 panchromatic film record remains classified at this publication in March 2012. However, the Director of National Intelligence declassified selected KH-9 panchromatic imagery products for the NRO to use during the turnover of Hexagon artifacts to the National Museum of the United States Air Force (NMUSAF). To illustrate the intelligence value of the Hexagon program, we included sixteen of those KH-9 imagery products as figures in a supplemental section at the end of the history.

The *Center for the Study of National Reconnaissance Classics* is a series of occasional CSNR publications whose purpose is to inform our readers about classic issues from the past. The books and monographs in the series most typically are histories, but they also could address lessons-learned topics, the legacy recognition of people and programs, insights into historically significant artifacts, or tutorials on the discipline of national reconnaissance. We issue the publications in the series on both an *ad hoc* basis, or in connection with a significant event. We are issuing a Gambit-Hexagon collection of histories in response to Director of the NRO Bruce Carlson's decision in June 2011 to declassify the programs and his subsequent declassification announcement on 17 September 2011. The Historical Documentation and Research (HDR) Section of the CSNR selected five classic histories of the Gambit and Hexagon programs:

- *A History of Satellite Reconnaissance—The Perry Gambit & Hexagon Histories* (by R. L. Perry)
- *The Gambit Story* (by F. C. E. Oder, J. C. Fitzpatrick, & P. E. Worthman)
- *The Hexagon Story* (F. C. E. Oder, J. Fitzpatrick, & P. E. Worthman)
- *Hexagon Mapping Camera Program and Evolution* (M. Burnett)
- *A History of the Hexagon Program—The Perkin-Elmer Involvement* (by R. J. Chester)

On 21 January 2012, the CSNR published the first volume in the Gambit-Hexagon CSNR Classics series, *A History of Satellite Reconnaissance—The Perry Gambit & Hexagon Histories*. We did this in support of the ceremony that marked the NRO turning over a collection of Gambit and Hexagon artifacts to the NMUSAF and their exhibit opening of these artifacts to the public. The opening of this exhibit represented the largest collection of satellite reconnaissance artifacts ever assembled and put on public display. That exhibit can serve as a companion resource to those who read the histories in this CSNR Classics collection.

Each of these histories offers a different perspective on the programs; the Perry Gambit and Hexagon histories are from the viewpoint of a former Air Force historian at RAND writing in response to tasking from the then NRO Program A (Air Force program); the Oder, et. al. Gambit and Hexagon histories are from the viewpoint of authors with program experience working under the sponsorship of the Deputy Director of the NRO; the Burnett Hexagon mapping system history is from the viewpoint of the Hexagon program office working under the direction of two Air Force officers in the program and the NRO Program A Director; and the Chester Hexagon history is from the viewpoint of Perkin-Elmer, which was an associate contractor for the Hexagon program.

All of the authors researched and wrote their histories during what some observers might describe as the height of the Cold War, from 1964 to 1985. This influenced them to react to and focus heavily on the threat from the former Soviet Union and its allies. Also, all of the authors had at least some degree of first-hand knowledge

about these programs, and in many cases, they had first-hand experience working in the programs. This gives you a window into what it was like to be a participant-observer in the development and operation of these film-return satellite photoreconnaissance systems during the Cold War.

Dr. James D. Outzen, the NRO Senior Historian and Chief of the CSNR's HDR section, is the editor for the Gambit-Hexagon CSNR Classics series. Dr. Outzen selected the five histories for this CSNR Classics series from the NRO Records Center and CIA archives that collectively best retell the impressive Cold War story about these programs. He has prepared a brief preface and introduction for each history to provide context and explain its significance.

When you read the histories you will note that some information is missing. Even though the Director of the NRO authorized the declassification of almost all the programmatic information about these programs, some information, because of its potential impact on other sources and methods, remains classified. Dr. Outzen usually let the redacted text stand on its own, but in some instances he has done some editing for readability. For some of the histories, Dr. Outzen has incorporated supplemental reference material into the publication.

Robert A. McDonald, Ph.D.
Director
Center for the Study of National Reconnaissance

Preface

Coinciding with the commemoration of the 50th Anniversary of the National Reconnaissance Office (NRO), the Director of the NRO, Mr. Bruce A. Carlson, publicly announced the declassification of the Gambit and Hexagon imagery satellite systems on 17 September 2011. This announcement constituted the NRO's single largest declassification effort in its history. The Gambit and Hexagon programs were active for nearly half of the organization's history by the time of the declassification announcement. Their history very much represents the NRO's history—one that is defined by supremely talented individuals seeking state of the art space technology to address difficult intelligence challenges.

The United States developed the Gambit and Hexagon programs to improve the nation's means for peering over the iron curtain that separated western democracies from east European and Asian communist countries. The inability to gain insight into vast "denied areas" required exceptional systems to understand threats posed by US adversaries. Corona was the first imagery satellite system to help see into those areas. It could cover large areas and allow the United States and trusted allies to identify targets of concern. Gambit would join Corona in 1963 by providing significantly improved resolution for understanding details of those targets. Corona provided search capability and Gambit provided surveillance capability, or the ability to monitor the finer details of the targets.

For many technologies that prove to be successful, success breeds a demand for more success. Once consumers of intelligence—analysts and policymakers alike—were exposed to Corona and Gambit imagery, they demanded more and better imagery. Consequently, the Air Force, who operated the Gambit system under the auspices of the NRO, entertained proposals for an improved Gambit system shortly after initial Gambit operations commenced. They received a proposal from Gambit's optical system developer, Eastman Kodak, for three additional generations of the Gambit system. Ultimately the Air Force settled on only developing the proposed third generation because the proposed second generation offered minimal incremental improvement and the fourth generation appeared technologically unachievable at the time. The third generation became known as Gambit-3 or Gambit-cubed while it was under development. Once it replaced the first generation, it simply became Gambit. The new Gambit system, with its KH-8 camera system, provided the United States outstanding imagery resolution and capability for verifying strategic arms agreements with the Soviet Union.

Corona was expected to serve the nation for approximately two years before being replaced by more sophisticated systems under development in the Air Force's Samos program. It turned out that Corona served the nation for 12 years before being replaced by Hexagon. Hexagon began as a Central Intelligence Agency (CIA) program with the first concepts proposed in 1964. The CIA's primary goal was to develop an imagery system with Corona-like ability to image wide swaths of the earth, but with resolution equivalent to Gambit. Such a system would afford the United States even greater advantages monitoring the arms race that had developed with the nation's adversaries. The system that became Hexagon faced three major challenges. The first was development of the technology, which was eventually overcome by the Itek and Perkin-Elmer Corporations. The second was bureaucratic, deciding how the CIA and Air Force would cooperate in building such a system because they each had strengths and weaknesses in the development of national reconnaissance systems. The third challenge was to secure the resources that were required to build the most complicated and largest reconnaissance satellites at the time. By 1971, the NRO overcame the challenges to successfully launch the Hexagon satellite and fulfill, or even exceed, expectations for unparalleled insight into capabilities of US adversaries.

At the time of the Gambit and Hexagon declassification announcement, the NRO released a number of redacted Gambit and Hexagon documents and histories on its public website. One of the histories is contained in this volume.

Hexagon Mapping Camera Program and Evolution was written in 1982 by Maurice G. Burnett at the request of the Director of the Air Force Program at the NRO, known as Program A. The Hexagon mapping camera flew on 12 of the 20 Hexagon missions. It proved to be a remarkably efficient and prodigious producer of imagery for mapping purposes. The mapping camera system was successful by every standard including technical capabilities, reliability, and capacity.

Hexagon Mapping Camera Program and Evolution is very unique among the histories of the Gambit and Hexagon programs for two reasons. First, Burnett provides a comprehensive history of the development of imagery satellites. The volume is very useful for individuals who want to become familiar with the development of these systems as well as the evolution of the Air Force's Program A at the NRO. Second, Burnett also provides a very detailed history of the mapping camera system including technical development as well as the operation of the mapping camera system.

Burnett prepared a history that is further strengthened by the visual content he chose to enhance the historical narrative. *Hexagon Mapping Camera Program and Evolution* includes multiple photographs, engineering drawings, and examples of satellite imagery. This content gives readers, especially those interested in the development of satellite imagery, another reason to read this volume.

Hexagon Mapping Camera Program and Evolution joins five other volumes of Gambit and Hexagon histories that the Center for the Study of National Reconnaissance is reprinting in conjunction with the program declassifications. Those other volumes include *The Gambit Story* and *The Hexagon Story* both written by Frederic Oder, James Fitzpatrick, and Paul Worthman, Robert Perry's histories of Gambit and Hexagon, a Perkin-Elmer history of Hexagon, and a compendium of key Gambit and Hexagon program documents. In total, this collection of Gambit and Hexagon publications provides the public with broad insight into previously classified programs. The volumes complement each other by providing details not found exclusively in any single program history volume.

At the time of this writing, KH-9 panoramic camera system imagery has not been declassified. I have included in a separate section of this publication a small number of KH-9 images that were released in conjunction with the Hexagon declassification.

I have chosen not to reprint pages that were redacted in their entirety in *Hexagon Mapping Camera Program and Evolution*. Those pages are: 157 and 158. We also did not reprint blank pages, which consist of pages 24, 32, 36, 38, 40, 42, 54, 62, 76, 84, 86, 90, 92, 94, 98, 102, 1-8, 1-10, 1-12, 1-16, 1-18, 1-22, 1-24, 1-26, 1-28, 1-30, 1-32, 1-36, 2-22, 2-48, 3-4, 3-10, 3-12, 4-4, 4-8, 5-10, 5-12, 6-4, 6-6, 6-10, 6-12, & 6-14. The unedited redacted Hexagon Mapping Camera Program and Evolution can be found in the declassified records section of NRO.gov for those interested in reviewing a document with the completely redacted and blank pages.

The Gambit and Hexagon systems became reliable means for addressing difficult intelligence challenges once they became operational. The Hexagon system, in particular, provided broad area imagery that was essential for understanding the strategic capabilities and arms control compliance of the Soviet Union and other Cold War adversaries. These national reconnaissance systems dutifully provided the nation reliable vigilance from above until the next generation of imagery satellites advanced US intelligence collection capabilities.

James D. Outzen, Ph.D.

Chief, Historical Documentation and Research
The Center for the Study of National Reconnaissance

CENTER FOR THE STUDY OF NATIONAL RECONNAISSANCE

The Center for the Study of National Reconnaissance (CSNR) is an independent National Reconnaissance Office (NRO) research body reporting to the NRO Deputy Director, Business Plans and Operations. Its primary objective is to ensure that the NRO leadership has the analytic framework and historical context to make effective policy and programmatic decisions. The CSNR accomplishes its mission by promoting the study, dialogue, and understanding of the discipline, practice, and history of national reconnaissance. The CSNR studies the past, analyzes the present, and searches for lessons-learned.

HEXAGON (KH-9) MAPPING CAMERA PROGRAM AND EVOLUTION

Prepared for:

THE DIRECTOR, PROGRAM A—NATIONAL RECONNAISSANCE OFFICE
(DIRECTOR OF SPECIAL PROJECTS, OFFICE OF THE SECRETARY OF THE AIR FORCE)

JOINT SI/TK/B (IMAGERY)
CLASSIFIED BY SISR VOL I & IPM
DECLASSIFY OADR

MCS HISTORY

WARNING NOTICE:
INTELLIGENCE SOURCES AND
METHODS INVOLVED (WINTEL)

**DISSEMINATION AND EXTRACTION OF INFORMATION
CONTROLLED BY ORIGINATOR (ORCON)**

NATIONAL SECURITY INFORMATION:
UNAUTHORIZED DISCLOSURE SUBJECT
TO CRIMINAL SANCTIONS

BIF-059W-23422/82
Handle Via
BYEMAN/TALENT KEYHOLE
CONTROL SYSTEMS JOINTLY

PUBLICATION REVIEW

This document was produced by the Hexagon Program Office for the Director of Special Projects, Office of the Secretary of the Air Force.

Prepared by: *[signature]*
Lt. Colonel Guy F. Welch
Deputy for Payloads

[signature]
Colonel Lester S. McChristian
Director, Hexagon Program Office

Approved by: *[signature]*
Maj. General John E. Kulpa
Director, Secretary of the Air Force
Special Projects

Because of the breadth of the document, custodial organizations are required to insure access to this document or any part thereof is limited to a "must know" basis and controlled by individual signature acknowledgement of each access. The list of which is to be maintained with the document. This document will not be downgraded or duplicated/copied in whole or in part without the express written approval of Program A, Director, Security and Policy and Director, Hexagon Program Office.

BIF-059W-23422/82
Handle Via
BYEMAN/TALENT KEYHOLE
CONTROL SYSTEMS JOINTLY

MCS HISTORY

ACKNOWLEDGEMENT TO AUTHOR/EDITOR

The Office of the Secretary of the Air Force, Special Projects, wishes to express its appreciation to Maurice G. Burnett for his outstanding efforts in providing the research, data compilation, organization, and presentation of this historical reference document.

Mr. Burnett has been continuously involved with the Reconnaissance Community for over 30 years. It is this wealth of experience and knowledge that made his particpation invaluable in the location and selection of material, interviewing of key personalities, and the identification of significant world events and technological advances. This will undoubtedly make this historical documentation of the evolution of satellite reconnaissance and mapping much more understandable and meaningful to future generations.

BIF-059W-23422/82
Handle Via
BYEMAN/TALENT KEYHOLE
CONTROL SYSTEMS JOINTLY

MCS HISTORY

FOREWORD

This report was prepared from data provided by the Secretary of the Air Force Special Projects Office (SAFSP), the Defense Mapping Agency (DMA), the Aerospace Corporation, and the HEXAGON Associate Contractors.

As the HEXAGON (KH-9) Mapping Camera Program was approaching its scheduled completion date of mid-1981, the Secretary of the Air Force Special Projects Office (SAFSP) in El Segundo, California, initiated a task to prepare a history of this highly successful program. It was initially envisioned that the report would cover only the Mapping Camera System (MCS), its development, operation and results. But during ensuing discussions between government and contractor personnel it was recognized that a history of the MCS alone would give future readers only a fragmented account of how the MCS fit into the overall picture of satellite reconnaissance—obviously we did not just suddenly acquire this sophisticated capability.

Of particular concern to Lt. Colonel Guy F. Welch (SAFSP) at the time, was the scarcity of historical data readily available for the indoctrination of newly assigned personnel to SAFSP. "Old timers" who might be capable of giving a well-rounded briefing (from first-hand experience) on the evolution of satellite reconnaissance were not easily accessible in 1980, over two decades after the development and first launch of an American satellite. And, though there were printed accounts covering certain periods in satellite development, there was seen a need to prepare a condensation of records and to include synoptic charts, schematics and photographs to serve better as an initial indoctrination tool.

For these reasons, it was decided that a supplement should be added to the MCS history to provide a synopsis of the evolution of satellite reconnaissance in general, and that in this particular case, the supplement should be placed ahead of the "original" main subject. The result then was to prepare the report in two parts, Part I covering the evolution of the satellite reconnaissance capability, and Part II, the Mapping Camera System (MCS). Part II also includes a description of the overall HEXAGON (KH-9) system, but only briefly, since a full history of the HEXAGON program is in preparation. It is the intention of the present SAFSP Director, Major General John F. Kulpa that histories of the HEXAGON program and other programs, which are still active, shall be completed at appropriate times and in adequate detail.

Four reports were used principally as reference sources in the preparation of Part I. Through extensive research, authors have presented from varying perspectives the evolution of missiles, satellites and government organizations during the period 1945 to 1960 in sufficient detail to develop the theme we were looking for. Hence, the following reports were drawn on extensively (in many instances verbatim) and are referenced throughout part one.

TOP SECRET/RUFF/GAMBIT/HEXAGON

BIF-059W-23422/82
Handle Via
BYEMAN/TALENT KEYHOLE
CONTROL SYSTEMS JOINTLY

"U.S. Military Space Programs: A Brief Analytical History, and Interaction with Operational Forces, 1945, 1975," B. W. Augenstein, August 1975 (UNCLASSIFIED).

"The U.S. Military in Space, Its Inheritance and Bequest," the United States Air Force Academy Military Space Doctrine Working Group, March 1982 (UNCLASSIFIED).

"A History of Satellite Reconnaissance," Volumes I through IIIB prepared by Robert Perry under direction of the NRO (TOP SECRET/BYE).

"CORONA Program History, Vol I Program Overview," produced by the Directorate of Science and Technology, Central Intelligence Agency, 19 May 1976 (TOP SECRET/BYE).

CONTENTS

Title Page	i
Publication Review	iii
Acknowledgement	iv
Foreword	v
Contents	vii
Figures	xli
Tables	xvi
Distribution	xvii
Acronyms	xviii

PART I — EVOLUTION OF SATELLITE PHOTOGRAPHIC SYSTEMS

Background	1
Early Missile Research	1
Pre-World War II	1
World War II Era	2
Post-World War II	2
Early Satellite Research	4
Interim Aerial Surveillance Methods	7
Government Organizations Evolved as Space Vehicle Development Continued	9
Space Program Complexity Deepens	13
Post-Sputnik "Catch Up" Actions—ARPA and NASA Established	14
Rekindling of USAF Interests in Space	16
Transfer of Space Activities to the Services	16
U-2 "Spy Plane" Incident Closes Russian Skies	17
Establishing the NRO and SAFSP	18
Space Activities 1958-1960	19
Early Satellite Systems	25
Preface to Early Satellite Systems	25
SAMOS Programs	33
E-1/F-1, E-2 Series	33
E-4 Mapping Satellite	47
E-5 Recovery Program	51
E-6 Recovery Program	59
CORONA Program	68
DISCOVERER Series	79
Army SECOR System	81
Army ARGON Mapping System	82
MURAL Camera	82

MCS HISTORY

1.5-Inch Terrain and Stellar/Terrain Cameras	82
LANYARD Camera	87
JANUS Camera	88
DISIC Mapping Camera	88
J-3 Constant Rotating Camera	88
Defense Weather Satellites	99
GAMBIT Program	111
The KH-11 System	118
Exhibits of Typical Imagery From PHOTINT Programs	119
References	138

PART II — THE HEXAGON (KH-9) MAPPING CAMERA SYSTEM

Introduction	i-3
1. The HEXAGON Satellite	**1-1**
Government Organizations	1-1
Associate Contractors	1-3
The Aerospace Vehicle	1-6
Satellite Basic Assembly Structure	1-13
Satellite Basic Assembly—Aft Section	1-13
Attitude Control	1-14
Orbital Adjust and Reaction Control	1-14
Electrical Distribution and Power	1-19
Telemetry and Tracking	1-19
Command and Timing	1-20
Lifeboat II	1-20
Search/Surveillance Cameras	1-25
The Mark 8 Satellite Reentry Vehicle (SRV)	1-29
The Mark V Satellite Reentry Vehicle	1-33
Introduction	1-33
System Description	1-34
Hardware Description	1-39
Mission Overview	1-42
2. The Mapping Camera Program Development	**2-1**
Introduction	2-1
MCS Development to First Flight	2-3
Events Leading to Contract Award	2-3
System Configuration Changes	2-8
Establishing First Launch Schedule	2-8
NPIC Statement of MCS Product Requirements	2-9
MCS Calibration Responsibilities and Procedures	2-9
MCS Block I Procurement	2-11
MCS Software Development	2-11
Incorporation of Wobble Roller	2-12
Early Plans for Block II Procurement	2-12
Proposal for MCS Improved Design	2-13

~~TOP SECRET~~/RUFF/GAMBIT/HEXAGON

Pressure Makeup System	2-14
Two Additional Mapping Cameras	2-15
MCS Testing to First Flight	2-16
Mapping Camera System Hardware and Test Cycle	2-19
Configuration	2-19
Physical Characteristics	2-20
Lens Systems	2-20
Mechanical Description	2-28
Terrain Camera	2-29
Stellar Camera	2-32
Terrain Primary Shutter	2-33
Rotary Shutter	2-33
Capping Shutter	2-33
Stellar Shutter and Light Baffle Subsystems	2-34
Mechanical Operations	2-34
Terrain Pressure Plate and Platen Press	2-34
Forward Motion Compensation	2-34
Terrain Film Transport	2-36
Frame Advance and Mechanical Operation	2-38
Other Rollers	2-38
Stellar Film Transport and Platen Press	2-39
Frame Advance and Mechanical Operation	2-39
Terrain Film Supply Subsystem	2-41
Stellar Film Supply Subsystem	2-41
Stellar and Terrain Takeup Subsystem	2-42
Terrain Thermal Shutter Subsystem	2-42
Mechanical Operation	2-42
Chutes	2-42
Light Baffle	2-42
Main Instrument System Electronics Assembly	2-43
Electrical Distribution and Power	2-43
System Harness	2-43
Physical Orientation	2-43
Test and Integration	2-45
Preflight Calibration	2-46
System Integration and Test	2-53
Flight Readiness Assurance Program	2-56
3. Mission Scenario	3-1
Operational Summary	3-1
Development of Requirements	3-2
Mission Requirements Validation	3-2
Defense Mapping Agency Validation	3-2
Director Central Intelligence (DCI) Validation	3-6
Mission Objectives Sent to Satellite Test Center	3-6
HEXAGON (KH-9) Data Base	3-7

MCS HISTORY

Satellite Test Center Operation	3-7
Purpose of the Satellite Test Center	3-7
Overview of STC Operations	3-8
STC Staffing	3-8
USAF Tracking Network	3-9
Associate Contractor Support	3-13
Operational Support Data Base	3-14
Operational Support Activities—Associate Contractors	3-15
Operations Data Flow	3-16
De-Orbit/Recovery Sequence	3-20
De-Orbit Preparations	3-20
De-Orbit	3-20
Reentry	3-21
Backup Timer Events	3-21
Post Flight	3-22
4. Processing/Duplication	**4-1**
Processing Site Production Work Flow	4-1
Receipt of Mission Film	4-1
Downloading/Presplice	4-2
Original Negative Processing	4-2
Optical Titling	4-2
Preliminary Evaluation	4-2
Densitometry/Duplication Route Determination	4-2
Printing	4-5
Duplicate Film Processing	4-5
Duplicate Copies Inspection	4-5
Final Original Negative Inspection	4-5
Acquisition Material	4-5
Stellar Camera	4-6
Terrain Camera	4-9
Distortion in Film Handling Systems	4-11
Duplication	4-12
Films	4-12
System Sensitometry	4-12
Physical Handling	4-13
5. Post Flight Analyses	**5-1**
Concept and Purpose	5-1
PFA Background	5-1
Mapping Camera PFA Organizations and Procedures	5-3
In-Flight Camera Calibration Procedures	5-8
6. Exploitation	**6-1**
Exploitation Systems Developed and/or Utilized	6-1
Exploitation of Imagery	6-15
Analytical Triangulation	6-15
Continental Control Network (CCN) Data Base	6-15
Photogrammetric Compilation	6-16

7. Operational Considerations and Statistics 7-1
    Highlights 7-1
    Mission Planning, Strategies, and Accomplishments 7-1
        Summary of Collection Interests for Mission 1213 MCS 7-3
    Mission Engineering Summaries and Major Changes 7-5
        Mission 1205 (SV-5) 7-5
        Use of SO-131 7-6
        Mission 1206 (SV-6) 7-6
        New Launch Schedule 7-7
        Mission 1207 (SV-7) 7-8
        Project 80 Study 7-9
        Mission 1208 (SV-8) 7-11
        Mission 1209 (SV-9) 7-12
        Mission 1210 (SV-10) 7-13
        Mission 1211 (SV-11) 7-14
        Star Sensor System (S^3) 7-16
        Ultra-Thin-Base (UTB) Film Implementation 7-17
        Mission 1212 (SV-12) 7-19
        Mission 1213 (SV-13) 7-20
        Mission 1214 (SV-14) 7-21
        Mission 1215 (SV-15) 7-24
        Mission 1216 (SV-16) 7-25
    Closeout of the Operational Phase 7-26

8. KH-9 Products and Imagery 8-1
    Medium and Small Scale Maps and Charts 8-1
    Large (1:50,000) Scale Topographic Line Maps 8-1
    Digital Data 8-1
        Digital Terrain Elevation Data 8-2
        Digital Feature Analysis Data 8-2
    Point Positioning 8-2

9. Summary 9-1

10. References 10-1

11. Acknowledgements 11-1

MCS HISTORY

FIGURES

PART I — EVOLUTION OF SATELLITE PHOTOGRAPHIC SYSTEMS

1-1	Advanced Reconnaissance System Evolution	12
1-2a	Standardized Agena as Viewed in 1959	23
1-2b	U.S. Space Vehicles as Viewed in 1959	23
1-3a	Number of Launches Each Year	30
1-3b	Orbit Duration of PHOTINT Systems	31
1-4a	WS-117L Vehicle System	35
1-4b	WS-117L Subsystems	35
1-5	E-1 Visual Reconnaissance Component Test Model	37
1-6	E-2 Visual Reconnaissance Payload	39
1-7	E-2 Payload-Functional Schematic	41
1-8a	SAMOS E-5 Recovery Camera Payload	53
1-8b	SAMOS E-5 Recovery Capsule	53
1-9	E-6 Reconnaissance Satellite Subsystems	61
1-10	Major Components of the CORONA J-3/CR (KH-4A) Launch Vehicle	69
1-11	CORONA Photographic Payload Profiles	70
1-12	C Triple Prime (C''') Camera (KH-3) in Test Stand	75
1-13	ARGON (KH-5) Army Mapping Camera	83
1-14	MURAL (KH-4) Twin Panoramic Camera System	85
1-15	Index and Stellar-Index Cameras	85
1-16	LANYARD (KH-6) Panoramic Camera System	89
1-17	Artist's View of the J-1 (KH-4A) Camera System With Dual SRV's	91
1-18	Major Components of the J-3 (KH-4B) Subsystem	93
1-19	CORONA Program Synopsis	95
1-20	CORONA Program Museum Display	97
1-21a	Early Weather Reconnaissance System	101
1-21b	DMSP System Evolution	101
1-21c	DMSP System Network	101
1-22	GAMBIT (KH-7) Configuration	112
1-23	KH-1 Camera System Imagery (CORONA C)	122
1-24	KH-2 Camera System Imagery (CORONA C')	123
1-25	KH-3 Camera System Imagery (CORONA C''')	124
1-26	KH-4 Mapping Camera System Imagery From Panoramic Camera (CORONA-MURAL)	125
1-27	KH-4 Mapping Camera System Imagery From Stellar/Index (S/I) Camera (CORONA-MURAL)	126
1-28	KH-4A Camera System Imagery (CORONA-JANUS J-1)	127

TOP SECRET/RUFF/GAMBIT/HEXAGON

BIF-059W-23422/82
Handle Via
BYEMAN/TALENT KEYHOLE
CONTROL SYSTEMS JOINTLY

1-29 KH-4B Mapping Camera System Imagery (DISIC) 128
1-30 KH-4B Pan Camera System Imagery (CORONA-JANUS J-3) 129
1-31 KH-5 Army Mapping Camera System Imagery (CORONA-ARGON) 130
1-32 KH-6 Camera System Imagery (CORONA-LANYARD) 131
1-33 KH-7 Camera System Imagery (GAMBIT-PROGRAM 206, CUEBALL) 132
1-34 KH-8 Camera System Imagery (GAMBIT-G-CUBED) 133
1-35 KH-9 Mapping Camera System Imagery (HEXAGON-MCS) 134
1-36 KH-9 Camera System Imagery (HEXAGON-PAN) 135
1-37 KH-11 Camera System Imagery, ▮▮▮▮ 136
1-38 KH-11 Camera System Imagery, ▮▮▮▮ 137

FIGURES

PART II — THE HEXAGON (KH-9) MAPPING CAMERA SYSTEM

Frontispiece—Mission 1216 (SV-16) Launch on 17 June 1980		i-1
1-1	Locations of Government and Contractor Facilities Supporting HEXAGON Program	1-3
1-2	The Aerospace Vehicle	1-7
1-3	HEXAGON Vehicle on Orbit	1-9
1-4	Satellite Vehicle Configuration	1-11
1-5	Satellite Basic Assembly Structure	1-15
1-6	Satellite Basic Assembly—Aft Section	1-15
1-7	Attitude Control	1-17
1-8	Orbit Adjust and Reaction Control	1-17
1-9	Electrical Distribution and Power	1-21
1-10	Telemetry and Tracking	1-21
1-11	Command and Timing	1-23
1-12	Lifeboat II	1-23
1-13	Search/Surveillance Cameras	1-27
1-14	Two-Camera Assembly	1-27
1-15	Mark 8 Recovery Vehicle Equipment	1-31
1-16a	Mark V Reentry Vehicle	1-35
1-16b	Mapping Camera Terrain and Stellar Takeup Assemblies	1-35
1-17a	Mapping Camera Module	1-37
1-17b	SRV Inboard Profile	1-37
1-18	Mission Flight Envelope	1-38
1-19	Operational Flight Envelope	1-38
1-20	Mark V SRV (Exploded View)	1-40
1-21	Mission Flow	1-43
2-1	Mapping Camera System	2-21
2-2	Mapping Camera Major Assemblies	2-24
2-3	12-Inch Metric Lens	2-25
2-4	10-Inch Stellar Lens	2-27
2-5	Stellar Lens Structure	2-27
2-6	MC Center Structure and Mechanical Interface	2-28
2-7	Terrain Camera Mechanical Arrangement	2-29
2-8	Data Block With Readout Overlay Superimposed	2-30
2-9	Terrain Camera Structural Elements	2-31
2-10	Stellar Camera Mechanical Arrangement	2-32
2-11	Terrain Camera Primary Shutter Arrangement	2-33
2-12	Stellar Shutter and Light Baffle Arrangement	2-35
2-13	Pressure Plate and Platen Press Arrangement	2-36
2-14	FMC Drive Arrangement	2-37
2-15	Terrain Film Transport	2-37

MCS HISTORY

2-16	Terrain Transport Film Flow	2-38
2-17	Stellar Camera Film Transport	2-40
2-18	Stellar Transport Film Flow	2-41
2-19	APSA and MC Coordinates and Assembly Identification	2-44
2-20	Mapping Camera Integration and Operational Flow	2-45
2-21	Distortion Boresight Test Unit (DBTU)	2-46
2-22	Air Force Celestial Calibration Site, Cloudcroft, New Mexico	2-47
2-23	Cloudcroft Facility	2-51
2-24	Cloudcroft Calibration Enclosure	2-51
2-25	Cloudcroft Operational Flow	2-52
2-26	Main Instrument Shipping Container	2-54
2-27	SV in High-Bay Test Area	2-57
2-28	SV Entering Vacuum Chamber	2-57
2-29	SV Going Vertical	2-59
2-30	SV in Position for MCS Film Loading and Vertical Tests	2-61
3-1	Operational Events	3-3
3-2	Mapping Camera Operations—92-N.Mi. Altitude	3-5
3-3	Operational Support Overview	3-8
3-4	USAF Tracking Network	3-11
3-5	OIWG Organizational Structure	3-14
3-6	Problem Management Flow Chart	3-16
3-7	Mapping Camera Operational Data Flow	3-18
3-8	Mapping Camera Operational Data Flow	3-19
3-9	Flight Profile	3-23
3-10	Typical Air Recovery Sequence	3-24
4-1	Processing Site Production Work Flow	4-1
4-2	Satellite Recovery Vehicle (SRV)	4-3
4-3	SRV Interfaced With Presplice Complex	4-3
4-4	Examples of Optical Titling Exposed During Development	4-5
4-5	Kingston Continuous Printers	4-7
4-6	Distortion Analysis System	4-7
4-7	Ontario Processing Complex	4-12
5-1	PFA Organizations and Data Flow	5-4
5-2	Visual Edge Match Equipment (VEM)	5-9
5-3	VEM Matrix Array	5-11
6-1	Replacement of Photographic Imagery Equipment (RPIE)	6-3
6-2	Universal Automatic Map Compilation Equipment (UNIMACE)	6-5
6-3	Automatic Reseau Measuring Equipment (ARME)	6-9
6-4	Analytical Stereoplotter Systems (AS-11)	6-11
6-5	Strip Vacuum Frame	6-13
7-1	Mission Statistics	7-2
7-2	HEXAGON Program Museum Display	7-27
8-1	1:50,000-Scale Map	8-5
8-2	Joint Operations Graphics (Air) Map	8-7
8-3	Joint Operations Graphics (Ground) Map	8-9
8-4	Comparison, Operational Films—Sky Harbor, Phoenix, Arizona	8-11
8-5	Tag-On of QX-801 Film—Dayton, Ohio	8-13
8-6	Experimental Photography—IR Color, SO-131	8-15

DISTRIBUTION

Distribution controlled
by ▮▮▮▮▮▮▮

ACRONYMS

ACIC	Air Force Aeronautical Chart and Information Center
ACS	Attitude control system
ADPACS	Automatic data processing and control system
AEI	Aerial Exposure Index
AFSC	Air Force Systems Command
AMODE	Ascent mode
AMS	Army Map Service
APSA	Auxiliary Payload Structure Assembly
APTC	Aerial Positioning Terrain Camera
ARDC	Air Research and Development Command
ARME	Automatic Reseau Measuring Equipment
ARPA	Advanced Research Projects Agency
ASA	American Standards Association
ATV	Altitude thermal vacuum
AV	Aerospace vehicle
BMD	Ballistic Missile Division
CIA	Central Intelligence Agency
C&S	Control and synchronization
CCN	Continental Control Network
COMIREX	Committee on Imagery Requirements and Exploitation
CSR	Category satisfaction ratio
DBS	Doppler Beacon Subsystem
DBSG	Data base subgroup
DCI	Director Central Intelligence
DDR&E	Director of Defense Research and Engineering
DDS&T	Deputy Director for Science and Technology
DFAD	Digital feature analysis data
DIA	Defense Intelligence Agency
DIU	Data interface unit
DISIC	Dual Improved Stellar Index Camera
DMA	Defense Mapping Agency
DMAHTC	Defense Mapping Agency Hydrographic Topographic Center
DoD	Department of Defense
DTED	Digital terrain elevation data
ECS	Extended command system
EDAP	Electrical Distribution and Power Assembly
ETL	Engineering Topographic Laboratory
FEAF	Far East Air Forces
FMC	Forward motion compensation
FRT	Frame reference time
FRAP	Flight Readiness Assurance Program
FV/H	Forward velocity divided by height
GOPSS	Geodetic Orbiting Photographic Satellite System
GRD	Ground resolved distance
GWC	Global Weather Center
HAP	Interagency High Altitude Photographic Program
HO	Horizon optics

IAG	Imagery Analysis Group
ICD	Interface control document
ICRS	Imagery Collection Requirements Subcommittee of COMIREX
IFD	In-flight disconnects
IOC	Initial operational capability
IPIN	Integrated Photogrammetric Instrumentation Network
JPL	Jet Propulsion Laboratory
LOL	Limited operating life
MC	Mapping Camera
MC&G	Mapping, Charting, and Geodesy
MC&G-WG	Mapping, Charting, and Geodesy Working Group
MCATS	Mapping categories
MCS	Mapping Camera System
MCS	Minimal command system
MCC	Mission control center
MCM	Mapping Camera Module
MISEA	Main Instrument System Electrical Assembly
MOB	Mission objective file
MPE	Mission performance evaluation
MPR	Mission performance report
MTD	Missile target data
NAVPAC	Navy Navigational System
NEC	Northeast contractor
NFIB	National Foreign Intelligence Board
NL	Number of looks
NRO	National Reconnaissance office
NRP	National Reconnaissance Program
NSC	National Security Council
NTB	National Target Base
OAS	Orbit adjust system
OLOPS	Off-line orthophoto system
ONC	Operational Navigation Chart
OTV	Orbital thermal vacuum
PAR	Product Authorization Request
PBC	Planetary block concept
PEL	Precision elastic limit
PET	Performance evaluation team
PFA	Post flight analyses
PHOTINT	Photographic Intelligence
PIER	Performance interim evaluation reports
PMS	Pressure makeup system
PMU	Pressure makeup unit
POM	Program objectives memorandum
PPDB	Point positioning data base
Program A	U.S. Air Force
Program B	Central Intelligence Agency (Cognizance)
PSD	Power spectral density
RADC	Rome Air Development Center
RAF	Royal Air Force

MCS HISTORY

RCS	Reaction control system
RDTE	Research, development, test, and engineering
RGP	Relative geometry point
RPG	Reference point graphic
RPIE	Replacement of photographic imagery equipment
RTS	Remote tracking station
SAFSP	Secretary of the Air Force Special Projects
SAMSO	Space and Missile Systems Organization
SBA	Satellite Basic Assembly
SBAC	Satellite Basic Assembly Contractor
SCF	Satellite Control Facility
SI	Stellar index
SIOP	Single integrated operational plan
SMAC	Simultaneous multi-camera analytical calibration
SAC	Strategic Air Command
SAO	Smithsonian Astrophysical Observatory
SOC	Satellite Operations Center
SRAM	Short range air missiles
SRM	Solid rocket motor
SRV	Satellite reentry vehicle
STB	Standard thin-base
STC	Satellite Test Center
SV	Satellite vehicle
TA	Technical advisor
TAC	Tactical Air Command
TAC	Trend analysis console
TAS	Technical advisor staff
TERCOM	Terrain contour matching
TOPOCOM	Topographic Command
UNAMACE	Universal Automated Map Compilation Equipment
UPDRAMS	Universal Photogrammetric Data Reduction and Mapping System
USAFE	United States Air Forces in Europe
USIB	U.S. Intelligence Board
UTB	Ultra-thin-base
UUTB	Ultra-ultra-thin-base
VOD	Vertical obstruction data
Vx/H	Orbital angular rate, in track
Vy/H	Orbital angular rate, cross track
WAC	World aeronautical chart
WACM	World aeronautical chart mosaic
WAG	World area grid

PART I
Evolution of Satellite Photographic Systems

BACKGROUND

As future readers look at the history of space exploration, they will probably be perplexed at the variety of booster vehicles and satellite systems that were developed in the 1950's and early 1960's, the genesis of the space age. Since this report will be used in part for orientation purposes, it was felt that brief accounts should be given of the many facets which were influential in shaping the events of those formative years. Although historical records vary as to emphasis and detail, there is common agreement on three points:

1. The technology to build space vehicles, admittedly of great challenge in the early days, nevertheless arrived much sooner than did the government organizations which were required to direct this radically new concept of vehicle development and deployment.

2. There were two significant events during the 1950's which served to overcome a general lethargic attitude toward space and satellite development.

    - In 1952 the United States exploded a thermonuclear device—several months later the Soviets exploded a similar weapon.
    - On 4 October and 3 November 1957, Russia scooped the U.S. by successfully launching Sputniks I and II into orbit.

3. The stimulus which finally brought priorities to focus on satellite reconnaissance came on 1 May 1960 when a United States U-2 high-altitude reconnaissance plane was shot down over Russia, resulting in the closing of Russian skies to overflight by airplanes.

EARLY MISSILE RESEARCH

Pre-World War II

The first guided missiles in the United States were built and tested from 1916 to 1918 during World War I, but were never used in combat. As propeller driven airplanes without pilots, these first guided missiles had "pre-set controls," which means that their target could not be changed in flight. They were equipped with automatic pilots and vacuum devices to drive the planes to their destinations, and explode the bombs they carried. Charles F. Kettering, the automobile engineer, and Elmer Sperry, who developed the gyroscope, helped design these early types of guided missiles.

The first command-type guided missile, or one that could be maneuvered in flight by remote-control command, was a radio-controlled airplane built by the United States Navy in 1924.

The first successful drone missile, called the QUEEN BEE, was demonstrated in Great Britain in 1935. This drone, a standard Navy training plane fitted with radio controls, was used as an antiaircraft target.

World War II Era

During World War II Germany made the first successful use of ground-to-ground guided missiles in combat. The Germans built a huge missile-research center in 1937 at Peenemunde on the Baltic Coast, and at this center developed and tested more than 20 types of missiles.

By 1944 bombing raids by allied airplanes had ruined many industrial cities in Germany. To help revive the spirits of the German people, German leaders announced the VERGELTUNGS-WAFFE EINS, or Vengeance Weapon One. The V-1 was 25 feet 4 inches long, carried one ton of explosives, and was powered by a pulse-jet engine developed by Paul Schmidt, a Munich engineer. The British called the V-1 the "Buzz-Bomb," because the loud noise of its engine announced its coming long before it exploded. The V-1 could go about 150 miles at a speed of 360 miles an hour.

Defense fighters shot down V-1's rather easily because of their relatively slow speed, but by the end of the war, buzz-bombs had killed thousands of persons in England.

On 8 September 1944, the German's began to use an even more terrifying rocket-propelled V-2 guided missile, produced under the direction of Count Wernher von Braun. Having the appearance of a giant wingless artillery shell, the V-2 was 46 feet long with four arrowlike fins at its tail. Carrying more than a ton of explosives it was a pre-set missile, like the V-1, but much more complicated having more than 30,000 parts. The V-2 was guided by an automatic pilot and had an electronic brain which shut off the rocket engine at the proper time to make the missile dive to its target. The people could not hear this missile coming since it traveled at over 3,600 miles an hour, much faster than the speed of sound. Having a range of about 200 miles, it was launched straight up and zoomed to a height of about 60 miles before diving toward the target.

In the United States many experiments were made with guided missiles, but only a few were put to use during World War II. The simplest guided missile was the Air Force's guided bomb called the AZON, a command-type air-to-ground missile. It was a standard 1,000-pound bomb which could be steered to the right and left by a radio operator in the bomber. Special control apparatus in the tail of the bomb consisted of gyroscopes, batteries, and a radio receiver. The AZON bomb was successfully used in 1944 against river locks and viaducts in Germany.

Another successful but far more complicated missile was the Navy's BAT. The BAT was a homing air-to-ship glide-bomb missile. While the BAT was still slung under the wing of an airplane, its radar nose was carefully pointed at an enemy ship. The radar was "locked-on" to the target and the BAT was released to glide and steer itself.

Although the world was still airplane oriented at the close of WW II, the knowledge gained by German specialists in missile development would soon be surfaced in countries where numerous German scientists and technicians were relocated after the war, predominantly the United States and Russia.

Post-World War II

Following the close of World War II, three of the armed services, Army, Navy, and Air Force each began research in the area of guided missiles. As a result, each claimed itself the legitimate heir to the responsibility of developing and organizing missile and space research.[1]*

---

*References are presented at the end of Part I (page 138).

MCS HISTORY

Recruiting German scientists and technicians who had worked on the German V-2 project, the Army commenced their research at Ft. Bliss near El Paso, Texas. Most of their early testing was conducted with captured V-2 missiles. Dr. Wernher von Braun, the leader of the German scientist team assisting the Army, was responsible for introducing many American scientists and engineers to the technology required to design and build missiles.

Much of the research conducted with the V-2's involved experiments with the upper atmosphere. For several years, von Braun's team launched V-2's and enhanced V-2's, until the supply of these captured missiles was nearly exhausted in early 1950. By this time, guided missile technology having advanced beyond the range of the White Sands Missile testing area, the Army decided to relocate von Braun's group to the Redstone Arsenal at Huntsville, Alabama. The advantage of this site was its relative nearness to the Eastern Test Range at Cape Canaveral in Florida.

The Korean War broke out in late 1950 and the missile development group at Redstone was asked to design a missile having at least a 500-mile range. The missile von Braun's team built was named the REDSTONE. Liquid fueled, this missile was initially flight-tested on 20 August 1953. Thirty-six more of these missiles were built and flight-tested from 1953 to 1958.

At the end of World War II the Navy also had discovered a use for the "newly" developed rocket technology. A Rocket-Sounde Research Branch was established within the Naval Research Laboratory (NRL) to produce rockets with the capability to research conditions in the upper atmosphere. Two rockets were produced by the Navy—the VIKING and AEROBEE. The VIKING was derived largely from V-2 technology and was powered by a 20,000-pound-thrust liquid oxygen-alcohol engine. The AEROBEE was originally powered by a 21,000-pound-thrust solid-fueled engine, but later versions were capable of being fueled with JP-4 jet fuel. The AEROBEE was such a successful sounding rocket that a great variety was produced. The VIKING did not share that same type of success. First launched on 3 May 1949 to an altitude of 50 miles, the VIKING became the forerunner of America's infamous VANGUARD booster.

The Air Force was the first service to begin work on long-range missile technology. The early work was not done on missiles, but on long-distance guided airplanes or rocket powered flying bombs. The successful development of the U.S. short-range flying bomb (the JB-2) before World War II ended convinced Air Force leaders that the winged missile could probably be deployed sooner than long-range missiles—still to be developed. Understandably, at this point the Air Force was still conditioned to think in terms of "winged" vehicles.

Following the end of the war, the Air Force began research on three pilotless flying bombs. The SNARK (of Northrop Corporation) was the first winged missile to have intercontinental range. Designed to fly up to 7,000 miles, it could carry a 5,000-pound nuclear warhead. The Air Force funded the SNARK program somewhat unenthusiastically and it did not become operational until 1958.

A second guided missile project was the Martin MATADOR. It was not built to fly so great a distance as the SNARK, but this shorter range and development time made the MATADOR an ideal missile for deployment in Europe. In 1955, a year after completing flight tests, the MATADOR entered operational service in Europe and the Far East.

The third and by far most important cruise missile was the NAVAHO of North American Aviation Corporation. Although this missile never got past early development and testing phases, its technology was used on almost all advanced missile and high-speed bomber projects throughout the three military services in the late fifties and early sixties.

As a true intercontinental ballistic missile (ICBM), the Convair Corporation had earlier designed the MX-774 for the Air Force. Funding, however, was terminated by the Air Force prior to 1953 and only token work continued.

One general argument against ICBM's was weight. The weight of atomic (fission) warheads in the immediate post-war years was so great that many Air Force leaders believed an ICBM was an impractical weapon. The weight and fuel required to deliver a warhead over a 5,000-mile distance made development and construction prohibitively expensive. And besides the weight problem, no one had yet developed a guidance system accurate enough to guide a rocket over the intercontinental distance and then hit within a few thousand feet of the target. These technological problems made easy the decision to cancel the MX-774 and instead to support cruise missiles. The advent of lightweight hydrogen (fusion) warheads in the mid 1950's, having megaton capabilities, would resurrect the idea of feasibile ICBM's for military purposes.

EARLY SATELLITE RESEARCH

Space interests in the post-World War II United States began with a May 1945 report (von Braun to H. S. Tsien) discussing German views on prospects and potentials of satellites, during an early interrogation. (As an aside, Tsien went on, after his return to China in the 1950's, to become the leading weapons expert in the Peoples Republic of China.) The Navy learned of the report, and interest started in the Bureau of Aeronautics for further study; Hyatt, Havilland, Berkner and Hall were responsible for the study request of December 1945 to the Jet Propulsion Laboratory (JPL) for satellite vehicle studies.

Army Air Force* interest in satellites was also evident in two reports in 1945:

- In a November 1945 Arnold Report "... the design of space ships ... is all but practicable today ..."
- In a December 1945 report on the occasion of a proposal for Atomic Energy Detection Systems "... including space vehicles, space bases, and persuasive devices ... therein."

Budget cuts in 1946 prompted the Navy to propose to the Air Force combined sponsorship of programs at a joint meeting (March 1946). In July 1946, the Navy assigned contracts to Aerojet, North American Aviation (NAA), and Martin for propulsion and vehicle engineering design work for satellites.

The Navy work showed the following interests:

- Discussion of hydrogen-oxygen propulsion.
- NAA work on a High Altitude Test Vehicle (HATV): pressurized structural tanks, etc., later on ATLAS; single stage concepts.
- Reasonably detailed design and layout studies for both Martin and NAA satellite vehicles with substantial payloads (up to 2,000 lb).
- Though only peripherally related, work on nuclear rocket and ramjet propulsion.

---

*Near the end of World War I, Brig. General William "Billy" Mitchell became Chief of Air Service for the U.S. Army. The Air Service was renamed the Army Air Corps in 1926. The Army Air Forces was formed in 1941. Then, on 18 September 1947, Congress created the United States Air Force (USAF) as an equal partner with the Army and Navy.

The Air Force's action was to assign a major study to Project RAND* to investigate the feasibility of artificial satellite vehicles. The conclusion was that a large rocket would have sufficient performance to place several hundred pounds of payload on orbit.

The Rand report, although done very quickly to stake out a claim for the Air Force in the space field, vis-a-vis Navy, contained some rather thorough preliminary scientific and engineering analyses of satellite feasibility. Concepts studied included multi-stage vehicles, meteor problems, reentry considerations, scientific applications, detailed trajectory analyses; military uses for assisting missile guidance, and for reconnaissance, weather surveillance, and communications; and the potential impact and significance of the satellite project were assessed and highlighted. Major documents coming from the Air Force studies included:

- Douglas Aircraft (Project RAND), May 1946, "Preliminary Design of an Experimental World-Circling Space Ship."
- Project RAND Summary Report, June 1946, "World-Circling Space Ship," RA-15001.

In February 1947, Rand published a multi-volume detailed study amplifying their prior work. As a result, in September 1947, the USAF requested an Air Material Command (AMC) evaluation of the Rand reports of February 1947. The December 1947 response of AMC verified feasibility, but had questions of utilization, and reflected doubts that funding would become available at the appropriate level. AMC suggested the establishment of a satellite project (to prepare specifications, requirements, and scheduling). AMC further suggested the priority of guided missile development, but proposed nevertheless starting on satellite component developments.

Subsequently, the Air Force Chief of Staff General Hoyt S. Vandenberg's policy statement (January 1948) constituted the first clear service statement of space program interest ("USAF . . . has logical responsibility for satellites . . ."). Although funding competition effectively devised development of satellites, the January 1948 policy was put into effect (by February 1948) by authorizing Rand to do research and to let subcontracts in field. However, the military worth of satellites was not yet fully recognized in the USAF. The Research and Development Board concurred in the USAF action (mid 1948); and the USAF became the only service authorized to expend funds on satellite vehicle studies.

In 1954, Rand published summary reports entitled Project "Feed Back."[1] These reports (edited by J. E. Lipp and R. M. Salter) covered the preceding 8 years of work, and with cognizance having been turned over to the Air Force, Rand made the following specific policy recommendations:

1. The earliest possible completion and use of an efficient satellite reconnaissance vehicle is of vital strategic interest to the United States.

2. The satellite operation must be considered and planned on a high policy level.

3. The project should be handled as sensitive matter as regards disclosure. Secrecy concerning the operation should be maintained, particularly during the period just preceding and immediately following launching.

4. The extent and nature of disclosures regarding the actual operation should be determined in the light of the general political situation.

---

*Progenitor of the Rand Corporation, but then a special element of the Douglas Aircraft Corporation.

5. The international, legal, and political implications of the operation should be carefully considered; defense against possible legal attacks from the Soviet side should be prepared.

6. Delayed transmission techniques, permitting receiving stations in territories under U.S. jurisdiction, are preferable to instantaneous transmission, because they would reduce the political vulnerability of the project.

The most promising first-use appeared to be reconnaissance by means of television. Reasonable comprehensive investigation of this scheme was carried out by Rand during 1949 and 1950 and a satisfactory utility was indicated. Study of auxiliary power plants for the satellite was undertaken by Westinghouse Electric Corporation for RAND during this time.

Investigations were conducted of certain critical elements of the reconnaissance satellite. These included:

1. Studies of the suitability of television for reconnaissance by satellites, made by the Radio Corporation of America for Rand on subcontract.

2. Studies of auxiliary power plants to supply electricity to vehicle-borne equipment, made by Allis-Chalmers Manufacturing Company, Bendix Aviation Corporation, Frederic Flader, Inc., and Vitro Corporation for the Atomic Energy Commission (AEC) at Air Force request.

3. A subcontract from Rand to North American Aviation, Inc., to study an attitude-sensing and control system for the orbiting vehicle.

4. A prime contract from the Air Force to North American Aviation, Inc., to study a takeoff guidance system to place the vehicle in orbit.

5. Various supporting studies by Rand and its consultants on political and psychological effects of satellite operations, weather analysis by satellite television pictures, properties of the upper atmosphere, effects of meteors on satellite vehicles, and component characteristics.

The characteristics of a satellite reconnaissance scheme, referred to at the time by the unclassified code name "Feed Back," was conceived as being an integrated means for obtaining initial pictorial reconnaissance of potential enemy territory. As such it included not only the television equipped satellite vehicle, but also ground facilities for handling and evaluating information gathered.

At first, the satellite would probably provide initial reconnaissance, determining the existence, approximate location, and general nature of targets and activities. Other promising reconnaissance applications were mapping and weather intelligence.

The need for secrecy in developing this capability was stressed due to the novelty of satellite reconnaissance and the recognition that counteraction against Feed Back could be effected in many ways—action against the vehicle, action both physical and political against the communication stations, and attempts at deception.

Although it had been proposed that the Navy work with Rand and the Air Force, in the summer of 1948 the Navy relinquished interests in satellite work because of the limitations in R&D budget allocations.

INTERIM AERIAL SURVEILLANCE METHODS

As this ground work for space vehicles was being laid in the late forties and early fifties, it was apparent that satellite reconnaissance, if to be possible at all, was at best several years in the future.

The real significance of the German V-2 scientists taken to Russia following World War II was brought home by the explosion of the first Soviet hydrogen bomb on 12 August 1953, just 9 months after America's first. Recognition of the need to know what was going on behind the Iron Curtain was evidenced by the forthcoming diversity of ideas and plans—some new and some revitalized—to conduct aerial surveillance by "non-conventional" means. Few of these, however, survived political or funding constraints.

One plan discussed but not pursued beyond some initial planning, was the idea that the critical need to obtain photographic coverage of Russia for mapping purposes might be accomplished in a one-shot operation through a specialized deployment of high-altitude airplanes.[3] The plan envisioned that a fleet of super aircraft could be stationed in Europe in sufficient numbers to span Russia (photographically) from north to south, being maintained on alert status awaiting a period of wide area high pressure when the entire U.S.S.R. would be virtually cloud free. The fleet would then cover the country in one sweep from west to east, landing then at friendly bases. This idea was born (and died) before the U-2 era, the U-2 being an airplane that would have been technically qualified for such a mission had it been available at an earlier date. There was one airplane undergoing tests at the time, however, that might have been worthy of consideration for this operation, the Northrop Flying Wing. Although the operational ceiling of the Wing was published as being 40,000 feet, there was speculation that the photographic reconnaissance version (YRB-49A) with four 5,000-pound-thrust Allison J-35-A-19 engines in the wing and two more suspended in pods below the wing, might be capable of operating at much higher altitudes. But even if airplanes with desired performance had been available to make the plan technically feasible, it is doubtful that any such overflight would have received political sanction.

Another plan to photograph Russia by unconventional means was initiated in the early fifties and did become operational, though for a very brief time. This approach to acquiring intelligence data over denied areas involved the use of high altitude balloons. The program, code name "GENETRIX," called for large plastic balloons carrying light-weight cameras and electronic equipment to be floated across the U.S.S.R.[4]

As to the operational concept, they were to be launched between 0° and 60° north latitude during the winter months to take advantage of the high altitude westerlies. The vehicles were to be capable of remaining above 60,000 feet for periods of 8 to 10 days, during which they should travel some 5,000 to 10,000 miles. Recovery sites located in the Northwest Pacific would locate and track the balloons as they came into range, and suitable aircraft would recover the payloads. The payload, consisting of two 450-foot rolls of $9\frac{1}{2}$-inch film and one roll of 16-mm film, would be packed and shipped to the U.S. for processing, indexing, and studying. It was estimated that successful operation of 2,500 vehicles would net 85% coverage of the area of interest.

Original research and development work in this system was started about 1950. By 1952, after unsatisfactory progress, the project was reorganized, given more financial support, higher priority, and transferred from Wright Air Development Center (WADC) to Air Force Cambridge Research Center (AFCRC) for continued development. Here the vehicle was developed as a weapon system under Project 119L, and the development time was cut by one year under a 1-A priority crash effort.

In February 1954, Colonel Richard Philbrick, Commander, Aeronautical Chart and Information Center (ACIC) at St. Louis, Mo., received a letter from Mr. Walter J. Levison of Boston University, stating that Boston University had a contract with AFCRC to develop the photographic system for Project 119L, and discussing possible photographic equipment to be used in the project. He asked coordination with Colonel Philbrick and ACIC in the design of the equipment. Final authority, outlining ACIC's responsibility in the project, was a top secret letter from Headquarters, USAF, dated 29 July 1954, assigning an additional function to ACIC. This letter called for ACIC to accomplish data reduction, i.e., processing, indexing, and photointerpretation on a crash basis; however, late in March 1955, the responsibility for processing film was transferred to SAC, mainly for budgetary reasons—since the SAC Reconnaissance Technical Squadrons had the equipment and personnel in place and were capable of doing the job, whereas ACIC would have to procure the equipment and train the personnel.

ACIC's answer to Mr. Levison's letter pertaining to camera configuration recommended, as first choice, two 6-inch focal length, 10.4 × 10.4-inch format cameras with tilts between 35° and 40°. An exposure interval of approximately 10 miles was suggested; based on an estimated speed of 50 knots at 80,000 feet, this would mean a time interval of 12 minutes between exposures. An azimuth device would be preferred but the rotation would be acceptable if split vertical cameras were used. The system that was recommended as second choice (which consisted of two 6-inch focal length 9 × 9-inch format, 35° split vertical cameras) was finally selected for the project.

The photointerpreter (PI) team was made up of members from the Army, Navy, CIA, RAF, SAC, TAC, USAFE and FEAF (Far East Air Force).

The cover story to account for the existence of the large balloons stated that the project was part of a worldwide meteorological survey presently being conducted by the USAF to secure vital high altitude scientific data in conjunction with the International Geophysical year.

The first vehicle was launched on 10 January 1956 between 0100-0200 GMT. On 6 February 1956, operations were suspended as a result of formal protests by the Russians. During this brief period, 512 vehicles were launched and 54 were recovered, i.e., by the "intended" recovery forces. 1,984,173 square miles of photographic coverage was obtained of which 322,304 square miles consisted of duplicate coverage. Of this total, 1,116,449 square miles fell within the Sino-Soviet area.

Although this program was short lived, there were many benefits derived from the product. New targets were located, and confirmation of intelligence on previously known targets was possible. Probably the most significant long-term benefit was that this program provided experience in processing data from random reconnaissance over a large area, thus giving the using community an insight into future data handling requirements.

But by far the most productive aerial photographic reconnaissance development to become operational in the mid fifties was the U-2 "Spy Plane." Designed by Clarence L. (Kelly) Johnson, one of the most renowned and imaginative aircraft designers in the world, the plane was built by the Lockheed Aircraft Corporation in Burbank, California in a location that became known as the "Skunk Works."

The U-2, with somewhat the appearance of a glider and having exceptionally high altitude capability, began flights over the Soviet Union in 1956 to monitor ICBM tests and to find out just how far the Russians had progressed toward developing a nuclear armed intercontinental ballistic missile (ICBM). As anticipated, the Soviets were able to track its flights by radar, but much sooner than expected—in fact, on the very first flight. A formal protest was filed by the Soviets; however, U-2 operations were resumed after a temporary standdown and the U-2 ranged over much of the world for nearly four years.

MCS HISTORY

GOVERNMENT ORGANIZATIONS EVOLVED AS SPACE VEHICLE DEVELOPMENT CONTINUED

Experiments in the construction of thermonuclear devices by the Atomic Energy Commission (AEC) in 1953 demonstrated that the weight of the warhead could be reduced and hence an effective ICBM could be built. These technological advances coupled with intelligence reports that the Soviets were already developing missiles with intercontinental range, helped to convince the new Assistant Secretary of the Air Force, Trevor Gardner, that the United States should increase its ICBM research efforts.

Trevor Gardner, because of the Soviet threat and the improving state of missile technology, formed the Strategic Missiles Evaluation Committee (SMEC) in late 1953. The SMEC or "Teapot Committee" as it was popularly known, was chaired by Dr. John von Neumann, a prominent American scientist. The committee's membership consisted of leaders of the scientific and technological community. Holding its first meeting in November 1953, the SMEC investigated all strategic missile programs then in progress in the United States and the Soviet Union. Forwarding its recommendations in early 1954, the Teapot Committee formally urged the creation of a separate military organization to conduct the research and development of long-range missiles. In addition, they suggested this new organization assume control of the ATLAS* project and develop an upgraded version of the underfunded project.

The organization created by the Air Force at the request of Trevor Gardner, was the Western Development Division (WDD). Located at Inglewood, California, the WDD was first commanded by Brigadier General Bernard Schriever. In late 1954, he became responsible for the development of America's first ICBM—the improved ATLAS. In 1955, ICBM development became the first national priority and Schriever's budget grew accordingly.

As a hedge against ATLAS failure or the possibility of unforeseen developmental delays, the WDD began work on the TITAN, a second ICBM, using systems that had been developed for the ATLAS. An Intermediate Range Ballistic Missile (IRBM) was also designed using ATLAS components; this was the THOR, a 1,500-mile-range missile weighing about 100,000 lb. Planned as a stop-gap measure to be deployed in Europe, the THOR was an "ICBM hedge" against the advanced Soviet ICBM technology. A final ICBM project, undertaken by the WDD before SPUTNIK was the design of the first solid fueled missile—the MINUTEMAN. This missile project succeeded largely through the efforts of Colonel Edward N. Hall. Because a solid-fueled missile offered the advantage of longer storage and short response time at a less expensive price, the government chose to build a 1,000 missile MINUTEMAN force in the early 1960's.

On 8 November 1955, at the recommendation of President Eisenhower's staff, the Army and Navy began a joint research project to develop an IRBM. The Navy, about a week later, created the Special Projects Office for missile research on the IRBM under Rear Admiral William F. Raborn. Then the Army, on 1 February 1956, renamed the guided missile organization at Redstone Arsenal to the Army Ballistic Missile Agency (ABMA). The missile these agencies decided to develop jointly was named the JUPITER.

The JUPITER was basically an upgraded REDSTONE missile. This design was entirely satisfactory to the Army because the technology of liquid fueled missiles was well advanced for that time, and development time of this IRBM would not be long. The Army had actually begun development on the JUPITER before jointly working with the Navy. The first mockup of the

---

*In 1951, the Air Force had revived the contract on MX-774 and redesignated it MX-1593, or more commonly the ATLAS.

JUPITER—the A Series—was launched beginning in September 1955. The first of the Third Series, the JUPITER-C, reached an altitude of 682 miles and a range of 3,400 miles on 20 September 1956.

The Navy, certain that a liquid fueled rocket would not be suitable for sea service, pulled out of the joint program with the Army, and on 20 November 1956, the Secretary of Defense, Charles E. Wilson, issued a roles and missions memorandum that completely stripped the Army of its long-range missile program. Dr. von Braun's group became limited to developing missiles with ranges of less than 200 miles.

The roles and missions memorandum, however, did not stop the Army from completing the development of the JUPITER IRBM. On 31 May 1957, an operational JUPITER traveled over 1,600 miles in a launch from Cape Canaveral making the JUPITER the first successful IRBM. In mid-1958, the ABMA delivered the first operational JUPITER's to the Air Force for use in Turkey.

Meanwhile, in response to Rand's Feed Back studies, in May 1954 the Air Force's Air Research and Development Command (ARDC) was directed to assume responsibility for study of applications of Project Feed Back. System Requirement No. 4, in March 1955, established approval of a clear effort to develop a reconnaissance satellite system.

By November, the Wright Air Development Center (WADC) at Dayton, Ohio had a small team in place, and study contracts had been let to RCA, Martin and Lockheed for further definition of time and technology requirements for satellite developments, under the nickname "Pied Piper." Two Air Force officers assigned to this small team, Q. A. Riepe and William G. King, Jr., then holding the ranks of Major and Lt. Colonel, respectively, later assumed key leadership roles in operational satellite programs. As a personal insight to attitudes and priorities of the time, Brigadier General King recalls that their charter at WADC was not very clear and that they initially received little notice by the important people at Wright Field.[5] But in fairness to the Wright Field leadership at the time, these were the days of the "Century Series" fighters—the B-58 Bomber had a big program office—the B-52 Bomber office dominated every meeting and action. Airplane development at the time was critically important and was pursued on that basis. In the priority of things of that day—satellites—("spelled how?? one or two L's") didn't command much attention or supervision. For the several people in the office, the facilities were relatively adequate. There were four desks and five people; one person was required to be on TDY at all times! Luckily, King said, there were contingents in Baltimore (ARDC Hq) and at Rome and at Boston who knew the ropes and were intent on getting a charter—or SOR or SOC or whatever the license was to get a line item in a budget—and of course they were successful. At this time, and during the years to follow, it was the personal enthusiasms of a few individuals that helped keep things moving.

Briefings being the WADC team's principal product, audiences were easy to come by, but real help in the way of requirements, etc., was missing—"What the hell can you really do with satellites??" But with General Schriever at WDD coming to almost everybody's notice, the team looked upon this missile oriented organization as a potential source of help that should be investigated—after all, satellites needed boosters. Shortly following a briefing given in Los Angeles in which the ARDC Commander General Donald Putt and other dignitaries were in attendance, the decision was made to transfer the program, the resources and the people to WDD.

The transfer to WDD had actually been considered as early as October 1954 by the ICBM Scientific Advisory Committee, which considered the possible interaction between satellite and missile proposals. In January 1955, the committee proposed to address the question:

Would working on the satellite vehicle interfere with the missile program? Although the committee expressed concerns on the interference of satellite work with the ICBM work in June 1955, recognition that ICBM-IRBM boosters were essential for satellites eventually prompted the program change from WADC to WDD.

Finally, on 29 October 1956, Lockheed was awarded the development contract for Weapon System 117L (WS-117L), the designator for the military satellite program that was now committed to the development and test of actual flight articles.

The key characteristics or "sales highlights" for WS-117L, which was also referred to by the official nickname "New Horizon," were:

- Complete target coverage
- Accurate specific target location
- Continuous target area surveillance
- Instantaneous warning of ICBM attack
- Nearly invulnerable to attack or counter measures*
- No air crews
- No (typical) overseas bases
- Invades no airspace†
- High data rate
- Economical per unit of data
- Fast response
- Growth potential.

The planning of WS-117L contemplated a family of separate systems and subsystems employing satellites for the collection of photographic, electronic, and infrared intelligence. The program, which was scheduled to extend beyond 1965, was divided into three phases. Phase I, the THOR-boosted test series, was to begin in November 1958 and had a primary objective of development/initial testing. Phase II, the ATLAS-boosted test series, was to begin in June 1959 with the objective of completing the transition from the testing phase to the operational phase and of proving the capability of the ATLAS booster to launch heavy loads into space. Phase III, the operational series, was to begin in March 1960 and was to consist of three progressively more sophisticated systems: the pioneer versions (photographic and electronic), the advanced version (photographic and electronic), and the surveillance version (photographic, electronic, and infrared).

The main tasks of the photographic reconnaissance satellites, that hopefully would be operational in the early 1960's, were to study the Soviet ICBM's in as much detail as possible—number of missiles, number and construction of missile sites, support buildings, equipment, and personnel—and to map the entire Soviet Union to provide targeting data for U.S. missiles. (Later, as mapping was accomplished, it was discovered that the positions of some cities shown on Soviet maps were deliberately falsified.) The nature of these requirements and the state of the art together suggested a multiple approach in selecting the payloads to be developed. Physical film recovery was considered best for high resolution and metric accuracy, but the art of recovering vehicles from space had still to be developed. Recovery of instrument packages from rocket flights had been very successful using parachutes when used with rockets attaining peak altitudes of less than 60 miles. However, bringing back politically sensitive payloads from orbiting

---

*By 1961, the official program management prediction on this issue was that during the 1960's the Soviets' would develop the capability to destroy or render useless a satellite on orbit.

†The legal aspects of space overflight had not been resolved to the same degree as penetration of air space.

satellites presented a problem of great challenge.* On the other hand, radio transmission, with resolution thought to be adequate for many of the area survey requirements, was attractive from the aspect of timely data return. The decision was made to develop both film recovery and radio transmission (readout) satellites as complementary programs, thus leading to the first two photographic reconnaissance programs: CORONA for recovery and SENTRY for readout (SENTRY was later renamed SAMOS).

The full scope development plan for WS-117L was endorsed in April 1956. Thus, despite suggestions of high level indifference to the notion of militarily useful satellite vehicles, the programs suggested by Rand over the previous ten years finally got under way in 1956. The major milestones in the evolution of an "advanced reconnaissance system" up to this point are summarized graphically in Fig. 1-1.

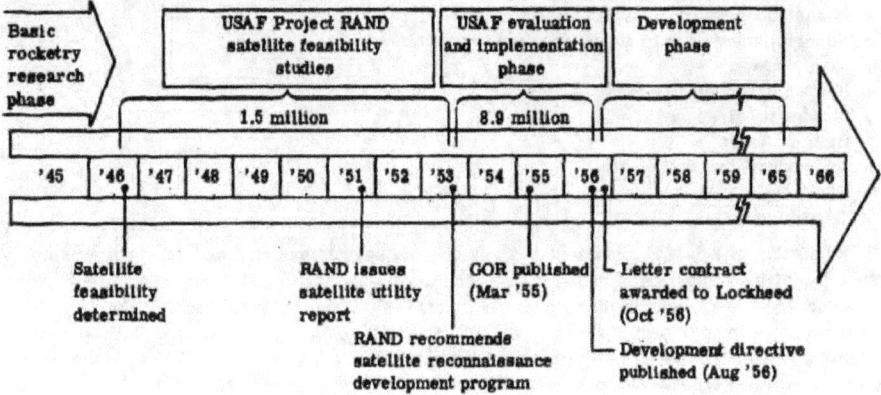

Fig. 1-1 — Advanced reconnaissance system evolution

---

*In 1951, 9 years before the first recovery of an orbiting body, Robert M. Salter, Jr. touched briefly, but with great insight, on this subject during an address on Engineering Techniques in Relation to Human Travels:[6] "It is physically possible to bring a satellite back without great additional source of power. This is not easy and would require considerable development in control equipment. In launching a satellite, a long, coasting (elliptical) trajectory is indicated, with a small additional kick provided to pull it into the orbit. The same kick in reverse will put the vehicle back into the original ballistic flight path, but the vehicle might burn on the way down. By using a carefully selected and maintained gliding trajectory it is believed possible to enter the atmosphere without disastrous skin temperatures and high landing speeds. In fact, terminal speeds slightly over sonic are indicated, at which point parts of the vehicle could be landed with parachutes. The main problem then, of the returnable satellite, is that it requires a very accurate control during the descent phase—automatic programmed control at the least—and possibly the continuously computed variety."

SPACE PROGRAM COMPLEXITY DEEPENS

The following years were critically important years in the military space program. Once the decision to undertake the basic WL-117L program had been made, program proposals began to proliferate rapidly. The U.S. military space program rapidly became very complex.

In August 1954, Congress had approved U.S. participation in the International Geophysical Year (IGY), and "launchings of small satellite vehicles" was recommended for such U.S. participation. Planned to begin on 1 January 1957 and run through 30 June 1958, the IGY was to be a worldwide scientific effort to gather data about the sun and the upper reaches of the earth's atmosphere. Both the United States and the Soviet Union had announced that each would orbit an artificial satellite to aid in the monumental research effort.

In early 1955, the Army and Navy had proposed a joint satellite effort, Project Orbiter. However, a major policy directive in May 1955 (NSC Dir. 5520) directed that no missile intended for military purposes could be used for IGY satellites by the United States. The directive supported President Eisenhower's "peaceful uses of space" concept.

Under the complex arrangements for the IGY satellite, the DoD was to supply the booster to put the payload into orbit, the National Academy of Sciences would determine the experiments to be placed on the satellite and the National Science Foundation would finance the venture. Since each of the services were in some way involved with missile technology, an ad hoc committee was set up to choose the booster.

Two of the three main contenders (the Army-Navy system based on REDSTONE, and the Air Force ATLAS based system) were in conflict with the NSC directive. The other contender was the Navy VANGUARD, not yet even under development. The underlying reason for choosing the VANGUARD was that it had no "military taint," a reason difficult to understand since a military service, the Navy, was in charge of the VANGUARD program.

In retrospect, the decision to go with the VANGUARD was a serious blunder. Von Braun's team had proven the Jupiter booster was capable of orbiting a satellite after their first test launch of 20 September 1956. To go with a completely new booster built by organizations not even experienced in designing long-range missiles or boosters only because VANGUARD was "virtuous" was a poor political and technological decision.

NSC Directive 5520 notwithstanding, ARDC was requested in August 1955 to establish a scientific satellite program integrated with WS-117L (as a vehicle to satisfy IGY requirements, by implication). In November 1955, after a short hold on the project, WDD was assigned responsibility for a plan to use WS-117L prototypes for scientific satellites. By January 1956, WDD responded with a proposal to orbit a 3,500-lb satellite by August 1958, using ARS (Advanced Reconnaissance System) items, and capable of conducting a number of specific scientific experiments. ATLAS C was to be the booster.

The WDD plan was not acted upon in 1956 by the committee because NSC 5520 still prevailed, but in February 1957, OSD requested a USAF estimate of the U.S. capability to build a "backup" scientific satellite for IGY use. The WDD response was that with the ATLAS program it might be possible to conduct one or two maximum risk launchings during 1958. Again, however, the decision was made not to tie the 117L to IGY needs. The ATLAS satellite (Project Score) was actually launched December 1958, equipped to broadcast President Eisenhower's voice in a Christmas message from space. But this was 14 months after Russia had stunned the world by successfully launching into orbit the world's first satellite, SPUTNIK I on 4 October 1957.

POST-SPUTNIK "CATCH UP" ACTIONS—ARPA AND NASA ESTABLISHED

Decisions during this time were made against the backdrop of this stupendous Russian space achievement. The impact on world opinion, U.S. public opinion, and congressional concern was almost immediate. The shock that a backward Eurasian power could leapfrog the once mightest technological nation to achieve a first in space was the general perception. Of more specific concern was that the Soviets had a space booster which demonstrated not only the ability to place objects in space, but also the capability to carry warheads to intercontinental distance. Any attempts to belittle or minimize the significance of the 184-pound SPUTNIK I satellite were dismissed when on the third of November a second satellite, SPUTNIK II, was launched by Russia. This satellite, weighing 1,120 pounds, placed an animal into orbit for the first time, the dog named Laika. The location of ICBM sites in Russia would now become a highest priority task for the intelligence community.

Prior to SPUTNIK, Headquarters USAF, together with Rand, AMC, ARDC, BMD, and WADC had evidenced widespread interest in astronautics and had acquired a fairly sophisticated grasp of its technology. On the other hand, at no level within the Government had there been a clear statement of the ultimate objective of a space program, or a systematic evaluation of the widely varying aims of the suggested projects. Clearly, government management must be structured with mandates to narrow and erase the "presumed" missile gap and space gap.

To meet this challenge the Secretary of the Air Force, James H. Douglas, called upon a committee of distinguished scientists and Air Force officers headed by Dr. Edward Teller to propose a line of positive action.

The first major organizational development came on 7 November 1957 when the President added to the existing organizational structure by appointing Dr. James R. Killian as Presidential Special Assistant for Science and Technology.

In the same month, the Secretary of Defense, Neil H. McElroy announced that to eliminate the possibility of interservice rivalry over space he intended to create a special projects agency to handle the research and development of satellites and other space related and advanced technologies for the DoD. General Schriever protested this decision to build a special projects agency because the Air Force already had the capability to conduct operations in space. Nevertheless, on 7 February 1958, McElroy established the Advanced Research Projects Agency (ARPA). The new organization was headed by Roy W. Johnson; and contrary to the wishes of the Joint Chiefs of Staff (JCS), he was authorized to direct the research and development projects within the Department of Defense that the Secretary might assign to it.

In practice, ARPA would then reassign the projects on a contractual basis to the military departments, other Government agencies, or civilian institutions. (Between 7 February and 1 October 1958, ARPA actually served as the national space agency.)

Also in November 1957, Senator Lyndon B. Johnson, the Senate Majority Leader, had opened hearings on the apparent American lag behind the Soviets in military science and technology. In the course of these hearings, one of the witnesses, von Braun, provided a solution to the apparent funding and advocacy problem that hampered the VANGUARD program. In his testimony to the Senate committee, he suggested the creation of a national space agency, with a separate budget, having the initial missions of putting a man in orbit and constructing a manned space station.

The Senate committee was impressed by von Braun's proposals. Indeed, Eisenhower's science advisor, Dr. Killian, had also visualized a space program conducted primarily by a single civilian organization. A civilian run space program would also complement President Eisenhower's statements made in January 1958, "that outer space be used only for peaceful purposes."

MCS HISTORY

With the thought of a single civilian space agency in mind, the Bureau of the Budget, in close cooperation with Dr. Killian and NACA,* drafted the legislation which later became the NASA Act. Signed into law on 29 July by President Eisenhower, the Space Act of 1958 declared that the general welfare and security of the United States required that adequate provisions be made for aeronautical and space activities. The agency was delegated the responsibility to conduct research on space and aeronautic activities except those primarily military in nature. Its charter was very clear, to catch up with the Russians in space and missile technology.

NASA was built from already existing organizations. At the core of NASA was NACA with all its research organizations. Dr. John P. Hagen, the Director of the Navy-run VANGUARD program with his 180-man VANGUARD operation and the Army's Jet Propulsion Laboratory (JPL) also became a part of NASA. A meteorological satellite program, the TIROS, and a modified Air Force launch vehicle, the ATLAS-CENTAUR, were also legislated to NASA jurisdiction. The final transfer requested by NASA was not approved until almost two years later. ABMA and von Braun's team of scientists and engineers—complete with all the facilities of the Redstone Arsenal—later became part of NASA on 1 July 1966.

In the spring of 1958, shortly after its activation, ARPA, acting as the national space agency, had organized its space projects into four programs: (1) Missile Defense against ICBM, (2) Military Reconnaissance Satellites, (3) Developments for Application to Space Technology, and (4) Advanced Research for Scientific Purposes.

In September 1958, shortly before the activation of NASA, ARPA redefined the Advanced Reconnaissance System and broke it down into separate projects with different designations. The reconnaissance aspect was renamed SENTRY. The vehicle tests, biomedical flights, and recovery experiments were grouped together as DISCOVERER. The infrared sensing system became MIDAS. In the last months of 1958, ARPA assigned those three projects to ARDC-AFBMD with the usual contractual arrangements. At the end of June 1959, the Air Force was still without a space program of its own, but was supporting a large part of the ARPA program and some NASA projects as well.†

---

*As an aeronautics research agency since World War I, NACA (National Advisory Committee for Aeronautics) had been very successful in its efforts to advance general and military aviation, and thus enjoyed a fine working relationship with the military while still retaining its civilian outlook.

† Space projects wholly or partly entrusted to AFBMD by mid-1959:

1. Discoverer
2. Sentry
3. Midas
4. 6-Hr Comm Sat
5. 24-Hr Comm Sat
6. Deep probes
7. Transit Nav Sat
8. Tiros Cloud Cover
9. Courier Passive Army Comm Sat
10. HETS
11. Mercury (MIS)
12. Willow
13. Special Testing
14. Centaur
15. Saturn
16. Manned Sat & Interceptor and Inspection
17. Geo-Astro-Physical Program
18. Hustler Engine
19. Aerojet 104 Engine
20. Delta
21. Vega

REKINDLING OF USAF INTERESTS IN SPACE

In the spring of 1959, widespread dissatisfaction with the progress made by the space program led to changes in organization that were of substantial consequence. On 13 April 1959, Headquarters USAF issued the equivalent of a charter that gave the Directorate of Advanced Technology authority to coordinate within the Air Staff all USAF space activities.[7]

Still another change occurred on 9 November 1959 when DCS/Development redesignated the Director of Advanced Technology as the Assistant for Astronautic Systems. The Chief of Staff approved the shift to "Assistant" status in December, but would not permit use of the term "Astronautic Systems."

In his new position, the Assistant for Advanced Technology had overall responsibility within DCS/Development for policy guidance and program direction in the broad areas of ballistic missiles or vehicles, ballistic missile warning and defensive systems, and vehicles and systems to operate in space, including those for detecting and tracking.

Simultaneously with these USAF organizational changes, differences among the military services came into the open. In the midst of these discussions ARPA recommended in June a MERCURY Task Force to assist NASA, and the Secretary of Defense requested JCS advice in assigning operating responsibilities for several projects, including MIDAS and SENTRY—the latter soon to be redesignated as SAMOS. In the months that followed, the services held their positions. The Army and Navy wanted a Mercury Task Force and a Defense Astronautical Agency to control the space systems. The Air Force objected to both.

TRANSFER OF SPACE ACTIVITIES TO THE SERVICES

In September, the Secretary of Defense made three important decisions. He disapproved the proposed Defense Astronautical Agency. He terminated the move for a MERCURY Task Force, but as a substitute selected Major General Donald N. Yates, USAF, Atlantic Missile Range commander, to "direct military support" for the project. And finally, McElroy reversed his established policy on ARPA by dividing the military space program among the three services. Under this arrangement, MIDAS and SAMOS were marked for the Air Force, although formal transfer did not follow immediately. Likewise TRANSIT, a more recently planned navigational project, would go to the Navy, and a NOTUS family of four communication satellites to the Army.

The actual transfer of SAMOS and MIDAS occurred in late November 1959. ARPA also relinquished Project DISCOVERER to the Air Force, something not mentioned in the September decisions.

The funding decision and allocation process for the space program for fiscal years 1958-1960 was very complex. Since SPUTNIK came early in the fiscal year 1958, appropriations were already in effect, and adjustments were difficult.

Although the Air Force had long entertained space plans up to October 1957, little hard money had been allocated to the projects. The budget for fiscal year 1958, which had come into effect three months before the advent of SPUTNIK, allocated $65.8 million to the Advanced Reconnaissance System (as compared to $13.9 million the previous year) to make up the lion's share of an approximate total of $70 million for space and near-space.

In January 1958, three months after SPUTNIK, when the Air Force presented its first systematic plans for a space program, Headquarters proposed that an extra $155 million to be added to the original $70 million to make a total of $225 million for fiscal year 1958. The emergence of ARPA changed all these hopes. Between March and October 1958 all the true space projects of the Air Force, together with funds, passed either to ARPA or NASA. Consequently, the USAF budget allocation to near-space and space projects for fiscal year 1959 fell to a mere $8.7 million. About half of this sum was allocated to DYNASOAR* and BRATS,† and the remaining $4.7 million to space studies and lesser items. For fiscal year 1960 the total fell again, this time to a paltry $2.2 million.

In the fiscal year 1959, however, ARPA reassigned nearly $300 million to ARDC for work on the Advanced Reconnaissance System and its subsystems, and for applied projects. In the same year, ARPA reassigned $64 million to the Army and $24.9 million to the Navy. But all was not to be smooth sailing from an organizational sense.

At the beginning of February 1960, USAF research and development covered the whole range of space subjects from exploratory research to system development; but it was being done under the sponsorship of ARPA, a fact that galled many at the time. ARPA enjoyed almost complete freedom in deciding which military requirements, stated by the services, would be pushed, combined, or ignored.

The projected removal of ARPA from the space field, announced in September 1959, served in large part to settle this issue; but it did not alleviate the tight control which continued to be exercised by OSD agencies. Balancing influence and decision making between civilian and military officials was also a problem within the OSAF-Air Staff complex itself. In October 1959, Secretary of the Air Force James H. Douglas directed that all space actions be taken "within the framework of the AFBMC (Air Force Ballistic Missile Committee)," concentrating decision making prerogatives in civilian hands and reducing air staff participation.

In February 1960, Dr. H. York approved the shift to the Air Force of a major segment of ARPA's space study and component development program.

The Air Force divided the space development effort in 1960 into three major areas. The first, pure studies, sought new ways of doing military jobs and outlined possible system approaches. The second included applied research for the purpose of developing techniques that provided essential ingredients for future systems. The third area, system development, was the final goal, the last step in the study-research-development process to meet requirements stated years earlier.

Throughout the year, planning and programming for the SAMOS Reconnaissance Satellite, the MIDAS Early Warning Satellite, and the DISCOVERER Research Satellite were lumped together. The three had grown out of a proposal partially outlined as early as 1946 and established as a system development 9 years later.

U-2 "SPY PLANE" INCIDENT CLOSES RUSSIAN SKIES

As mentioned earlier, the U-2 "Spy Plane" had been ranging over much of the world since 1956. The advantages to be gained from overflights were dramatically emphasized by the intelligence the U-2's were bringing back. As the need for information grew, the flights into Russia became longer and longer until it was decided to make one way trips across the Soviet Union from ▬▬▬▬▬▬▬▬▬▬▬▬▬▬▬▬▬▬▬▬▬. But on the first of the one-way flights attempted, the U-2 piloted by Francis Gary Powers was shot down by a surface-to-air missile near Sverdlovsk on 1 May 1960.

---

*DYNASOAR (Dynamic Soaring): Proposed manned vehicle employing boost-glide principles.
† BRATS (Ballistic Research and Test Systems): Formerly the Ballistic Systems Research and Supporting System (BALWARDS).

Upon this politically delicate event, President Eisenhower declared the Soviet Union off limits to overflight by airplanes. Although the U-2's had been overflying communist countries for several years, a large fraction of Eurasian landmass had yet to be photographed even once. To the general public the President's decree may have seemed to end all chances for continued American aerial surveillance of the Soviet Union. Yet, in reality, terminating the U-2 overflights was to bring stronger emphasis on the attractiveness and urgency of developing the satellite reconnaissance concept.

ESTABLISHING THE NRO AND SAFSP

During the Spring of 1960, Congress became intimately involved in SAMOS-MIDAS progress. In light of the U-2 incident, Congress called for the rapid development of both space systems, and voted sums of money far in excess of the administrations's request for fiscal year 1961. Nevertheless, uncertainties and indecision on the technical and budgetary aspects of SAMOS and MIDAS continued to affect planned operational dates for these two space programs.

In mid-1960, there was concern over the delays that were resulting from multiple layers of management. The urgency surrounding the proliferation of space programs seemed to call for a new management approach.

As various government organizations searched for a way to eliminate some of the multiple-role frustrations of past decades, it was decided that the best approach to getting a streamlined operation in being was to have direction for such a plan emanate from the President.

Under panel discussions involving President Eisenhower's Science Advisory Committee and the Air Force, an administrative arrangement was seen whereby various echelons of management would be bypassed leaving program development directors reporting directly to a new National Reconnaissance office (NRO) within the office of the Secretary of the Air Force. It was visualized that the organization should have a clear line of authority, and at top level direction should include OSD and CIA, and not just the Air Force.

Following several meetings and discussions in preparation for the National Security Council (NSC) briefing on this subject, the panel made their recommendation in a formal briefing on 25 August. The President and the Secretary of Defense Thomas S. Gates, Jr. agreed that this should be the arrangement, i.e., that the line of command should be directly between the program development directors and the office of the Secretary of the Air Force. From instructions which followed, the Secretary of the Air Force issued orders on 31 August establishing:

a. A Director of the SAMOS Project at AFBMD as a field extension of the Secretary's Office, responsible to and reporting directly to the Secretary, and

b. An Office of Missile and Satellite Systems (SAFMS) within the Secretary's staff, to assist him in discharging his responsibility for the direction, supervision, and control of the SAMOS Project.

Brigadier General Richard D. Curtin was designated as Director of SAFMS. Brigadier General Robert E. Greer was designated as Director of the SAMOS Project,* with additional duty as Vice Commander for Satellite Systems, AFBMD, ARDC, with duty station at ███████ ███████ California.

---

*The SAMOS Project was later renamed SAFSP (Secretary of the Air Force Special Projects).

SPACE ACTIVITIES 1958-1960

Despite organization and policy confusion during preceding years, by 1960 the United States had made noteworthy contributions to space science by orbiting satellites in major fields of interest: scientific; communication; weather; navigation; and reconnaissance. Some successful launch events in these specific fields are summarized for brevity:[*]

a. The first U.S. satellite, EXPLORER I, launched on 31 January 1958, discovered Van Allen radiation in space.

b. The second, a 6-inch ball, VANGUARD I launched into a 400-plus mile orbit on 17 March 1958, discovered that the earth is pear shaped.

c. Project SCORE on 18 December 1958 broadcast the first voice transmission from space, a Christmas message from President Eisenhower.

d. VANGUARD II, on 17 February 1959, was the first satellite to send weather information back to earth.

e. On 13 April 1959, the second vehicle in the DISCOVERER Program (later to become the CORONA Program) was successfully launched and orbited for 17 revolutions; however, a timing malfunction precluded successful recovery of the Satellite Recovery Vehicle (SRV).

f. On 7 August 1959, EXPLORER VI mapped the Van Allen Radiation.

g. The first navigation satellite, TRANSIT 1B, was launched on 13 April 1960.

h. Then came TRANSIT IIA on 22 June 1960, carrying Canadian-built instruments to measure radio interference in space. This satellite also carried GREB I the first SIGINT satellite. This was a combined Navy/NSA effort.

i. On 12 August 1960, ECHO I, the first passive communication satellite was launched.

j. The very first recovery of an object from an orbiting satellite was made by the United States on 11 August 1960. This was the SRV from DISCOVERER XIII, the thirteenth vehicle in the DISCOVERER exploratory launch series which had begun on 21 January 1959.

These successes were due largely to the ingenuity of individuals in the armed services and Government organizations, and the industrial scientists who devised miniaturized instruments to fit the small payload capacity of then available rocket boosters.

In addition to orbiting earth satellites, the United States and Russia were both shooting rockets into outer space. Early in 1959, first Russia and then the United States launched probes that escaped earth's gravity. Russia's LUNIK I and the U.S. PIONEER IV, aimed at the moon, zoomed past the moon but went into orbit around the sun as the first man-made "planetoids." In September 1959, Russia launched LUNIK II, and 33 hours after the launch the probe crashed to the moon's surface.

---

*A broader view of early U.S. and Russian launches is provided in Table 1-1. This list, extracted from a TRW Space Log, has been limited to the period deemed appropriate at this point in the report. Also, Fig. 1-2a and 1-2b show examples of boosters, as viewed in 1959, and the Lockheed Standardized AGENA which has been used extensively in satellite programs.

Russia's LUNIK III, launched 4 October 1959 photographed the dark side of the moon.

PIONEER V, launched 11 March 1960, transmitted information on conditions more than 22,000,000 miles in space.

Then, on 20 July 1960, demonstrating a new method of weapon deployment, the Navy's POLARIS was successfully launched from a submerged nuclear-powered submarine.

Man in space did not come until 1961. On April 12, 1961 Russian cosmonaut Yuri Gugarin made a single orbit around the earth. The American astronauts, Alan Sheppard, on 5 May 1961, and Virgil Grissom, on 21 July, rocketed to a height of 117 miles in 300-mile-long suborbital flights. Then Russian cosmonaut Gherman Titov orbited the earth 17 times on 6, 7 August 1961. The first American astronaut to orbit the earth, John H. Glenn, Jr. made three orbits on 20 February 1962.

Table 1-1 — TRW Condensed Log of 1957-60 Space Projects

NAME	INITL DESIG.	PROJ. DIR.	LAUNCH DATA				INITIAL ORBITAL DATA				STATUS
			Date	Site	Vehicle	WEIGHT	Period	Perigee	Apogee	Incl.	
Sputnik 1	1957 A2	USSR	Oct 4, 1957	Tyuratam	A	184	96.2	141	588	65.1	Decayed 1-4-58; first artificial satellite, transmitted 21 days
Sputnik 2	1957 B1	USSR	Nov 3, 1957	Tyuratam	A	1121	103.7	140	1038	65.3	Decayed 4-14-58; carried dog Laika, transmitted 7 days
Vanguard TV 3	None	USN	Dec 6, 1957	ETR	Vanguard	3	-	-	-	-	Failed to orbit; lost thrust after 2 seconds
Explorer 1	1958 A1	USA	Jan 31, 1958	ETR	Jupiter C	31	114.3	224	1584	33.3	In orbit; transmitted until 5-23-58, discovered Van Allen belt
Vanguard TV 3 backup	None	USN	Feb 5, 1958	ETR	Vanguard	1	-	-	-	-	Failed to orbit: control system malfunction
Explorer 2	None	USA	Mar 5, 1958	ETR	Jupiter C	32	-	-	-	-	Failed to orbit: unsuccessful fourth stage ignition
Vanguard 1	1958 B2	USN	Mar 17, 1958	ETR	Vanguard	3	134.3	405	2462	34.3	In orbit: transmitted "pear-shaped" earth data until 5-64
Explorer 3	1958 Γ1	USA	Mar 26, 1958	ETR	Jupiter C	31	114.7	117	1739	33.5	Decayed 6-28-58; radiation, micrometeoroid data until 6-16-58
Vanguard TV 5	None	USN	Apr 28, 1958	ETR	Vanguard	22	-	-	-	-	Failed to orbit: third stage ignition malfunction
Sputnik 3	1958 Δ2	USSR	May 15, 1958	Tyuratam	A	2926	105.8	140	1168	65.2	Decayed 4-6-60; variety of scientific data returned up to decay
Vanguard SLV 1	None	USN	May 27, 1958	ETR	Vanguard	22	-	-	-	-	Failed to orbit: improper third stage trajectory
Vanguard SLV 2	None	USN	June 26, 1958	ETR	Vanguard	22	-	-	-	-	Failed to orbit: premature second stage cutoff
Explorer 4	1958 E1	USAF	July 26, 1958	ETR	Jupiter C	38	110.1	163	1372	50.1	Decayed 10-23-59; mapped Project Argus radiation until 10-6-58
Thor-Able 1 (Pioneer)	None	USAF	Aug 17, 1958	ETR	Thor-Able	84	-	-	-	-	Lunar probe failed; initial lunar attempt, first stage failed
Explorer 5	None	USA	Aug 24, 1958	ETR	Jupiter C	38	-	-	-	-	Failed to orbit: upper stages fired in wrong direction
Vanguard SLV 3	None	USN	Sept 26, 1958	ETR	Vanguard	22	-	-	-	-	Failed to orbit: insufficient second stage thrust
Pioneer 1	1958 H1	NASA	Oct 11, 1958	ETR	Thor-Able	84	-	-	70,717	-	Decayed 10-12-58; failed to reach moon, sent 43 hrs of data
Beacon 1	None	NASA	Oct 23, 1958	ETR	Jupiter C	9	-	-	-	-	Failed to orbit: third stage ignition unsuccessful
Pioneer 2	None	NASA	Nov 8, 1958	ETR	Thor-Able	67	-	-	-	-	Lunar probe failed: third stage ignition unsuccessful
Pioneer 3	1958 Θ1	NASA	Dec 6, 1958	ETR	Juno II	13	-	-	63,580	-	Decayed 12-7-58; failed to reach moon, provided radiation data
Score	1958 Z1	ARPA	Dec 18, 1958	ETR	Atlas B	8750	101.5	115	914	32.3	Decayed 1-21-59; first comsat, tx tapped messages for 13 days
Luna 1	1959 M1	USSR	Jan 2, 1959	Tyuratam	A-1	797	450 days	.9766AU	1.314AU	0.01°	In solar orbit: lunar probe, passed within 37/28 mi of moon
Vanguard 2	1959 A1	NASA	Feb 17, 1959	ETR	Vanguard	22	125.9	347	2064	32.9	In orbit; transmitted 19 days, satellite wobble degraded data
Discoverer 1	1959 B1	ARPA	Feb 28, 1959	WTR	Thor-Agena A	1300	90.0	114	657	90.0	Decayed 3-5-59; first polar orbit, no recovery capsule
Pioneer 4	1959 N1	NASA	Mar 3, 1959	ETR	Juno II	13	398 days	.9871AU	1.142AU	1.30	In solar orbit: lunar probe, passed within 37,306 mi of moon
Discoverer 2	1959 Γ1	ARPA	Apr 13, 1959	WTR	Thor-Agena A	1600	90.5	152	225	90.0	Decayed 4-26-59; capsule ejected on orbit 17, lost in Arctic
Vanguard SLV 5	None	NASA	Apr 13, 1959	ETR	Vanguard	23	-	-	-	-	Failed to orbit: second stage damaged at separation
Discoverer 3	None	ARPA	June 3, 1959	WTR	Thor-Agena A	1600	-	-	-	-	Failed to orbit: Agena fired, no satellite signals received
Vanguard SLV 6	None	NASA	June 22, 1959	ETR	Vanguard	23	-	-	-	-	Failed to orbit: second stage propulsion malfunction
Discoverer 4	None	ARPA	June 25, 1959	WTR	Thor-Agena A	1600	-	-	-	-	Failed to orbit: insufficient second stage velocity
Explorer 6	None	NASA	July 16, 1959	ETR	Juno II	92	-	-	-	-	Failed to orbit: destroyed by range safety officer
Explorer 6	1959 Δ1	NASA	Aug 7, 1959	ETR	Thor-Able	143	768	157	26,366	47.0	Decayed before 7-61; first earth photo, radiation data to 10-6-59
Discoverer 5	1959 E1	ARPA	Aug 13, 1959	WTR	Thor-Agena A	1700	94.1	135	456	80.0	Decayed 9-28-59; capsule orbited, decayed 2-11-61
Beacon 2	None	NASA	Aug 14, 1959	ETR	Juno II	10	-	-	-	-	Failed to orbit: first stage, upper stage malfunctions
Discoverer 6	1959 Z1	ARPA	Aug 19, 1959	WTR	Thor-Agena A	1700	95.2	131	528	84.0	Decayed 10-29-59; capsule rejected orbit 17, recovery failed
Luna 2	1959 Ξ1	USSR	Sept 12, 1959	Tyuratam	A-2	860	Flight time: 34.0 hours				Impacted on moon: first probe to hit the moon
Transit 1A	None	ARPA	Sept 17, 1959	ETR	Thor-Able	265	-	-	-	-	Failed to orbit: third stage ignition malfunction

Table 1-1 — TRW Condensed Log of 1957-60 Space Projects (Cont.)

NAME	INT'L DESIG.	PROJ. DIR.	LAUNCH DATA			WEIGHT	INITIAL ORBITAL DATA				STATUS
			Date	Site	Vehicle		Period	Perigee	Apogee	Incl.	
Vanguard 3	1959 H1	NASA	Sept 18, 1959	ETR	Vanguard	100	130.2	317	2329	33.3	In orbit: radiation, micrometeoroid data until 12-11-59
Luna 3	1959 θ1	USSR	Oct 4, 1959	Tyuratam	A-1	614	16.2 days	25,257	291,438	76.8	Decayed 4-20-60: photographed moon's far side for 40 min
Explorer 7	1959 I1	NASA	Oct 13, 1959	ETR	Juno II	82	101.2	346	676	50.3	In orbit: magnetic field, solar flare data until 8-24-61
Discoverer 7	1959 K1	ARPA	Nov 7, 1959	WTR	Thor-Agena A	1700	94.5	95	519	81.6	Decayed 11-26-59: poor stabilization, capsule not ejected
Discoverer 8	1959 λ1	USAF	Nov 20, 1959	WTR	Thor-Agena A	1700	103.7	120	1032	80.6	Decayed 3-8-60: capsule overshot recovery area on orbit 15
Atlas-Able 4 (Pioneer)	None	NASA	Nov 26, 1959	ETR	Atlas-Able	372	—	—	—	—	Lunar probe failed: payload shroud broke away after 45 sec
Discoverer 9	None	USAF	Feb 4, 1960	WTR	Thor-Agena A	1700	—	—	—	—	Failed to orbit: premature first stage cutoff
Discoverer 10	None	USAF	Feb 19, 1960	WTR	Thor-Agena A	1700	—	—	—	—	Failed to orbit: destroyed by range safety officer
Midas 1	None	USAF	Feb 26, 1960	ETR	Atlas-Agena A	4500	—	—	—	—	Failed to orbit: second stage failed to separate
Pioneer 5	1960 A1	NASA	May 11, 1960	ETR	Thor-Able	95	312 days	.8061AU	.995AU	3.35	In solar orbit: solar system data to 22.5M mi until 6-26-60
Explorer S-46	None	NASA	Mar 23, 1960	ETR	Juno II	35	—	—	—	—	Failed to orbit: apparent upper stage ignition malfunction
Tiros 1	1960 B2	NASA	Apr 1, 1960	ETR	Thor-Able	263	99.2	430	468	48.3	In orbit: first retest, sent 22,952 photos up to 6-17-60
Transit 1B	1960 Γ2	ARPA	Apr 13, 1960	ETR	Thor Able Star	265	95.8	232	463	51.3	Decayed 10-5-67: initial navsat, transmitted until 7-12-60
Discoverer 11	1960 Δ1	USAF	Apr 15, 1960	WTR	Thor-Agena A	1700	92.3	103	375	80.1	Decayed 4-26-60: capsule ejected orbit 17, recovery failed
Echo A-10	None	NASA	May 13, 1960	ETR	Delta	132	—	—	—	—	Failed to orbit: second stage attitude control malfunction
Sputnik 4	1960 E1	USSR	May 15, 1960	Tyuratam	A-1	10,009	91.3	194	229	65.0	Decayed 9-5-62: Vostok prototype, recovery failed 5-19-60 as cabin went higher orbit, cabin decayed 10-15-65
Midas 2	1960 Z1	USAF	May 24, 1960	ETR	Atlas-Agena A	5000	94.4	299	321	33.0	In orbit: data link quit 2nd day
Transit 2A	1960 H1	USN	June 22, 1960	ETR	Thor Able Star	223	101.7	385	665	66.7	In orbit: navmad navigation, geodetic data until 8-62
Solrad 1	1960 H2		June 22, 1960	ETR		42	101.5	382	657	66.8	In orbit: first sub-satellite, returned solar data until 4-61
Discoverer 12	None	USAF	June 29, 1960	WTR	Thor-Agena A	1700	—	—	—	—	Failed to orbit: second stage attitude instability
Discoverer 13	1960 Θ1	USAF	Aug 10, 1960	WTR	Thor-Agena A	1700	94.1	157	431	82.8	Decayed 11-14-60: first recovery, from ocean on orbit 17
Echo 1	1960 I1	NASA	Aug 12, 1960	ETR	Delta	186	118.2	941	1052	47.2	Decayed 5-24-68: first passive comsat, relayed voice, TV signals
Discoverer 14	1960 K1	USAF	Aug 18, 1960	WTR	Thor-Agena A	1700	94.5	113	502	79.6	Decayed 9-16-60: first mid-air capsule recovery, on orbit 17

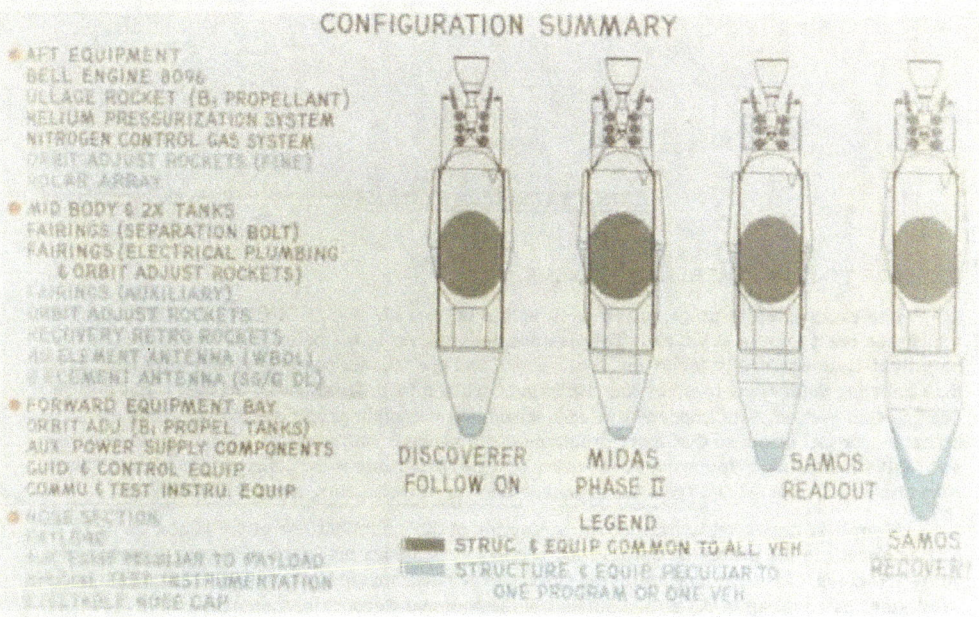

Fig. 1-2a — Standardized Agena as viewed in 1959

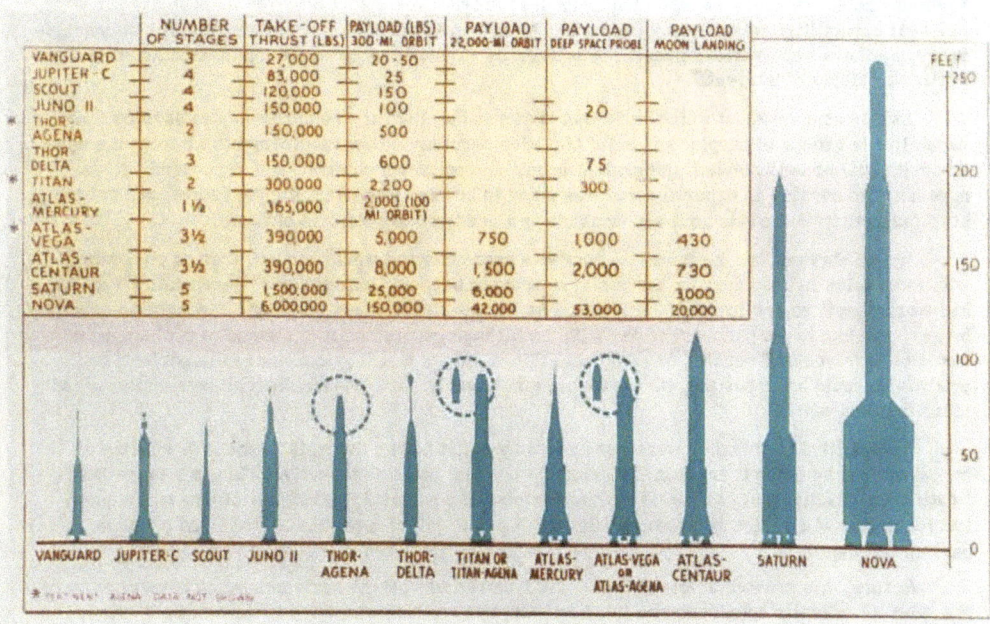

Fig. 1-2b — U.S. space vehicles as viewed in 1959

EARLY SATELLITE SYSTEMS

PREFACE TO EARLY SATELLITE SYSTEMS

The early satellite programs covered in this section (SAMOS, CORONA, GAMBIT, etc.) are addressed one program at a time. This is done deliberately in the belief that this approach will help first-time readers to follow the development and operational periods of each program, rather than bringing them along together and switching back and forth from one to another. But before reading this section, it is important to understand that the early programs did not come along in a neat sequential order so that their problems could be dealt with one at a time. As the charts and tables immediately following show, many of these programs were going on concurrently and were therefore interrelated from management aspects of technology, security, and budget.

Also, it is important to have some perception of the complications of the times we are talking about. In the preceding background section, it was seen that gradually over a period of years progress was made in simplifying Government management for the satellite reconnaissance programs, culminating in the establishment of the National Reconnaissance office.

Yet, this "ideal" arrangement did not mean that it would be easy to fulfill two prime objectives: first (with some programs already in development) to determine the kinds of reconnaissance systems the nation really needed, and second, to provide some sort of cover or security for what our real capabilities were. Many people had by now seen the potential of satellites for reconnaissance, so there were various pressures brought by various groups to get business started in their particular fields of interest.

In this science about which little was known at the time, it was necessary to develop a new capability in groups of people, and build facilities peculiar to new requirements wherein the new devices could be built, tested, integrated, launched, controlled on orbit, and recovered. It was necessary to develop an expertise that was going to be very important for the future, and yet to keep the activity in bounds so there wouldn't be a needless expenditure of funds.

It was obvious that the potential here was enormous and would require trained personnel, with motivation to keep them in the business so as to become "experienced" personnel. If all we had were people on one program, that would be inadequate, because the programs were to become bigger. We had to get people familiar with the technology, trained in the kinds of problems that would be involved and working on some real things, things that they considered important, and potentially could be important, in other words that could produce technological developments and ideas that had merit.

Competitive approaches were seen as a way to get better products in these new fields of endeavor, and by doing it early in the game it would not be all that costly. This had as its justification the development of experience, the training of personnel, testing out of new techniques and ideas, but without the requirement that it all had to jell at a particular time and produce a particular result.

In turn, this created a very difficult and frustrating working environment. There were very few who were really cognizant of a total activity, and persons who were involved in one particular facet of the business were in many cases completely unaware that there was another activity proceeding along similar lines. Thus, when decisions were made to modify, stretch out, or, in the extreme case, cancel a program that was showing great potential, it appeared to the uninformed that management was making irrational decisions. Such cases will be seen in this section, e.g., the LANYARD (KH-6) was a backup to the GAMBIT (KH-7) and was cancelled when the KH-7 performed satisfactorily on its first mission, and in another case, when the SAMOS E-6 program could no longer provide cover for the more desirable GAMBIT (KH-7) because of configuration differences, the E-6 became the pawn.

But competition was not always the most productive approach, particularly when it was in the form of interservice rivalry. You will read about two early mapping cameras which came about as the result of Army and Air Force arguments over roles and missions. These systems, the Army ARGON (KH-5) and the Air Force SAMOS (E-4), starting about 2 years before the NRO was established, took on some of the interservice complications in their evolutions. Had there been some sort of a national set of objectives in position earlier, a program could have been created that would have been better and would have provided both the Army and Air Force a more efficient way than the development of a competitive approach.

In the following discussions of early satellite programs, the intermix of program names and KH numbers, together with the overlapping time frames, would most certainly be difficult for first-time readers to follow without some measure of background and reference. To that end, a summation of pertinent program/system data is provided at this point for orientation and reference (see Tables 1-2 and 1-3). Also, immediately following, is a graphic study of the learning curve associated with these programs (refer to Figs. 1-3a and 1-3b). The ultimate goal, of course, was total operational success, mission after mission, but total success was not to come instantaneously. While one community was focusing on the high priority payloads, other circles were concentrating equally on other subsystems and functions which were also vital to program success, e.g., airframe, propulsion, vehicle internal electrical systems, guidance and flight controls, ground-space communications, and ground systems performance (including the launch complex, tracking stations, telemetry ship, Hawaiian Control Center and Recovery Force). In the early phases, systems were launched frequently and missions were short (1 to 4 days). It was the opinion of the BMD Commander, Brig. General Osmund J. Ritland, that the way to achieve early success in the DISCOVERER Program was through frequent flights. He believed that the real payoff in learning "what to do" would be proven in flight tests, and that in this initial phase of developing a space capability, it was better to suffer some embarrassments on end results than to have a stretched out "no action" program. As these developmental flights were in process, however, he was deeply concerned that more publicity was being given to failures than to the successful "firsts" being achieved, such as: placing in orbit an earth oriented and stabilized 1,700-pound satellite; achieving a maneuverable satellite using airborne programs and even ground-commands in orbit; and setting up and operating a world-wide communications and control organization, including recovery task forces, where no previous programs had even plowed the ground. But catastrophic failures became less and less frequent, and virtually disappeared as experience in space technology grew, as shown in Fig. 1-3b. From technological advances in both hardware and software, missions were stretched longer and longer, thus fewer launches were required each year to acquire the desired coverage.

For those who would like to go beyond the brief accounts of the early satellite systems presented in this document, two sets of reports are highly recommended: "A History of Satellite Reconnaissance," Volumes I through IIIB prepared by Robert Perry under direction of the NRO; and "CORONA Program History, Volume I, Program Overview," produced by the Directorate of Science and Technology, Central Intelligence Agency. These historical documents were used extensively in preparing the program summaries that follow.

TOP SECRET/RUFF/GAMBIT/HEXAGON

Table 1-2 — Satellite Photographic Systems—Overview

Reconnaissance Systems	Type		Sponsors			Security		Development/Operational Periods
	Readout	Film	USAF	CIA	U.S. Army*	Covert	DoD	1956–1985
SAMOS (E-1, 2, 3, 4, 5, 6)	E1, 2, 3	E4, 5, 6	X				X	
CORONA (KH-1, 2, 3, 4, 4A, 4B, and 6)		X	X	X		X		
ARGON (KH-5)		X			X	X		
GAMBIT (KH-7)		X	X			X		
GAMBIT³ (KH-8)		X	X			X		
HEXAGON Program (KH-9)		X	X	X		X		
Mapping Camera System		X	X		X	X		
MOL/DORIAN (KH-10)		X	X			X		
KH-11	X			X		X		
Defense Weather Satellites								
Program 417/DSAP/DMSP		X					X (Special Access)	

*Army Mapping Service prior to 21 November 1962. DIA MC&G Directorate 21 November 1962 to 1 July 1972. Defense Mapping Agency (DMA) 1 July 1972 to present.

Legend: Development, Operational, Projected

TOP SECRET/RUFF/GAMBIT/HEXAGON

Table 1-3 — Early Satellite "Photographic" Payloads

Program/System Designators	K/H Number	Mission Series	Development/ Operational	Camera Types Primary	Camera Types Auxiliary	Focal Length	Operating Altitude, nm	Resolution (avg GRD), ft	Film Size	Film Supply, ft	Launches Attempted	SRV's Recovered
SAMOS												
E-1	N/A	2100	1956-Feb 1961	Readout		6-in.	280	100	70-mm	1,800	3	N/A
E-2	N/A	2100	1958-Sept 1961	Readout		36-in.	100-120	20 est.	70-mm	4,500	1	N/A
E-5 (1A)	N/A	None launched	1958-Jan 1962	Film recovery, frame (mapping)		6-in. terrain (3-in. stellar)	90	150 est.	9.5-in.	4,000	0	0
E-3	N/A	2200	1959-Mar 1962	Film recovery, panoramic		66-in.	155	5-8 ft est.	5-in.	205-500	3	0
E-6 (666 BJ; Program II, 201, 723)	N/A	2400	1961-Jan 1963	Film recovery, twin panoramic		36-in. (two)	100-125	10 est.	6.6-in. Two Strands	6,512 6,613	5	0
CORONA												
C	KH-1	9000	1959-Sept 1960	Film recovery, Single panoramic		24-in.	110	50	70-mm	3,000	10	1
C'	KH-2	9000	1960-Nov 1961	Film recovery, Single panoramic		24-in.	110	35-40	70-mm	7,000	10	5
C'''	KH-3	9000	1961-Jan 1963	Film recovery, Single panoramic		24-in.	110	25	70-mm	7,900	8	4
MURAL	KH-4	9000	1961-Dec 1963	Film recovery, twin panoramic		24-in.	95-100	30	70-mm	15,800	26	20
			1962-Sept 1963		Stellar/Index (S/I) (one)	38-mm terrain 80-mm stellar		200 est. N/A	75-mm 35-mm	150 75		
			1963-Oct 1969	Film recovery, twin panoramic, dual SRV		24-in.	95-100	8-10	70-mm	32,000		
JANUS (J-1)	KH-4A	1000	1963-Sept 1969		Stellar/Index (S/I) (two)	38-mm terrain 80-mm stellar		200 est. N/A	70-mm 35-mm	500 250	52	52

Table 1-3 — Early Satellite "Photographic" Payloads (Cont.)

Program/System Designators	KH Number	Mission Series	Development/ Operational	Camera Types Primary	Camera Types Auxiliary	Focal Length	Operating Altitude, nmi	Resolution (avg GRD), ft	Film Size	Film Supply, ft	Launches Attempted	SRV's Recovered
CORONA (cont.) JANUS (J-3) Constant Rotator (CR)	KH-4B	1000	1965–June 1972	Film recovery, twin panoramic, dual SRV		24-in. (two)	85–100	6–6	70-mm	32,000	17	11
ARGON (A) (DAFF)	KH-5	9002	1965–Sept 1971		Stellar/Index DISIC (one) [2]	3-in. terrain (two) 3-in. stellar (one)		125	3.5-in. terrain 35-mm stellar	3,000 2,000		
			1960–Aug 1964	Film recovery, Army Mapping camera		3-in. terrain 3-in. stellar (one)	185	250 sat. N/A	5-in. terrain 35-mm stellar		12	6
LANYARD (L)	KH-6	8000	1963–Aug 1963	Film recovery, high resolution panoramic		66-in.	100	8	9-in.	8,000	3	2
PROGRAM 35; 417	N/A	5000	1961–Oct 1962	Readout, weather reconnaissance satellite [3]	Stellar/Index (S/I)	36-mm terrain 80-mm stellar	450	1.3 nautical miles	70-mm 80-mm	500 150	5	N/A
GAMBIT (G) (Program 206; Coe Ball)	KH-7	4000	1960–June 1967	Film recovery, high resolution strip		77-in.	85–95	1.5	9.5-in.	3,000		
			1963–June 1967		Stellar/Index (S/I) (one) [7]	36-mm terrain 50-mm stellar		200 sat. N/A	70-mm 35-mm	500 150	31	34

Notes:

1. The initial frame cameras were "Index" (down-looking) only. These cameras, flown first on mission 9031, 27 Feb 1961, were used until the Stellar/Index (S/I) cameras were deployed.
2. DISIC was the acronym for Dual Improved Stellar Index Camera.
3. Two SRV's each launch.
4. Progenitor of DMSP (Defense Meteorological Satellite Program).
5. Two satellite reentry vehicles effective mission no. 23 and subsequent.
6. Stellar/Index (S/I) camera effective mission no. 7 and subsequent. The S/I's flown initially on the G program were manufactured by Itek. Later, KE produced their own S/I, this camera was known as the APTC (Astro Positioning Terrain Camera). Of the payloads flown between 1966 and 1973, 41 of the APTC design were flown as integral subsystems on GAMBIT payloads.

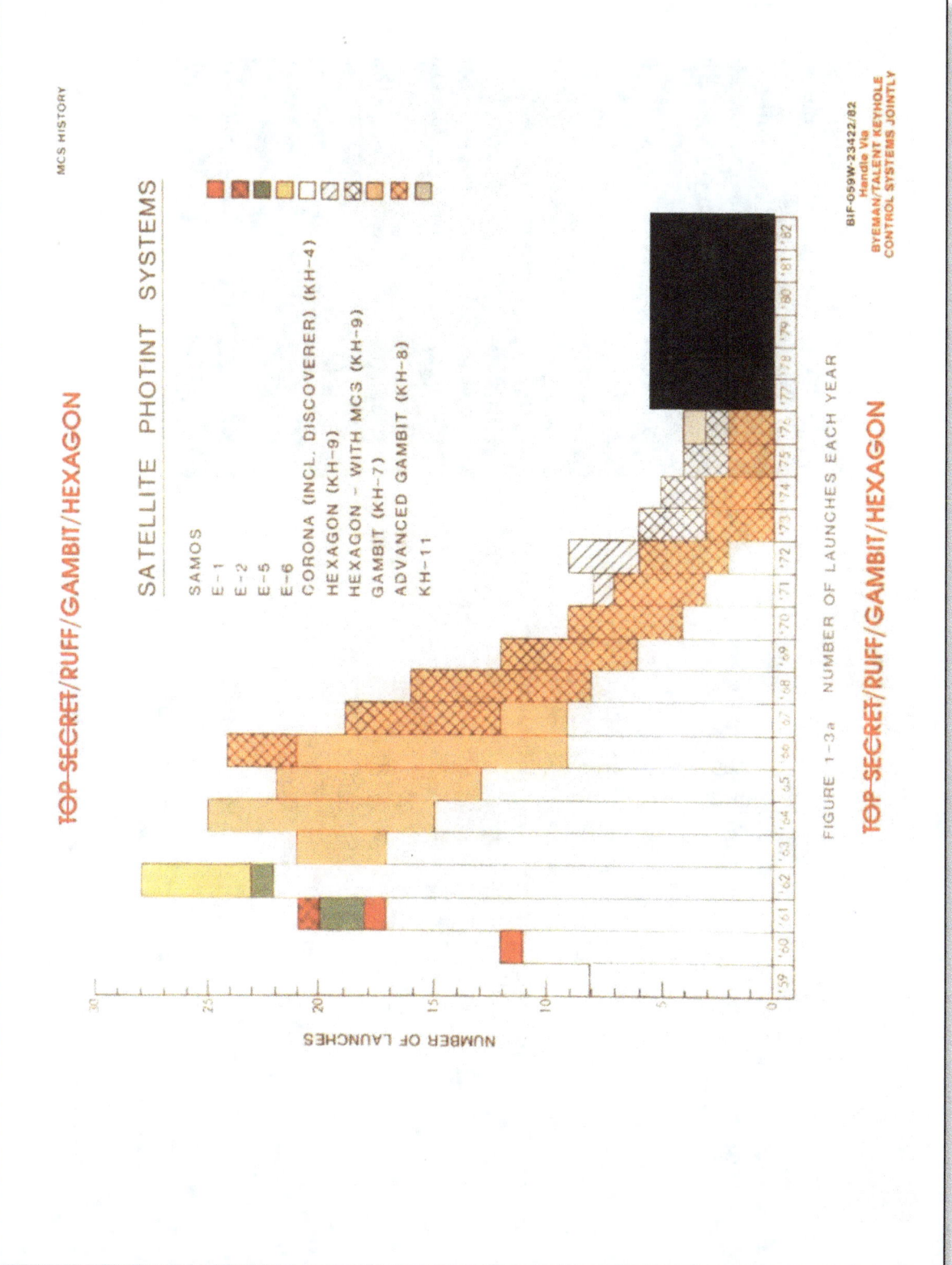

FIGURE 1-3a NUMBER OF LAUNCHES EACH YEAR

FIGURE 1-3b ORBIT DURATION OF PHOTINT SYSTEMS

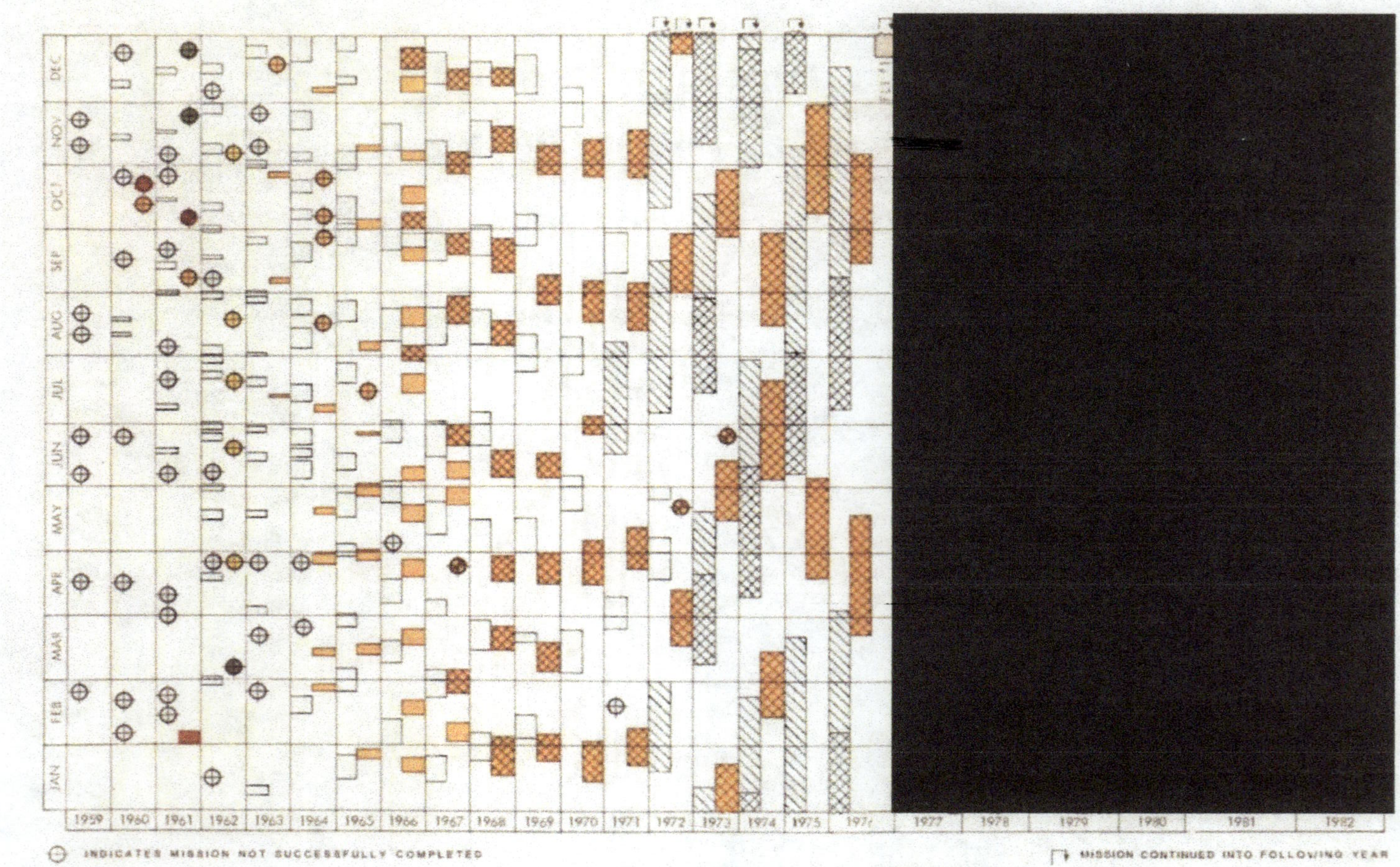

MCS HISTORY

THE SAMOS PROGRAMS

As late as March 1958, WS-117L embodied concepts of a "pioneer" system built around a 6-inch (focal length) lens, and an "advanced visual" system embodying a 36-inch lens as the basic data gathering devices. Both infrared and electronic collectors were being considered by that time, but the chief emphasis remained with visual modes.[9]

The pioneer of the visual reconnaissance "E"* subsystems (Fig. 1-4a) in the WS-117L family (Fig. 1-4b) was intended to provide in-camera definition approaching 100 lines per millimeter, based on an f/2.8 lens in combination with a very fine grain film. Orbital operation was predicated on the assumption that the camera system would function for 5 minutes during each pass over the "area of interest" and that on subsequent orbits three receiving stations within the continental United States would "read out" the intelligence thus acquired. (The stations were to be located at Fort Stevens, Oregon; Ottumwa, Iowa; and New Boston, New Hampshire. Offutt Air Force Base was to be the satellite operations control center.) It seemed probable that an efficient processing and dissemination complex would permit at least 10 percent of the derived intelligence to reach the central analysis station within 1 hour of its receipt and the remainder within 8 hours. The Strategic Air Command wanted an eventual "near real time" system, of course, hoping to use it for attack warning as well as general intelligence. Each of several vehicles to be aloft simultaneously was to have a useful time on orbit of 10 to 30 days, limited principally by battery life. The initial system (E-1) (Fig. 1-5) was designed to permit identification of ground objects measuring 100 feet on a side. The "advanced" E-2 (Figs. 1-6a and b) was to produce images that would permit "visual resolution" of objects 20 feet on a side and was to have a potentially long orbital operating life—assuming the availability of either solar or nuclear power sources. Functions and characteristics of the E-2 system were:

Functions:
1. Photographs ground from 300-mile altitude with payload camera
2. Processes photographic film in processor
3. Converts information in photographic image to electrical signal in readout equipment
4. Provides data to telemetry link
5. Provides a pressurized housing with trunnions for oblique and stereo operation

Characteristics:

Ground resolution	20 feet
Ground width coverage	17 miles
Maximum obliquity angle	26 degrees
Stereo convergence angle	34 degrees
Ground coverage per day	55,000 square miles
Film consumption per month	10 pounds
Operating life	4 months
Readout bandwidth	6 megacycles/sec
Weight	950 pounds
Length	67 inches
Base diameter	55 inches
Average power consumption	50 watts

---

*The letter designator assigned individual WS-117L subsystems had the following basis: Subsystem A—airframe; B—propulsion; C—auxiliary power; D—guidance and control; E—visual reconnaissance; F—electromagnetic reconnaissance (Ferret); G—infrared reconnaissance (later—Midas); H—communications; I—data processing; J—geophysical environment; K—personnel; L—biomedical recovery. The E designators ultimately ran from E-1 through E-6; the F designators through F-4.

One key to a useful readout system was a raw data processing subsystem which would include the equipment, techniques, and procedures to transform recorded raw data into intelligence, and to disseminate it to using agencies. Ground receiving stations, therefore, would identify, record, and retransmit information to an "Advanced Reconnaissance System Intelligence Center" (predictably dubbed "ARSIC"). The Intelligence Data Processing Subsystem ("IDPS"—later Subsystem I) was to be capable of performing all functions needed to transform the raw data into useful intelligence: processing, screening, interpretation, collation, evaluation, indexing, storage and retrieval, analysis, display, dissemination, and presentation.

The orbital vehicle (the upper stage and payload sections) was to be 19 feet long and 5 feet in diameter, was to carry a 2,680-pound payload and, including 5,080 pounds of propellants, would weigh 9,300 pounds at launch. The somewhat loosely defined operational concept of March 1958 anticipated that ultimately each of several E-2 satellites simultaneously on orbit would have a useful life of one year and be capable of providing 17-foot ground resolution.

Spot surveillance of selected targets rather than general reconnaissance was the objective of the development program. Surveillance of this nature was intended to provide advance warning of an imminent attack, a concept emphasized by application of the name SENTRY to WS-117L in June 1958. Unhappily, concept had little relevance to reality. Although a camera and readout system that could actually resolve 17 feet on each side would be capable of locating and identifying intercontinental missile sites, the total satellite system was incapable of such precision. Moreover, within the existing state of the art, the capacity of the system to scan and transmit images to ground stations was severely limited.

Even though electronic transmission of photographs to ground receiver degraded definition, the chief objection to readout was that relatively little coverage could be provided each day.

The readout technique that had evolved by 1958, and which was refined but not radically changed during the next two years, embraced a strip camera subsystem loaded with 4,500 feet of 70-millimeter film. The film moved past a slit aperture which served as a shutter, at a rate determined by image motion compensation settings. (The "slit" was actually a line scribed through the aluminum coating on a glass plate.)

Once exposed, the film was pressed against a chemically impregnated web at intervals over a period of approximately 16 minutes. The pre-soaked web contained all the necessary developing and fixing ingredients. After completing the processing stage, the developed film went to a storage section—a series of loops which held it in readiness for later scanning and transmission.

The readout mechanism consisted of a revolving drum line scan tube, a scanner lens system, a light collector lens system, a photomultiplier tube, and a video amplifier. An electron beam which focused on the phosphor-coated inner surface of the revolving drum was emitted through an optically flat window, the light beam going through a scanning lens that was moved vertically by a motor-driven cam. The lens moved the spot of light across the width of the processed film as the film moved laterally through a readout gate. The beam motion had the shape of a square wave, permitting continuous top-to-bottom travel rather than returning to a zero point for each scan operation. That portion of the beam which passed through the film was collected by another lens system capable of relaying 75 percent of the transmitted light to a photomultiplier tube which transformed the light energy into electronic signals. After passage through a video amplifier, those signals were relayed to the satellite's communication equipment section for transmission to the ground stations (see Fig. 1-7).

Image motion compensation, exposure control, and focus factors were set by command from a ground station. Attitude recording, a key factor for interpretation, was provided through inscription of a binary code on the edges of the film.

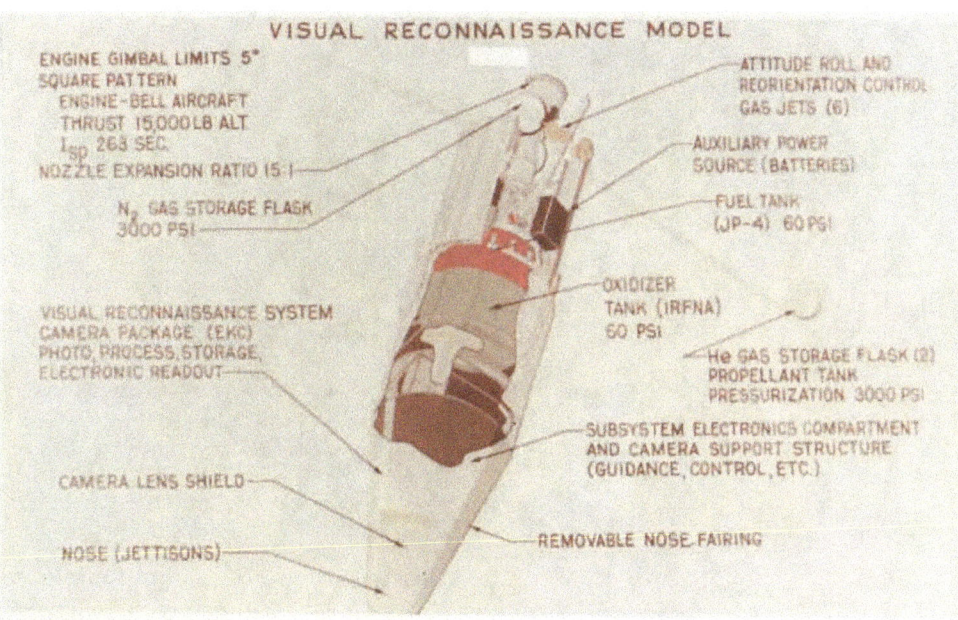

Fig. 1-4a — WS-117L Vehicle System

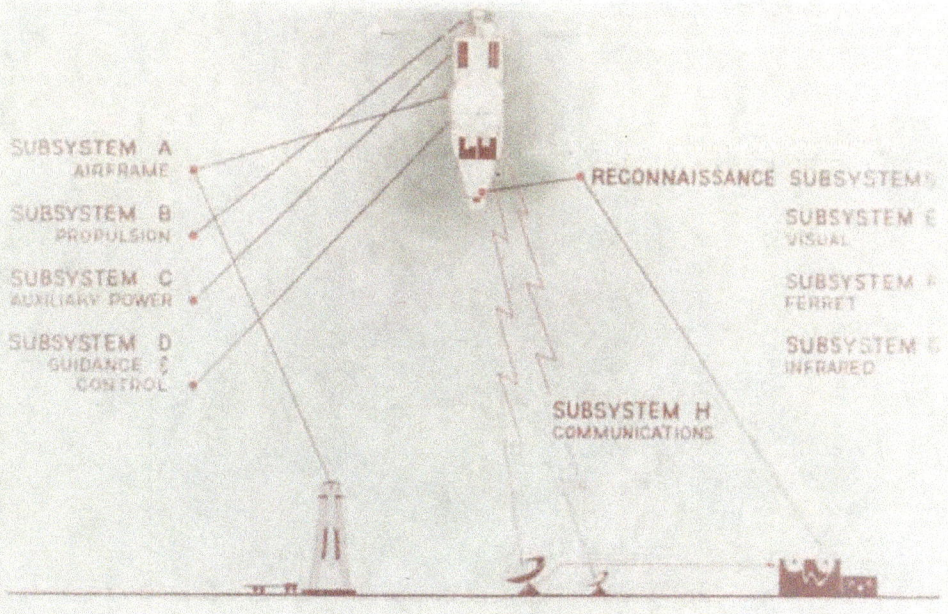

Fig. 1-4b — WS-117L Subsystems

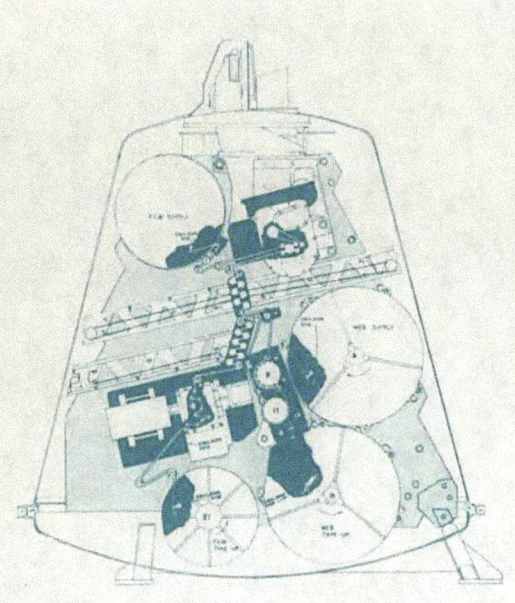

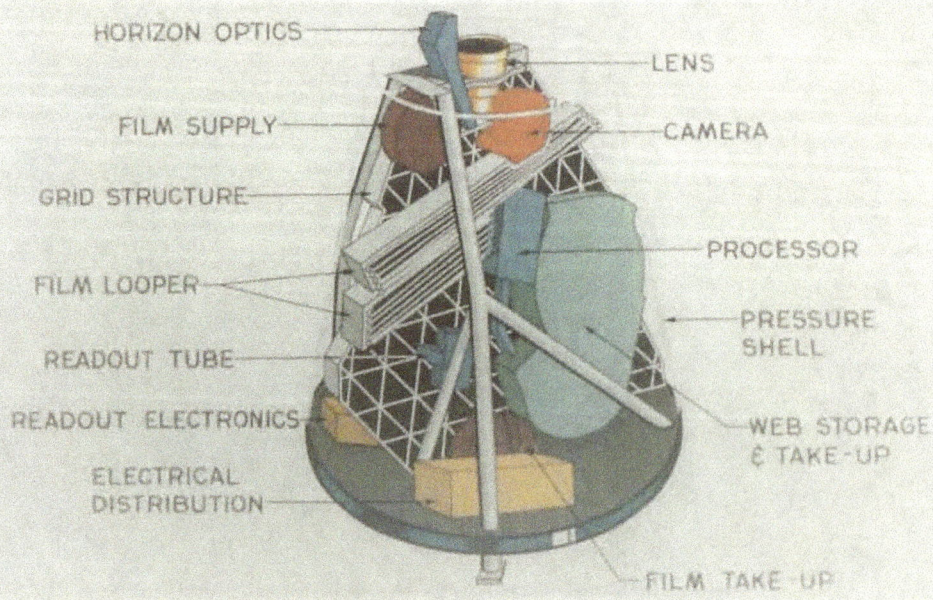

Fig. 1-5 — E-1 Visual Reconnaissance Component Test Model

TOP SECRET/RUFF/GAMBIT/HEXAGON

BIF-059W-23422/82
Handle Via
BYEMAN/TALENT KEYHOLE
CONTROL SYSTEMS JOINTLY

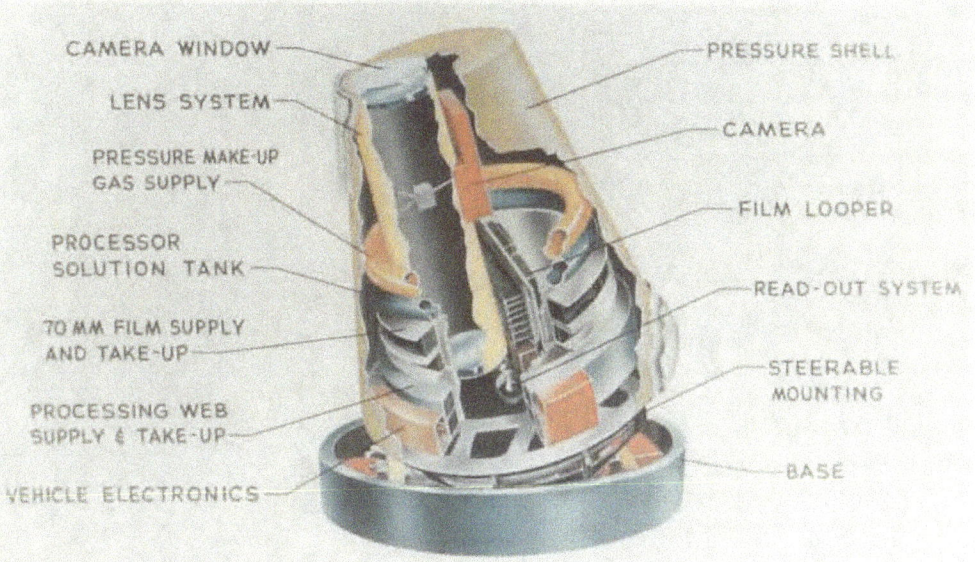

Focal length	36 inches
Film width	70 mm
Picture width	2 inches
Limiting resolution	250 lines/mm
Aperture ratio	f/4
Aperture diameter	9 inches
Exposure time nominal	1/100 second
Spectral range	0.500 to 0.720 micrometers
Slit width	0.0027 to 0.085 inch
Weight	83 pounds

Fig. 1-6 — E-2 Visual Reconnaissance Payload

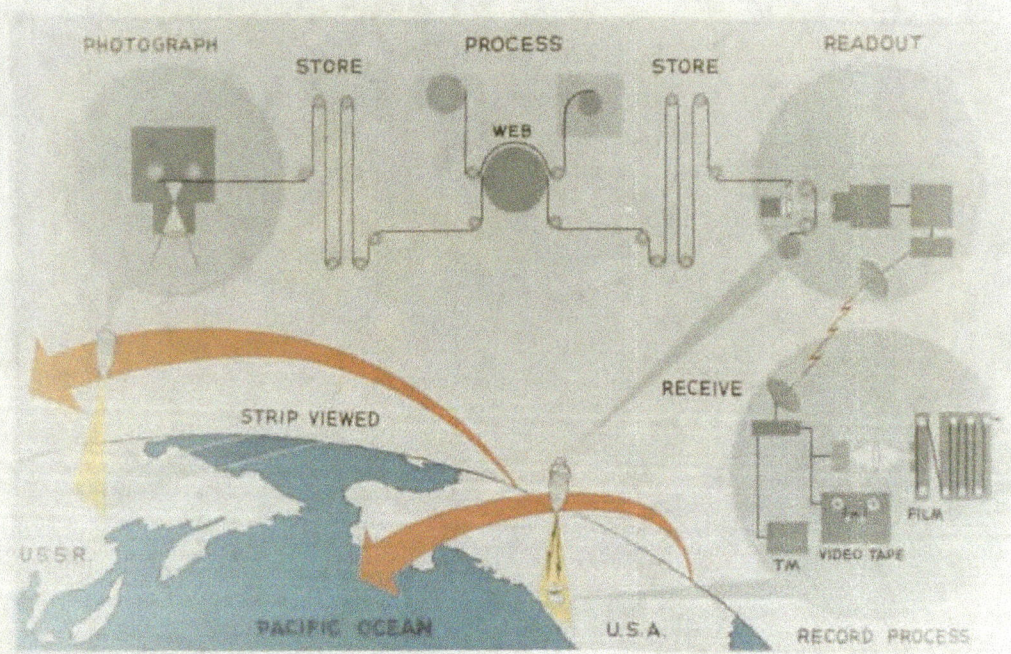

Fig. 1-7 — E-2 payload-functional schematic

The process, though complicated, could be performed by existing or available techniques and equipments. Limiting technical factors were the speed and width of the scanning beam, governed by bandwidth (megacycles per second) considerations. Unless traveling wave tubes could be incorporated in the system—and nothing suitable was available either in 1958 or three years later—the usable bandwidth was but 6 megacycles per second. Even though E-1 and E-2 systems were designed to limit their coding to white, black, and one gray scale, the scanning beam could travel across the width of film only once each second. The beam spanned only one-tenth of an inch of film during each transit. A complete scanner-beam pass—bottom-to-top-to-bottom—required 2 seconds. The transmission, readout, and reconstruction process transformed the signals from each such path into an 18-inch strip of 35-millimeter film in the ground station. Seven such strips, halved and realigned to conform to the pattern of the original film, could be reassembled into a single print measuring 9 inches along each edge.

Long before flight trials could be attempted, the limitations of the readout technique were well appreciated. On the assumption that a ground station could receive fully useful information for 8 minutes during each of five daily passes of the satellite within its reception range, it was apparent that each station could accept no more than 62 individual frames representing 16,740 square miles of target area each day. (An early CORONA system could scan 1.5 million square miles each day.)

Such considerations influenced the transformation of basic requirements in the period between March and September 1958. By that later date, Air Force headquarters had clearly indicated its desire that "consideration" be given to the use of a recoverable satellite in order to achieve maximum accuracy, information content, reliability of receipt of collected data, and reuse where economically feasible. Nevertheless, the stated objectives of the program were focused on early warning of attack, the collection of general intelligence, and support of the nation's emergency war plan. The use of satellite reconnaissance to assist in determining the war potential of the Soviet Union remained a secondary goal. Readout, even with its acknowledged limitations, still seemed the best means of satisfying the requirements.

Administration of SENTRY through the closing months of 1958 was complicated by the fact that the Advanced Research Projects Agency (ARPA) had custody of program funds and exercised a directive authority over the technical content of the effort. Although ARPA Director R. W. Johnson in mid-December 1958 had approved a new three-phase approach that included film recovery as well as Ferret and readout payloads, the research agency continued to press for the inclusion of an electrostatic tape readout system (later the E-2). Indeed, ARPA came to advocate cancellation of all other visual programs in favor of reliance on electrostatic tape methods.

Lockheed stated the case for an E-3 in a proposal dated 29 July 1959. Apparently proceeding on the reasonable premise that ARPA's predilection for readout would prevail over Air Force preferences, Lockheed painted the theoretical advantages of the E-3 in highly attractive colors. The contractor noted that E-2 technology was based on pre-1959 concepts and that the "recent addition" of a requirement for 5-foot visual reconnaissance had prompted attention to state-of-the-art improvements. In Lockheed's opinion (at least, in its 29 July 1959 opinion!), "an all-electronic approach would provide the highest possible performance in the earliest time period at minimum cost." Noting that the "technical feasibility" of electronic tape systems had been proven under Aeronautical Research Laboratory (Wright Air Development Center) contracts, Lockheed cited the availability of 100-line-per-millimeter definition (12,200 television lines for a 61-millimeter-square format!) and equivalent sensitivity of ASA 145 [standard reconnaissance film had an aerial exposure index (AEI) of 2 to 5], and a readout system substantially more simple than that of the E-2. The image was to be recorded on photoelectric-sensitive electrostatic tape, read out by deflecting the modulation of an electron beam to scan a portion of the tape, and the view signal amplified and then applied as a modulating signal for transmission to ground stations. A bandwidth of 12 megacycles per second was required (tubes had to be developed also) to provide a readout time of 8.7 seconds per frame. In such terms, ARPA's interest was entirely understandable.

Following further deliberations over requirements and funding, resulting in, among other things, the approval of the E-5 recovery system, the Air Force Ballistic Missile Committee took a hand, instructing BMD to submit a program that emphasized photography rather than Ferret subsystems and which clearly concentrated on recovery rather than readout data retrieval methods. One immediate consequence of the redirection was to eliminate some effort covered by existing contractors. Included in the termination package that BMD and Lockheed worked out early in December were all of the very advanced readout programs—E-3 and F-4 (a Ferret package development comparable to the E-3 in complexity and technical uncertainty).

When the SAMOS project finally became the direct responsibility of the Secretary of the Air Force, it included three photographic subsystems and one ground based subsystem that stemmed directly from the original WS-117L program. Others were pending approval, but only the E-1 and E-2 readout systems, Ferrets F-1 and F-2, and the E-5 recovery system were funded and in a hardware stage. Associated with the airborne was the ground-site complex of receiving, processing, storage, and dissemination equipment that was known as Subsystem I (eye).

As it happened, the period of SAMOS reorientation during the spring and summer of 1960 coincided with the climax of E-1 development, which had begun in 1956. Eastman delivered the first camera payload to Lockheed on 15 April and Lockheed completed its system test of the Agena plus payload on 3 June. The payload included both an E-1 and F-1 Ferret subsystem. By September, the Agena-payload complex was mated with an Atlas booster at Point Arguello, and on 11 October it was launched amid considerable fanfare that included elaborate press conferences and a large audience of cameramen.*

The launching went well enough to please photographers, but program people were less than happy. The umbilical connection failed to release at launch and the hefty push of the Atlas booster tore away the nitrogen fill line—complete with couplings to the Agena—when the hoses reached their physical stretch limits. Although the Atlas operated perfectly and the separation of the Agena from the first-stage booster occurred as programmed, nitrogen had been venting freely into the atmosphere through the entire boost period and the tanks were, for practical purposes, empty. Attitude stabilization depended on gas—and there was no gas. The Agena's engines ignited while the vehicle was improperly aligned for injection into orbit—and the flight was over. Investigation revealed that test base personnel had failed to install a half-inch assembly that should have joined the umbilical to the quick-disconnect fittings.

Between January and August 1960, redirection of the SAMOS program had caused elimination of five of the original eleven scheduled readout flights; only three E-1/F-1 and three E-2 payloads had been authorized. One of these was eliminated in early November, leaving only five readout payloads in the launch schedule.

---

*A few days prior to the scheduled launch date General Greer had taken the precaution to assemble a small group of his immediate staff—management, security, and technical—for a pseudo press question and answer session.

During the period between the first and second E-1 launches, Subsystem I was cancelled. Large among the reasons for ending the development was cost, the technical difficulty of development having been sadly underrated. For an input of nearly ▓▓▓▓▓ the Air Force had obtained a semi-obsolescent lot of partially developed equipment tied to an abandoned concept. In the final analysis, Subsystem I had been designed to satisfy the pre-1960 requirements for early attack warning; the shift of emphasis from surveillance to reconnaissance in July 1960 had doomed the development, although full appreciation of that circumstance was not widespread.

Although there could be heard strong recommendations to cancel all readout programs at this point and place full emphasis on recovery, empirical evidence remained the best basis for a judgment on program validity and, in the absence of flight data on which to base a finding, the cancellation of the E-1 program might prove difficult to defend.

By early January the second vehicle was on the stand and had been checked out. At that point a new complication arose. Negotiations with the Soviets for release of two imprisoned members of an RB-47 crew were approaching a climax; the newly installed Kennedy administration was extremely anxious to establish an early record of diplomatic achievement by getting agreement to the crew's return. In mid-January, with the E-1 on the pad and ready, General Greer received urgent encrypted instructions to delay the scheduled launch by some plausible subterfuge until the freed crew could actually be returned to American custody.

Following release of the Air Force crewmen, the second SAMOS vehicle (2102) was launched, on 31 January 1961. Like the first, in October, it carried a composite E-1/F-1 payload. This time the orbital vehicle was placed into a stable orbit having a period of 95.2 minutes, perigee of 260 miles, an apogee of 311.6 miles, and a nominal life expectancy of 1,130 days. The vehicle successfully relayed information to the readout station ▓▓▓▓▓

On 3 February, Colonel William G. King, SAMOS Program Manager at SAFSP, took the first analysis of flight results to Dr. J. V. Charyk, Undersecretary of the Air Force. An assembled photograph was available which indicated that the ground resolution of the system was roughly what had been anticipated, about 100 feet. Although the hand-processed pictures were relatively good in terms of original system requirements, the system itself did not promise much in the way of eventual utility. There seemed little justification for altering the premise of the previous 8 months—that the E-1 would be tested to prove out the feasibility of the in-flight processing, transmitting, and readout equipment. Charyk agreed that the relative success of the 2102 vehicle was sufficient proof of E-1 system feasibility; he approved General Greer's recommendation that the third of the programmed E-1 flights be cancelled and the equipment stored for some possible future application.* They also agreed to let the scheduled E-2 flights remain in the program for the moment, although again it was apparent that once a set of returns had been received there would be no real justification for further continuance of the E-2 program.

---

*There was one additional, almost afterthought, aspect to the E-1 program. In April 1961, representatives of the National Aeronautics and Space Administration (NASA) contacted Dr. Charyk's office to ask permission to examine and use E-1 technology in their own programs. It seemed possible for a time that the physical products of the E-1 development might actually find their way into a moon vehicle and, later, actually did in NASA's Lunar Orbiter. One stimulant was the obvious parallel between E-1 equipment and techniques and the devices used by the Soviets to photograph the back surface of the moon in October 1959.

On 19 April, the third E-2 vehicle-payload was cancelled. Another significant change came in July when a succession of payload, tracking net, and booster difficulties forced postponement of the scheduled launch of the first E-2.

On 9 September 1961 the initial attempt to orbit an E-2 payload ended in an awesome launch pad explosion. Loss of electrical power caused the Atlas to drop back on the pad less than 2 seconds after lift off. The E-2 payload was destroyed in the resulting blast and fire. (The Atlas failure was caused by a delay of 0.2 second in disconnecting the umbilical that carried the signal from external to internal electric power.)

The remaining E-2 flight test vehicle (2121) faced a problem of crowded launch pad schedules. After weighing the prospect of a major malfunction and the clear evidence that basic subsystem performance had been adequately demonstrated in the single E-1 flight, Charyk and Greer decided not to launch the second E-2 vehicle. On 30 September the contractor was instructed to remove it from flight readiness processing and put it in bonded storage. For all practical purposes, such action concluded the original readout oriented SAMOS program.

The E-4 Mapping Satellite (Program 1A)

> Note: It was disappointing that a record search did not produce any photographs or artist concepts of this system.

Development of a mapping and charting satellite had been a cherished Air Force dream for at least two years before SAFSP inherited the mantle of satellite reconnaissance responsibility—but progress had remained in the dream category![19] The requirement had been defined in September 1958, although considered abstractly even earlier. By the following January the notion had been translated into a proposal for a recoverable capsule system capable of taking pictures with high geometric fidelity and correlating them with the products of a stellar image recording camera. Called E-4, the proposed system was considered a companion to the E-5 surveillance system then being defined. Although the Ballistic Missiles Division and the Air Research and Development Command heartily favored starting development, even seeming to prefer the E-4 to the E-5, the Air Staff was never more than lukewarm. In part because highly influential intelligence officers withheld firm support, the E-4 took shape as a somewhat tenuous development which was in direct competition with a proposed ARPA-sponsored interim mapping system and with the ARGON system being covertly developed under Army auspices. In May 1959, ARPA directed the Air Force to cancel work on the E-4 mapping camera program. Most contracts were dropped the following month, although the Photographic Laboratory at Wright Field continued to fund related camera developments without calling much attention to the effort. The cancellation came, somewhat disconcertingly, on the day that Lockheed finished the initial version of the development plan.

On 18 October 1960, Major C. E. James of the newly organized SAMOS Washington office (SAFSS) met with Dr. Charyk to discuss geodetic and mapping satellites. He brought the Undersecretary up to date on the status and prospects of ARGON and then explained that the Air Force had a camera known as the "412" (actually the applied research development undertaken upon formal cancellation of the E-4 camera) which represented the logical follow-on to ARGON. Two were scheduled for completion by early 1961 and long-lead-time provisions had been made to purchase seven more. In James' opinion, the system represented the best that the existing state of the art could provide. He advised Dr. Charyk that the camera system could be readied for flight in an E-5 capsule by August 1961. (By using the considerably greater thrust of an Atlas booster, the E-4 avoided design compromises inherent in the Thor-boosted ARGON.)

The E-4 had other attractions. It promised new avenues for the future, seeming to be adaptable to evolution toward a long term objective defined in September 1960 by the National Security Council.* Moreover, an E-4 program under SAFSP auspices would eliminate any need for continuing the cumbersome ARGON management complex, which then included the Army Mapping Service, the National Photographic Interpretation Center, the Central Intelligence Agency, and the West Coast ARGON office. Finally and most important, E-4 promised better results than ARGON.

Convinced, Charyk authorized BMD-WADD (Wright Air Development Division) organizations to plan for early inclusion of the 412 camera in the SAMOS program. For the moment, he withheld any authorization to schedule use of the 412.

---

*Although the evidence is not entirely clear, it would appear that a discussion of mapping satellites during the September meeting of the National Security Council touched off Charyk's interest.

Although Charyk and his staff were relatively enthusiastic about the prospects of the E-4, neither General Greer nor Colonel King looked on it so favorably. Conceding the feasibility and general desirability of an E-4 system,* they nevertheless questioned the wisdom of substituting a mapping satellite for any of the E-5 payloads then on schedule.

Charyk, who thought less highly of the E-5, directed in December that the mapping camera be integrated in the total SAMOS effort as soon as possible and that the existing contracts be expanded to provide for three flight cameras, two test articles, and four follow-on models. (That total matched the figure of nine that Major James had described as "available" two months earlier). Flight hardware (Agenas and equipment) for three flights was to be purchased or transferred from other sub-programs. The booster problem was to be solved by using Atlas boosters made available by the decision to fly F-2 (Ferret) subsystems atop Thors, and the matter of inserting E-4's into the tight schedule of E-5 and E-6 flights was accomplished by slipping the entire sequence of shots.

Instructions and guidance along such lines came into General Greer's complex gradually, over a period of several weeks. Late in December 1960, Greer concluded that the net effect of redirection involving the E-4 and the F-2 had been to create two additional SAMOS technical programs. He cautioned Charyk that "nothing comes free in this business." Manpower and dollar increases were inevitable if the directions were carried out. The E-4 program promised to be particularly costly, he warned, since the implication of earlier directives was to give the E-4 precedence over both the E-5 and E-6. Greer was certain Charyk had not intended that result, and he was also sure that Charyk had not fully analyzed the cost impact of modifying Agenas for the F-2 and E-4 configuration.

After weighing the various considerations, Undersecretary Charyk in February 1961 decided that he wanted an E-4 but that it would have to be developed and tested within the limits of existing funds. He continued to insist, however, that rescheduling boosters and launches would permit the E-4 to progress without grossly affecting any of the search or surveillance payload programs.

---

*The objective of the E-4 development was a system capable of giving position accuracies of 500 feet or less. Based on the usual Atlas-Agena B combination, the recovery capsule was nearly identical to that of the E-5, having a 72-inch diameter and being 84 inches long. The terrain camera had a 6-inch focal length; the stellar camera a focal length of 3 inches. The customary gas reaction jets were to control attitude during a five day mission with an apogee of 178 nautical miles. Ground resolution could be, under good conditions, on the order of 150 feet, assuming a 90-mile perigee over the target area. The usual near-polar orbit was planned. The f/5.6 lens of the mapping camera was considered by reconnaissance camera experts to be the "best available today" for photogrammetric purposes. It had an axial resolution of 60 lines per millimeter with a distortion of 10 micrometers—which reduced to 2 micrometers upon calibration. Some 4,000 feet of 9.5-inch film would be carried and retrieved. Shutter speeds could be varied over a range from 1/50 to 1/800 second. Fiducial and reseau edge markings on the film were to be provided, based in part on a timer with an accuracy of 0.001 second. The f/2.5 stellar image camera would produce 4.5 by 4.5-inch film frames, exposing each frame for 4 seconds to provide an elongated tracer of star images on a total of 2,000 feet of film. Each mission could theoretically provide high quality photographs of about 50 million square miles of Sino-Soviet territory.

The term "Program 1A" was generally substituted for "E-4" as a means of obscuring project intentions. That subterfuge was also an element in the more widespread effort to remove all reconnaissance satellite effort from general view.

By early April 1961, the E-4 had acquired relatively firm configuration characteristics and had made the transition from proposed effort to funded procurement.* An effective working relationship between the Aeronautical Systems Division (ASD) and SAFSP was created in the following way: Colonel King and Major Maurice G. Burnett met with Brigadier General David M. Jones, Vice Commander, Wright Aeronautical Development Division (WADD) to formally brief him on the E-4 program and the proposed involvement of Aeronautical Systems Division personnel. This resulted in appointing a team, headed by Captain David G. Coleman and Mr. Leonard Crouch plus a few hand-picked specialists in the areas of requirements. The number of program-briefed personnel at ASD remained extremely small—perhaps ten or twelve—in effect, only General Jones and those who were actually working on the program. Through special arrangements, funds were transferred to ASD so that the "team" would have no funding restrictions on their operations. The team members were essentially removed from all other activities for the duration of the requirement.

Both the technical and the financial details had received Charyk's specific approval following a general presentation of 7-8 March, although the West Coast group remained rather "bearish" on the whole issue.

The ambitious E-4 program conceived during the SAMOS reorganization of August-September 1960 began to lose stature the following spring. On 28 March 1961, Deputy Secretary of Defense Roswell Gilpatric confirmed Air Force responsibility for development and operation of all defense department reconnaissance satellites, but also made the Army responsible for establishing and managing "a single geodetic and mapping program to meet Defense Department requirements." Within two weeks, the Army's chief of staff had contacted his Air Force counterpart, General T. D. White, to request nominations to an "integrated three-service" group to plan for mapping satellites . . . under Army cognizance. The first meeting was held early in May, and it was immediately apparent that the Army saw the Gilpatric directive as a mandate for establishing a new major research and development effort in satellite mapping and geodesy. The Air Force inevitably disagreed. The only product of the meeting was a decision to collect requirements statements from all three services.

The next meeting, on 11 May, was called on short notice but found the Air Force more determined than ever that reconnaissance satellite research and development should not be parceled out according to camera focal length.

---

*Lockheed was the system integrating contractor, under letter contract ▮▮▮▮▮ issued on 6 April 1961. The original work statement covered systems engineering and vehicles for three flights plus long-lead items for five more. Fairchild Camera and Instrument Corporation (FCIC) had payload development responsibility under letter contract ▮▮▮▮▮, issued by the Aeronautical Systems Division (ASD) Reconnaissance Laboratory on 25 March. (Fairchild actually accepted six days later.) Funds were initially released to ASD on 10 March. A total of ▮▮▮▮▮ was set aside for "Program 1A," the coverage extending through four fiscal years until fiscal 1964, but the bulk of that amount falling due in fiscal year 1963. Schedule called for initial launches in March, June, and September 1962 with the first of the five supplemental payloads going into orbit in April 1963.

On instructions from Charyk, the Air Force representative refused to discuss research and development in satellite geodesy, characterizing it a matter for secretarial resolution. Typically, the tri-service committee was unable to agree on anything significant, adjourning on the note that what was immediately needed was a commonly accepted definition of geodesy, some agreement on targeting requirements, and a standard viewpoint on data processing requirements.

Nevertheless the lines had been drawn, and under the rules outlined by Gilpatric, the E-4 program had become quite vulnerable. Yet had the matter remained one for resolution by a tri-service committee, Charyk and Greer might well have flown the E-4 before any decision could be taken. However, in late May 1961, the mapping satellites issue had passed to the Director of Defense Research and Engineering (DDR&E) for resolution, and the E-4 became but one of three proposed systems. Early in June, Gilpatric authorized continuation of the procurement of four cameras in the E-4 (Program 1A) configuration but instructed Charyk to suspend plans for buying and launching boosters and spare vehicles. There still was hope of course, that a decision to confirm proposed flight schedules would follow completion of an evaluation—but the hope was rather faint. Charyk therefore directed that all E-4 activity not essential to completion of four basic payloads (including accessories) should be halted. He subsequently modified the "complete stop" order to permit Lockheed to work on capsule engineering essential to creation of an "appropriate" interface between capsule and payload and to ensure compatibility of the payload with the capsule environment, but even then the Lockheed work was carefully limited.

For another six months, payload development continued at a slow pace and on a low key. It appeared to be progressing remarkably well, on the whole, a situation that most observers credited to the ability and industry of the immediate program officers (Captain Coleman and Mr. Crouch at ASD) and Major Burnett in Colonel King's Payload Division, consisting at the time of Majors Edward Conway, ▓▓▓▓▓▓▓▓▓ and Burnett. No firm decision on the future of the program had yet emerged from DDR&E, and Charyk seemed content not to push the issue. In October, he discussed mapping satellites with Dr. E. A. Fubini of DDR&E and got approval of a plan to bring E-4 payloads to a state of flight readiness and hold them there, the objective being to provide the least possible delay between a launch decision and an actual launch. He told SAFSP to begin putting together engineering and cost details for a "hold" program. General Greer's people, though reasonably optimistic about the promise of the E-4 camera and the functional effectiveness of the system, were not particularly encouraging—estimating that it would cost ▓▓▓▓▓▓▓▓▓ to orbit all four payloads, and ▓▓▓▓ for one—not counting launch and recovery charges.

Charyk, who had preserved the E-4 program through a succession of administrative moves and who had somehow managed to keep it alive in the face of a formal Gilpatric directive that denied his authority to do so, reacted angrily to the cost and time figures. "It appears that SAFSP does not want to do this job" he told Brig. General Richard Curtin, Director, Office of Space Systems. "The system is obviously gold plated and fat. It is necessary that the program be scrubbed down to the hard core and re-estimated."

Though unpalatable, the figures nevertheless proved to be well founded. By the end of the year, Charyk was apparently resigned to the fact that there was no prospect of early flight for the E-4 payload then approaching completion. Early in January 1962, he advised Gilpatric that as they were completed, the E-4 payloads were being stored in a readiness-in-9-months flight condition, and that a decision to fly would require the provision of substantial funds for launch, boosters, and space vehicle costs. And there the E-4's remained.

The E-5 Recovery Program

In April 1959 ARPA gave specific approval to the recovery (E-5) proposal. Funding difficulties then were responsible for cancellation on 23 June 1959.[11]

The issue, somewhat oversimplified, was essentially whether readout or recovery techniques should be employed to satisfy the 5-foot resolution requirement defined by the intelligence community.

Pressure from the Air Force to reinstate the E-5 resulted in ARPA's authorizing the award of an E-5 camera development contract on 4 September 1959 to "protect schedule." Five days later, on 9 September, ARPA formally authorized reinstatement of E-5 subsystem development, including the capsule.

A total of seven vehicle flights were programmed, two "diagnostic" vehicles being added in August 1960.

The E-5 (Figs. 1-8a and 1-8b) had remained relatively stable in terms of design details, having the following design characteristics:

Focal length	66-inches	Orbital life	1 mo.
Aperture	f/5.0	Weight (on orbit)	5,766 lb
System resolution	100 l/mm	Base diameter	6 ft
Orbital altitude	138 nm	Recovery capsule	
Ground resolution	5 ft	Weight	1,525 lb
Strip width	60 nm	Length	13.5 ft
Location accuracy	1 nm		
Film size	5-inch width		
	265 to 500-foot length		

Additionally, the E-5 was a stereo system. The camera had been developed by the Itek Corporation in Lexington, Massachusetts under subcontract to Lockheed, the system contractor. Each camera consisted of a sunshade and mirror, a window, an eight-element lens (with a temperature tolerance of but 1 degree), and camera body terminating in a 5-inch curved film plane with a 3-second pan cycle, a complex (64-step) exposure selection, and a complex film takeup system.

Along with the E-5, the emergence of the E-6 and GAMBIT systems raised the issue of whether all three recovery systems should be carried to completion. They had several overlapping qualities. Lockheed had total responsibility for E-5 and for the rapidly withering E-1 and E-2 satellite programs, and had prime responsibility for CORONA, but was no more than a vehicle supplier in the E-6 program. Lockheed, therefore was vitally interested in having the E-5 remain attractive. E-5 was then considered to be a logical successor to CORONA—still generally treated as an interim system with slight growth potential—although E-6 was a more promising candidate.

As with the E-1 and E-2, part of the discontent with E-5 arose from the fact that it did not represent the latest in satellite reconnaissance concepts and techniques. Even though development was not well under way until September 1959, the basic proposals embodied in E-5 dated from 1958, and considerable advances in optics, vehicle stabilization, and camera mode technologies had marked the ensuing two years.

Lockheed, aware of waning confidence in the prospects of E-5, proposed accelerating the program toward an April 1961 diagnostic flight and a subsequent launch rate of one satellite each month. An early demonstration could dispel doubts of the systems' usefulness. It would, if successful, provide a high-resolution recoverable system at least a year in advance of the first E-6 and some two years sooner than the first GAMBIT satellite, a consideration that could not well be ignored in an atmosphere of program urgency. Further, Colonel King and General Greer were realistically aware that E-6 and GAMBIT might represent the only insurance against program disaster.

Acceleration of a sort was approved for the E-5 effort before the close of 1960—in December, Charyk had authorized early diagnostic flights of degraded E-5 cameras to get telemetry data, prove out payload operation, and demonstrate the feasibility of capsule recovery in the E-5 configuration.

Along with the urgency of developing flight hardware, came naturally an urgency to accelerate development of test equipment and procedures. One essential element in the test flow was sub-system and system testing under simulated space environment.

With Itek, the camera contractor, on the east coast, and Lockheed, the prime/systems contractor, on the west coast, and neither contractor having on hand the necessary equipment, two vacuum chambers would have to be provided from scratch, one to be installed at each facility for sub-system and system testing.[12] The requirement was to design and build two identical vacuum chambers (10^{-4} mm Hg) with diffusion pumps. These chambers, 12 feet in diameter and 20 feet long, although dwarfs alongside the subsequently developed HEXAGON thermal/vacuum chambers, were considered "large" in the 1958 time frame. The logistics were seen at once as a problem, since the three companies considered capable of building the chambers were on either the east coast or the west coast.

The contract to build the chambers was won by the High Vacuum Equipment Corporation in Hingham, Massachusetts. The plan was to assemble and test the Sunnyvale chamber in Hingham, disassemble and ship by freighter through the Panama Canal to the Bay area, off-load and deliver (shipping time was up to 3 months). As recounted by Dana Jones, Itek's mapping camera contracting officer (then also in the Itek contracts department) there was, in retrospect, a certain amount of humor associated with the vexing task of getting the Sunnyvale chamber to its destination.

The chambers were late, and "oversize" trucking companies were contacted for estimates, since the three months via freighter was then totally unacceptable. One company said that it could be done in seven days. Upon reaching agreement to go this route, two I (eye) beams were welded to each side of the chamber to provide 8-inch ground clearance, one axle was placed in back and a hook was welded in front to form the "trailer". To make the whole thing interesting and relieve some of the tension, "anchor" pools were formed for one dollar each to pick day and time through the front gate at Sunnyvale. The route was the Massachusetts Turnpike, New York Thruway to the Ohio Turnpike, then pick up southern Route 66 through Los Angeles and on up to Sunnyvale.

The driver, for $20, agreed to call each night and report his progress which was plotted by flags on a map showing his route. The overall plan appeared, on the surface at least, to be well laid out. The trucking company had agreed to provide two additional drivers and automobiles with flashers to accompany the truck across the country. Since rules and regulations would vary in the states that would be crossed, permits would be obtained by the company as appropriate for each state and mailed to the post offices in the town nearest the next state border to be crossed. Then as the group approached the border, one of the escort drivers would speed ahead to the post office, pick up the permit at the general delivery window (hopefully) and be back at the border by the time the truck and other car arrived.

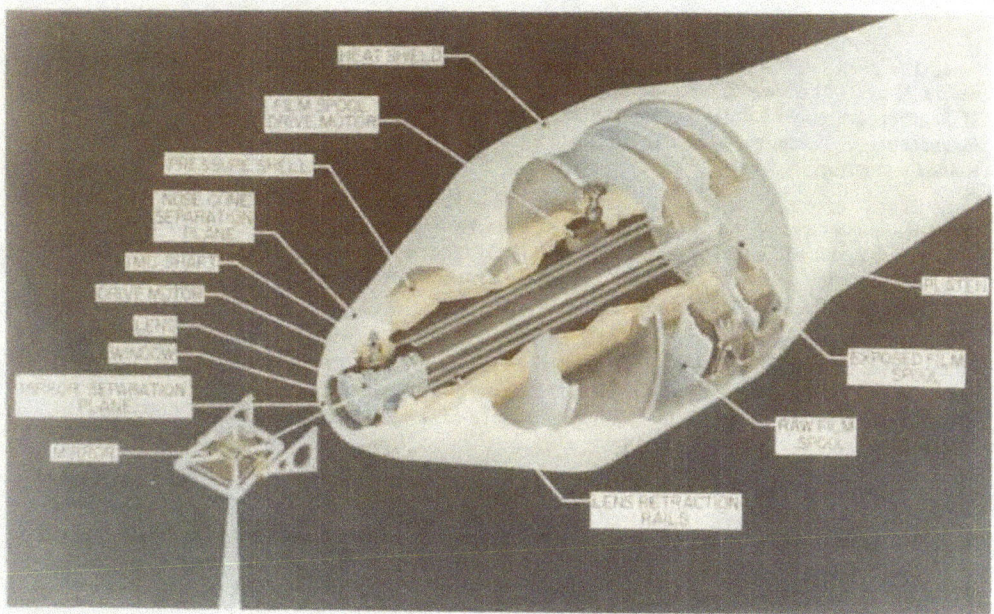

Fig. 1-8a — SAMOS E-5 Recovery Camera Payload

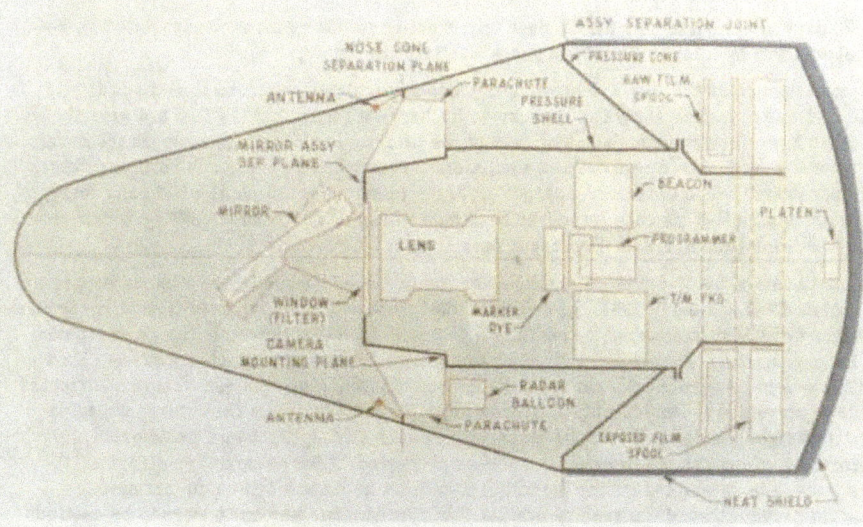

Fig. 1-8b — SAMOS E-5 Recovery Capsule

The crew left the Boston area one Thursday evening at 6 o'clock, estimating the Ohio border within 24 hrs. At 8 o'clock the following morning the driver called—he was in Springfield, Massachusetts, only 90 miles west of the starting point. The "hook" weld had broken on the Massachusetts Turnpike and he had been held up all night. Thirty six hours later he was re-welded and going.

That afternoon, Friday at 4 o'clock, he called from the entrance to the New York Thruway—they wouldn't let him on. At 4:55, after DX-A2 Priority ratings and "Highways for National Defense" threats, and a promise to call Nelson Rockefeller if the Secretary of Transportation could not work the problem, three New York State Troopers took the caravan through New York at 70 mph—got rid of him fast before the weekend.

At the entrance to the Ohio Turnpike he was weighed—the estimated 10 tons was actually 20 tons! He was told to get another axle. Three days later he pulled out on the highway and was pulled over by a Trooper asking to see the registration for the new axle. No registration! His next call for help was from the local jail. Four more days were lost before he was freed and under way again.

He became part of a parade in Indiana, and lost four more days because of floods and washed out bridges in Arkansas.

Lt. Colonel James Seay, General Greer's Director of Procurement at SAFSP, flew to Boston on a Sunday demanding that the rig be in Sunnyvale within four days. He was allowed to put his $1.00 bet on the date and time the rig would pull in the front gate. He picked twelve days, and almost won.

Walter Tyminski, one of Lockheed's vice presidents, was irate because one wall had been removed from Building 104 to receive the chamber, and during the long wait the temporary plastic wall which had been strung up didn't keep his office warm.

The truck went through Los Angeles on a Sunday without a permit. The driver figured that the policemen wouldn't stop anything so outrageous since it obviously had to have a permit.

So, 26 days after leaving for the "7-day" trip it pulled into the gate at Sunnyvale. The $400 anchor pool was won by a lucky guy whose guess was just one day off.

But any impression that the E-5 had thus become more highly regarded than the still embryonic E-6 was dispelled early in February with Charyk's ruling that the E-6 had priority over any other E-series development. The crux of the priority issue was not so much the development status of E-5 as that E-6 represented a solution to requirements for gross coverage, which carried higher priority than the specific target coverage mission for which E-5 had been designed. Further, confidence in E-5 success had never been high within SAFSP, and GAMBIT—which promised far better resolution than E-5—had begun development by February 1961.

The character of the E-5 test program had gradually been changed by the various program decisions of late 1960 and early 1961. In February 1961, that evolution received formal recognition in the statement of a test philosophy, essentially a determination that the early flights would contain very large quantities of instrumentation and would have limited functional objectives. Particular attention was to be devoted to reentry phase instrumentation since the sea-recovery-oriented E-5 capsule represented a considerable departure from the pattern set by Discoverer capsules—relatively lightweight and designed for air catch. Operations during flight test would gradually progress from the simple to the complex as success permitted. (For example, no steering manuevers were to be attempted during the initial E-5 flight because a failure in that mode probably would prevent test of the reentry system.) In essence, the E-5 tests were to be cautious research and development investigations rather than attempts to operate fully functional prototypes.

MCS HISTORY

Vehicle and payload problems indicated the launch schedule should be relaxed so the chances of mission success would not be endangered. Another factor intervened to ensure a relaxation of E-5 launch schedules. On 9 September the E-2 launch disaster had severely damaged the pad at Point Arguello. This caused the program office to slip the initial launch date to 12 December. (Vehicle 2203 slipped from 7 December to 18 January 1962 and 2204 slipped to 22 February 1962.)

On 17 October, General Greer directed Lockheed to make every effort to launch 2202 by 2 December rather than 12 December. The contractor reacted by shaping a "hard core" group of key personnel into a task force with a 24-hour, 7-day-per-week assignment: meet program objectives. Engineers and launch crews were shifted from the Midas program to provide the necessary work force.

The effort was extraordinarily successful. At 1245 hours on 22 November, 12 days in advance of the most optimistic schedule proposal in October, 2202 was launched from Pad 1. Every effort had been made to ensure a successful launch, including special provisions for "super clean" propellant tanks and x-ray checks of questionable transistors. But 247 seconds after lift-off, the Atlas lost pitch attitude control and shortly thereafter another programming error caused permature engine shutdown. That combination of errors caused the Agena to stabilize in a tail-first attitude after separation. When the Agena engines were ignited, the vehicle promptly de-boosted into the Pacific.

Taken together with the record of contractor problems, the launch failure had immediate repercussions. After hearing presentations on the status of the program and discussing its prospects with General Greer, Charyk on 4 December directed that all work on the E-5 program be halted except that in support of 2203 and 2204 launches.

Vehicles 2203 and 2204 differed from their predecessor in having a more comprehensive (ultra-high frequency) command and control system and more intricate telemetry. The camera was somewhat more refined as well.

In the midst of termination proceedings, and while the program office was trying to sort out the residue of a complex program, 2203 reached launch readiness. It climbed free of Pad 2 at Point Arguello at 1145 hours on 22 December, after two days of delay for the correction of minor defects. The launch was successful, and although there was a fault in the Atlas propulsion cutoff system, the net effect was to put the Agena in an orbit with a period 4.5 minutes longer than planned.

Once on orbit, the payload began its scheduled operation. At first all seemed well, and there were clear telemetry indications that the camera had functioned, but either the frame counter failed or the camera shut itself down earlier than scheduled. That was not too serious, even if undesirable. But a faulty command activated the reentry sequence on the sixth pass, and through a combination of errors, the payload, after separating, went into a new and higher orbit.

The dead Agena, relieved of its cargo, continued to circle the earth somewhat below the capsule. Because the reentry command had activated all systems in the capsule portion, the recovery battery was dead by the time it was needed to ignite squibs and activate the drag parachute. Further, the retro-rockets had been ignited during the unplanned manuever sending the capsule into its higher orbit, so any reentry would be entirely ballistic.

The Agena fell back and burned up somewhere south of Borneo on 31 December. Tracking stations calculated that the capsule would encounter enough atmospheric resistance to bring it down about 9 January. Air recovery would be impossible because of the complete absence of retro-rocket and parachute phases, but it was conceivable that the reentry vehicle might survive reentry

BIF-059W-23422/82
Handle Via
BYEMAN/TALENT KEYHOLE
CONTROL SYSTEMS JOINTLY

forces and impact where the payload could be recovered. In the course of Pegasus reentry experiments during September and October 1961, one reentry test vehicle had survived a ballistic return from an altitude of nearly 200,000 feet after its parachute failed to deploy.

E-5 program people bled the Spacetrack centers for whatever information they could obtain on the course and probable decay of the satellite. During the second week of January 1962 the tracking stations reported that the capsule had passed over the northernmost tracing screen but had not been picked up by the radars of the next belt southward.

Lieutenant Colonel V. M. Genez, SAFSP Operations Director, immediately contacted the 6594th Aerospace Test Wing in Sunnyvale, California, activating an earlier plan for the contingency recovery of decaying capsules that might enter intact. There was early indication that the payload had come down in northwestern Canada, so a C-119 carrying Lieutenant Colonel Lon Berry and a recovery crew flew into Great Falls, Montana, stopping there to get Canadian permission for a search along a specific path. The Royal Canadian Air Force wanted to know why. Colonel Berry explained that the USAF hoped to find part of a satellite. After several hours of delay, a direct phone call from Washington ordered Berry and the C-119 back to California. No reason was given.

It later developed that the area of the proposed search was along one of the Strategic Air Command's most heavily used polar patrol routes. Canadian authorities suspected that a B-52 had accidentally released a nuclear weapon and that the Air Force wanted to recover it surreptitiously. The issue was not of the sort that promised quick resolution, so the search party was ordered home. Later a pair of U-2 aircraft flew along the suspected reentry path, photographing the terrain in hopes there might be some sign of the capsule. Nothing turned up, and the affair ended on an inconclusive note.

The third and final E-5 vehicle was launched on 7 March 1962 at 1410 hours after an extended series of aborted countdowns. The Agena auxiliary power system and the command and control subsystem of 2204 had been substantially modified to reflect experience with the first two E-5's. Nevertheless, problems with the Agena, the Atlas, the guidance programmer, and various switches had delayed the launch since 22 February. Despite that omen, the launch and orbit injections were "near normal". For the first 13 passes, all went reasonably well. Then the New Hampshire tracking station improperly transmitted reentry sequence commands. The vehicle assumed and maintained reentry attitude, and over a period of several passes expended most of its attitude control gas. In part, the sequence of misadventures resulted from failure of a Fairchild timer. A recovery attempt on pass 17 failed because of another tracking station error, and by pass 21 all control gas had been exhausted. The only remaining recourse was to trigger the reentry system while the vehicle was in an appropriate reentry attitude. But instead of reentering, the capsule went into a higher orbit, much like its immediate predecessor.

More than a year later, in July 1963, the satellite's orbit had decayed to the point of imminent reentry. As the heavy heat shield still was attached, there seemed a chance that it would survive. Greer's staff, aided by computers and operations of the Aerospace Corporation, calculated the probable reentry path and impact point. They concluded that the satellite would impact toward the center of the Arabian Sea. Since any possibility of parachute deployment had passed months before, and since the shock of striking cold sea water after an uninhibited ballistic reentry almost certainly would breach the satellite casing, there seemed no possibility of retrieval. No recovery was attempted. All the available data suggested that the capsule had actually come down in the predicted impact area. Like both its predecessors, nothing more was heard of it.

After the failure of 2203, the E-5 program disappeared from organizational charts. No final report was written. On 1 March 1962, even before the last E-5 launching, Colonel King had been transferred to a new assignment and the remainder of the program office had been dispersed. As E-5, the program was thereafter of interest mostly to antiquarians and historians.

But the Itek camera, and the E-5 requirement, tenuously held to life notwithstanding the lack of program success. Charyk's decision to cancel the E-5 program had been taken on Monday, 4 December. Two days later ▮▮▮▮▮ of Itek proposed to Charyk that tests be run on Itek and Perkin-Elmer lenses to determine whether an improved lens might be substituted for the original in the still-pending 2204 flight. A comparison began early in January.

While arrangements for that work were in train, ▮▮▮▮▮ suggested to General Greer that advances in the camera and satellite technology since the start of E-5 should be adapted to a new reconnaissance system based on the Itek E-5 camera. After refining the original idea, he formally submitted it on 19 December 1961.

What ▮▮▮▮▮ proposed was combining a single re-engineered E-5 camera with the DISCOVER-CORONA capsule, a Thor booster, and a modified Agena.* The evolving camera, known as LANYARD is described further in the CORONA section, since it was developed and flown under the CORONA program management arrangement.

---

*Although Itek's record in E-5 development was scarcely faultless, the failures of the system had all originated in Atlas and Agena subsystems mostly peculiar to the original E-5 design. CORONA had a much better record by 1961, and Itek's reputation for camera development was quite respectable.

The E-6 Recovery Program

Note: At various times of no particular consequence, the E-6 program was officially known by other titles: Program II, Program 201, Program 698BJ, Program 722. The term most commonly in use in 1963 was "BJ". For the purpose of this account, and in the interest of narrative continuity, the identifier "E-6" is used throughout (see Fig. 1-9).[13]

Through the spring and summer of 1960, while matters of project structure and program objectives were being debated at various levels between the project office and the White House, the sixth and last of the SAMOS camera systems to receive formal designation was also taking shape.

By April 1960, CORONA had experienced its eighth successive failure (DISCOVERER IX) and was entering a limbo of engineering overhaul that would postpone further trials for two months. Early in May the U-2 incident abruptly halted use of the only other reconnaissance system available to take pictures over the Soviet heartland. The E-5 satellite system then in development was so designed that it would return relatively narrow film strips, each covering only about 15 by 53 miles along the ground. Moreover, it was still many months from its scheduled first trial.

Late in May and early in June suggestions were heard that a completely new photo-recovery system should be developed. On 5 July the United States Intelligence Board (USIB) issued a revision of satellite reconnaissance requirements, emphasizing the need for locating Soviet ballistic missile sites and calling for a search camera system capable of resolving objects 20 feet on a side before the end of 1962.

Until early July, the Air Force Ballistic Missiles Division (BMD) expressed a preference for some relatively minor modifications of the E-5 system rather than a new development. A 12 July BMD development plan revision, however, featured a proposal for a new camera payload (designated E-6) to be combined with a new recoverable and maneuverable re-entry body. On 11 August, BMD issued still another development plan which proposed an E-6 system generally conforming to the USIB statement of requirements. Featuring a panoramic camera with a 20-foot or better resolution, 8 days on orbit, and a highly precise recovery system, it was intended to provide broad coverage of those areas serviced by the Soviet railway network.

Even earlier, on 27 July, Colonel Paul J. Heran, then Director of Operations at the 6594th Test Wing, had been named to head a source selection board which was to evaluate contractor proposals for an E-6 system. Requests for proposals were dispatched to a selected list of contractors (from which Lockheed had been excluded) on the day the development plan was issued, 11 August. During the period of pre-proposal briefings, the SAMOS project was formally assigned-- to the Office of the Secretary of the Air Force, acquired a military chief, General Greer, and a secretariat-level overseer, Dr. Charyk, and in its revamped form received Presidential endorsement. The basic performance requirement was also modified to include 10-foot resolution ("or better") and 5 days on orbit.

The source selection board considered the E-6 to be a backup to the E-5 system, with assured recovery over land being more important than rigid adherence to the photography specifications. E-6, of itself, had to be "useful and usable even if the thing it's backing up also works." By implication, the E-6 had to differ from existing or programmed solutions to the reconnaissance problem, otherwise it would be duplicative—and undesirable.

The system Charyk described to President Eisenhower was composed of a precise land recovery subsystem (with air pickup a possible alternative) integral with a photographic subsystem that included a 24 to 36-inch panoramic camera. First flight, assuming progress consistent with that outlined in the development plan, was planned for January 1962. Seven flights, possibly augmented by two diagnostic tests, were on the proposed schedule.

The choice of subsystem contractors had, for practical purposes, been completed before the end of October, by which time the new SAMOS office structure had also been clarified. The source selection board recommended awarding the camera payload contract to Eastman Kodak and the recovery subsystem contract to General Electric.

Several factors, however (land versus water recovery, the role of the Aerospace Corporation in systems engineering, the relationship of SAMOS to ARDC programs) kept the source selection board in session until early December. The board had found no alternative to using Lockheed's Agena as the upper stage to inject the E-6 payload vehicle into orbit, and Lockheed thus became part of the contractor team. (Technical integration of the payload, upper stage, and recovery subsystems, however, was reserved for General Electric rather than Lockheed, which had that responsibility for all other SAMOS payload systems and for CORONA.)

On 14 December the board chairman, Colonel Heran, formally advised the BMD commander, Major General O. J. Ritland, that General Electric and Eastman Kodak had been chosen to develop recovery and camera subsystems, respectively. The maneuverable reentry aspect of the original requirement had been reduced to an applied research program aimed at the eventual design of a "terminally guided lifting type vehicle".

Even though the land recovery objective of the program defined in August had been substantially reduced in importance by December, the expectation that Martin's glide-control reentry technique would eventually be combined with the E-6 camera system remained a basic program concept through the early months of 1961. Fears for the possible loss of a SAMOS satellite over unfriendly territory, with repercussions perhaps more extreme than those of the U-2 incident, prompted continued concern for positive control of recovery modes and for the improvement of reentry accuracy.

Delays in completion of the source selection process had forced a slippage in the original program deadlines. During the last days of 1960, a technical direction meeting conducted by Aerospace produced revised milestone goals: delivery of the payload vehicle to Vandenberg Air Force Base and the first flight-ready Agena B to the missile assembly building by 20 November 1961, availability of the assembled vehicle on the pad by 18 December, and first flight by 1 February 1962. It was a schedule that seemed wildly optimistic in the light of earlier space program achievements—13 months from program approval (source selection) to first flight.

Some of the configuration details of the E-6 were decided less by engineering logic than by the need to camouflage GAMBIT. During the early months of the E-6 program it seemed essential not only to hide the GAMBIT technical effort under a screen of E-6 activity, but also to make the orbital vehicle portions of the two systems resemble one another in outward appearance. Thus, in theory, a GAMBIT could be launched without alerting many people to its real nature. Final evidence of the futility of the "look-alike" undertaking came after the technical evolution of GAMBIT continued with the result that GAMBIT rapidly assumed an appearance and character completely distinct from that of the final E-6 configuration.

Problems relative to tracking nets/communications and recovery had to be resolved. The formal decision to use Johnston Island as the descent and recovery zone was not made until late February and it was another month before a program office survey group could actually visit the

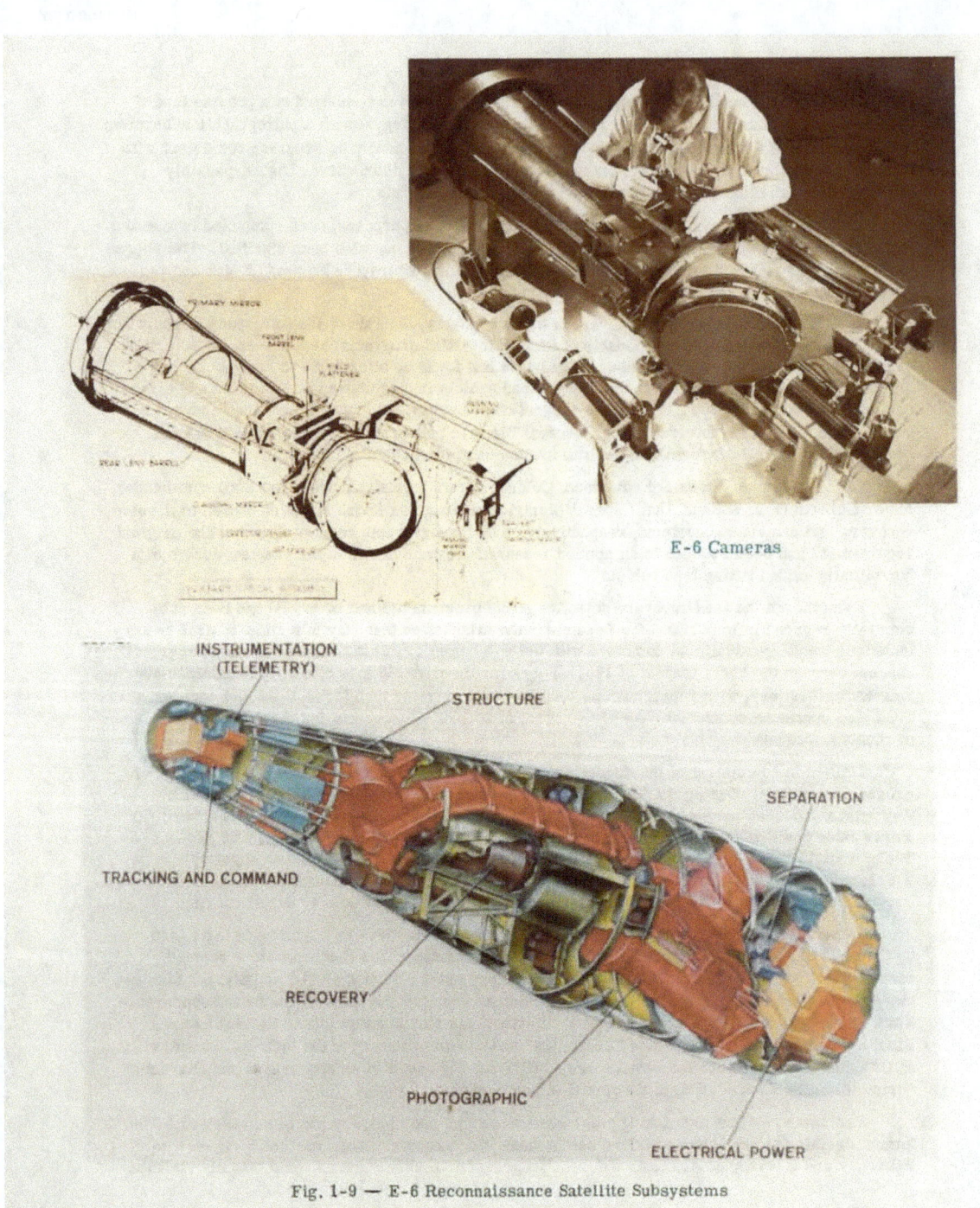

E-6 Cameras

Fig. 1-9 — E-6 Reconnaissance Satellite Subsystems

site and estimate needs. One of the last major technical redirections that could be incorporated before the program progressed so far that each change meant a significant delay was the 16 February 1961 deletion of air-catch considerations from the recovery subsystem. As with the E-5, the E-6 would depend on de-boost, aerodynamic deceleration, and water impact (and floatation) for its recovery mode. Sheer bulk was a principal deterrent to aerial recovery; the reentry body was 12 feet and 3 inches in length with a maximum diameter of 8 feet and 4 inches.

Although alternative modes of reentry and recovery operations were considered later, by March 1961 the basic techniques of E-6 launch, orbit, and recovery had been decided. The operation would begin with launch of the Atlas-Agena combination from Point Arguello and its control (in Atlas sustainer and vernier phases) by Atlas radio guidance. At Atlas burnout, the satellite vehicle (Agena B, camera section, and recovery vehicle) would coast to apogee, at which point the Agena B would deliver the impulse required to place the satellite combination in a preselected orbit within the Agena's guidance and control tolerance. Orbit insertion would take place at approximately 125 nautical miles altitude.

After insertion, the orbit would be defined from telemetry return, angle track data, and radar track information. The required orbit correction would be computed from track and rate radar derivations, and introduced as velocity changes provided by Agena re-burn. The final orbit correction system relied on a hydrogen peroxide propulsion unit contained in the camera section.

Photographic coverage normally would begin on the eighth orbit. The photographic subsystem was built around a pair of 36-inch (focal length) cameras (for stereo coverage) with horizon recording for attitude control.

Upon completion of the photographic portion of the mission, de-orbit requirements would be calculated from ephemeris data and sent to the orbiting vehicle. The Agena B would then be oriented to the proper attitude by its gas jets, and de-orbit thrust impulses would be applied to acquire the desired de-orbit trajectory.

The recovery vehicle would separate from the Agena B by retro-thrust derived from the orbit correction nozzles and would then be re-oriented to the desired reentry attitude by the nitrogen jets provided for reaction control. Pre-orientation of the Agena was intended to make the de-orbit technology relatively uncomplicated. Reliance on gas jets for spin-up was intended to eliminate the possibility of an unstable spin arising from unbalanced solid rockets.

Use of a parachute recovery system in combination with the recovery vehicle (based on General Electric's RVX-2) presumably provided a safe rate of descent plus adequate ablative protection for the recovery payload through the aerodynamic heating zone to the point of recovery.

Tracking, telemetry, and command equipments were contained in the recovery vehicle. Such devices had to be compatible with the Mod III track and command systems at the Atlantic and Pacific Missile Ranges; the S-band tracking radars at Hawaii, Kodiak, and Vandenberg; and the VHF and UHF telemetry receivers and command transmitters at various sites in the western hemisphere. During on-orbit operation, the satellite vehicle was controlled through time-coded binary signals transmitted by the Verlort tracking link. The satellite itself had a memory circuit adequate for the storage of commands necessary for both vehicle and payload operations during orbit. In actuality, some of the more precise circuitry required for command of the payload portion was essential to the GAMBIT system rather than the broad-swath E-6 camera, but for obvious reasons that fact was not widely known.

The original plan of an initial launch by December 1961, followed by six additional launches at 40-day intervals (and including two diagnostic launches from the Atlantic Missile Range, if necessary) had by early 1961 been changed to reflect a 9 March 1962 first launch target date. The entire slippage, at that point, had resulted from an August 1960 decision to permit prospective bidders more time than originally contemplated to develop their proposals.

The early objective of controlled land recovery became less than an integral part of the total program after 9 March 1961, when Undersecretary Charyk reduced the Martin effort to a study-through-mockup activity more slowly paced and less fully funded than initially proposed. The Martin Company's work statement was rewritten in April to reflect the changed emphasis and thereafter had no significant influence on the basic program.

While the program was being troubled by financial problems, the principal effort nonetheless was applied to remaining on schedule in the development, fabrication, and test aspects. The first key date was Kodak's delivery of a payload mock-up to General Electric, completed on schedule: 21 April. The first three flyable recovery vehicle cassettes reached General Electric before the end of June; in August, thermal environment tests of prototype lenses began; and on 18 September, the first drop test of a recovery vehicle (from a B-52 at Kirtland Air Force Base) ended in success. On 10 October, 1961 Colonel Heran assured Undersecretary Charyk that by all available indications the first launch would take place when scheduled: 9 March 1962. On the day of this report, Heran learned that the initial water-drop test of the reentry vehicle had also been successful, both in parachute deployment and in flotation characteristics. At the end of the month, recovery site facilities were complete.

This optimism was short-lived. A succession of technical problems combined to delay the launch date. Finally, at 1056 hours (local time) on 26 April 1962, the Atlas-Agena carrying E-6 number one climbed away from its launch pad, leaned toward the south, and vanished from the sight of observers at Vandenberg. At the proper time the Agena separated, the booster fell away, and the programmed injection into orbit began. Propulsion and guidance proved excellent. The orbit was near perfect, no adjustment was necessary. Telemetry signaled a possible failure of the camera window shields to open, and there was a clear indication of excessive use of control gas to maintain proper vehicle attitude, but it appeared that at least one of the cameras had operated as planned throughout the mission. The other of the camera pair showed no sign of functioning after orbit number seven. During the attitude adjust maneuver immediately before de-boost, however, the plume of the ullage rocket* impinged on the Agena's rocket exhaust nozzle and caused an unprogrammed pitch up, and the vehicle failed to enter through the proper "window". It could not be recovered.

Launch of the second system occurred on 17 June. Again the launch and orbit placement phases were "near normal" and the photographic subsystem functioned adequately, but premature exhaustion of attitude control gas forced a call down attempt during orbit 10 rather than during orbit 18, as planned. Again the de-boost phase was ineffective. The attitude control system of the Agena malfunctioned, a power failure prevented separation of the reentry vehicle from the Agena, and they re-entered as a unit. Because of that circumstance, the deceleration parachute did not deploy and the satellite completed a free-fall trajectory, impacting about 750 nautical miles further down range (north) than planned. The hard impact ruptured the recovery capsule, which sank before ships or planes could locate it.

---

*Ullage rockets are small solid propellant rockets attached to the Agena and are fired just prior to ignition of the Agena engine after its separation from the Thor. This is to ensure that the liquid Agena propellants are pushed against the bottom of the tanks so that proper flow into the pumps will occur.

Following corrective measures, the third flight, on 18 July 1962, produced another excellent orbit. A succession of difficulties of varying magnitude plagued the vehicle thereafter. The S-band beacon operated with marginal effectiveness throughout most of the mission and failed completely during orbit 18. The forward (main) camera failed to advance after the 10th orbit, the film cutter refused to function, and on revolution 18, during de-boost, the Agena secondary propulsion system again refused to ignite. Without ullage, the main engine could not fire, so no de-boost increment was available for the reentry operation. Again there was no recovery.

Changes introduced as a result of the third failure of the recovery system included redesigning circuits to isolate the secondary propulsion system from the solid ullage rockets and improving the pre-flight inspection of the circuitry. With these changes, flight number four began on 5 August 1962.

In what had by that time become an established pattern, the launch and injection operations resulted in an orbit within two percent of "perfect". No orbit adjust was needed. On-orbit telemetry was quite satisfactory, although some S-band peculiarities were noted in retrospect. (They caused a minor error in prediction of the impact point.) Steering gas consumption was normal and the command system performed with desirable efficiency. The camera payload, unhappily, developed some defects. Telemetry returns showed the main camera to be "operating" through pass number seven, but the film transport remained non-functional throughout the entire mission. The rear camera operated through revolution number six, after which the film transport failed. However, there was a clear indication that at least 1,500 feet of film had been properly exposed with the rear camera.

During the reentry and recovery phase, defects again appeared. Individual incidents of the de-boost sequence came in proper order, but the Agena imparted only 1,450 feet-per-second de-boost velocity instead of the programmed 1,600 feet-per-second. Nevertheless, the reentry sequence continued as scheduled until the vehicle emerged from the ion-sheath blackout. One second later, primary telemetry failed. Although telemetry signals briefly resumed after a lapse of 16 seconds, there was no indication of parachute operation and recovery aircraft in the impact zone were unable to secure a clear bearing on intermittent beacon signals which persisted over the next 40 minutes. Both electronic and visual search continued for four hours after presumed impact, but there was no sighting. A helicopter search over the next 24 hours produced nothing more tangible.

Analysis of the fragmentary telemetry indicated that excessive heating, principally in the aerodynamic wake of the reentry vehicle, had caused a failure in the parachute deployment circuitry.

The relatively rapid succession of flight tests—and mission failures—had not proceeded in a management vacuum, nor had work on improvement of the central E-6 configuration ceased. In the area of system improvement, two items were of particular interest during the months between April and October 1962. One was improved retrieval, either water to air or air catches. The second was the addition of an index camera which would more adequately pinpoint the location of sites photographed by the stereocameras, this addition was to be incorporated on the tenth and subsequent E-6 vehicles. However, unsuccessful attempts to develop a satisfactory water-to-air recovery system between March and October 1962 resulted in this idea and the index camera both being placed in the long-term category.

During the ensuing months, system configuration and costs received close scrutiny. The possibility of adapting the E-6 payloads to a thrust-augmented-Thor (TAT) launch vehicle and a DISCOVERER (CORONA) recovery capsule was examined but considered unjustified for several reasons. Use of the TAT would force "almost complete redesign and packaging" of the E-6 system, would reduce the quantity of film by at least one-half, and would essentially constitute a new program with all the complications inherent in such a procedure. Its affect would be to substitute a new launch system for one which had worked quite well.

During a complete program review in September 1962, costs and strategies for continuing the program were examined. Charyk directed that work on all vehicles additional to the nine originally programmed be halted. He further directed that three of the remaining five payloads be scheduled for flight in accordance with a philosophy of taking enough time to ensure a "maximum probability of success," and with intervals between the flights sufficient to permit complete analysis of all data from the previous flights and allow the incorporation of necessary changes.

Should these efforts not result in satisfactory demonstration of the E-6, cancellation of the program would be inevitable, but could not be straightforward. The E-5 effort had ended in termination by January 1962. With the last E-6 flight, the known "cover" for both CORONA and the still untested GAMBIT would vanish. Another casualty of E-6 program termination would be the known justification for the existence of General Greer's organization—SAFSP; only those with access to the cover programs appreciated that the E-6 effort was but a minor part of a major activity being managed from the fourth floor suite of offices in the "SSD complex" along El Segundo Boulevard in Los Angeles. CORONA program managers were also concerned that announced cancellation of E-6 might expose the CORONA effort. If the original objectives of SAFSP establishment were to remain valid, E-6 cancellation (should it finally occur) had to be accomplished by new camouflage for the covert programs, a logical explanation for continuation of SAFSP as an organization, and, ideally, a new overt program to cancel in case of a political decision to halt "open" support of satellite reconnaissance. One of the chief reasons for continuing the E-6 in its original form had been to permit the public cancellation and the clandestine continuation of other satellite reconnaissance activity, should international events so dictate.

Thus quite apart from considerations of technology, the launch of the fifth E-6 vehicle promised to be of considerable significance.

Following indepth reviews, modifications, and inspections, representatives of General Electric, Lockheed, Aerospace Corporation, and the program office made a final appearance before General Greer to assure him again that they had a very high degree of confidence in the chances of mission success.

Launch of the fifth E-6 occurred on 11 November 1962. System operation to the point of reentry was in many respects even better than during any of the earlier missions. Lift-off and orbit injection again resulted in establishment of a near perfect ephemeris. The only possible malfunction, suggested by telemetry but unconfirmable, was failure of hatch removal. The command system functioned without disorder and the photographic subsystem transported 3,400 feet of exposed film. De-boost sequencing was near perfect, and the reentry vehicle appeared to be performing without any error until it entered the blackout zone. Thereafter, events roughly paralleled those of flight four. There was some indication of parachute deployment, derived principally from telemetry indications that descent had lasted longer than would have been the case with a free-falling reentry body, and again one aircraft reported 16 minutes of indistinct beacon signal reception following impact, but none of the search craft sighted the vehicle, no further signals were reported, and at dark on the evening of 12 November the search ended. The fact that a recording station heard both SOFAR bombs detonate indicated to recovery team personnel that the vehicle had broken up on impact or sunk shortly thereafter.

MCS HISTORY

Investigation of the possible causes of the latest casualty continued and proposals for alternate configurations were resumed, but on 31 January 1963, Charyk formally notified General Greer that all proposals for further orbit tests of the E-6 payload had been disapproved.*

Thus with the cancellation of E-6 and LANYARD, none of the original E systems of 1960 survived in any form, yet the requirements that had caused their generation remained. At the end, the experience of E-6 payload development was to have a considerable influence on subsequent developments that led, by 1966, through the S-2 search system proposals to the eventual HEXAGON program (the S-2 system is addressed in Part II).

---

*At this time, few knew of CORONA, and fewer still were aware that the cancelled E-5 had reappeared in a different form as LANYARD. But some knew, and knowing were tempted to quip, quietly and privately, that it was a wise man who knew his own payload, that E-6 might have been cancelled, but it was equally possible that General Greer and Colonel Heran had found a way to stuff the E-6 cameras into something else and weren't telling.

THE CORONA PROGRAM

Note: On the facing page (Fig. 1-10) a sketch of the CORONA launch vehicle is shown for the purpose of orientation. On the following page, photographic payload profiles and characteristics are shown (Fig. 1-11 and Table 1-4) to provide the reader with a visual reference to associate with the payloads flown during the course of the CORONA Program.

WS-117L had been undertaken as a classified project. Although its existence was concealed, all findings were reported to Congress. The press soon began publishing stories on the nature of the program, correctly identifying it as involving military reconnaissance satellites, and referring to it as "Big Brother" and "Spy in the Sky". The publicity was of concern because of the sensitivity to the subject of overflight reconnaissance. It was decided therefore, that those portions of WS-117L offering the best prospect of early success would be separated from WS-117L. This would be designated CORONA and placed under a joint CIA-Air Force management team, an approach that had been so successful in covertly developing and operating the U-2. Air Force management, particularly Major General Bernard A. Schriever, the Commander of the Western Development Division, and Col. Frederick (Fritz) C. E. Oder, the Program Director for WS-117L, contributed greatly to the CORONA decisions in this time frame, as did Dr. Edwin Land of Polaroid Corporation, Dr. James A. Killian, the Special Assistant to the President for Science and Technology, and Brig. General Andrew J. Goodpaster, the President's Staff Secretary at the White House.[24]

The nucleus of a team was constituted as the Development Projects Staff under the direction of Richard Bissell, who was Special Assistant to the Director, Central Intelligence (DCI) for Plans and Development. Bissell was designated as the senior CIA representative on the new venture. His Air Force counterpart was Brig. General Osmund J. Ritland, who as Colonel Ritland, had served as Bissell's first deputy in the early days of Development Projects Staff and was then Vice Commander of the Air Force Ballistic Missile Division. Bissell recalls that his early instructions were extremely vague: the subsystem was to be developed out of work accomplished under WS-117L; it was to be placed under separate covert management; and the pattern established for the development of the U-2 was to be followed.

The splitting off of CORONA from the WS-117L program was accomplished by an Advanced Research Projects Agency (ARPA) directive of 28 February 1958 assigning responsibility for the WS-117L program to the Air Force and ordering that the proposed WS-117L interim reconnaissance system employing Thor boosters be dropped.

The ARPA directive ostensibly cancelling the Thor-boosted interim reconnaissance satellite was followed by all of the notifications that would normally accompany the cancellation of a military program. This was followed, as one would expect, by indignation on the part of all prospective contractors. Subsequent to the cancellation, only a very limited number of individuals in the Air Force and participating companies were cleared for Project CORONA. These people were informed of the procedures to be followed in the covert reactivation of the cancelled program.

Although CORONA was removed from WS-117L and placed under separate management as a covert activity, the original intent was to disguise its real purpose by concealing it as an experimental program carrying the name, DISCOVERER. The program was represented as a scientific program whose findings would be of value to many related programs. This permitted overt procurement of the necessary boosters, second stages, and hardware associated with the biomedical cover launches. It also provided an explanation for the construction of launch and ground control facilities. Only the program components associated with the true photographic reconnaissance mission had to be procured covertly.

BIF-059W-23422/82
Handle Via
BYEMAN/TALENT KEYHOLE
CONTROL SYSTEMS JOINTLY

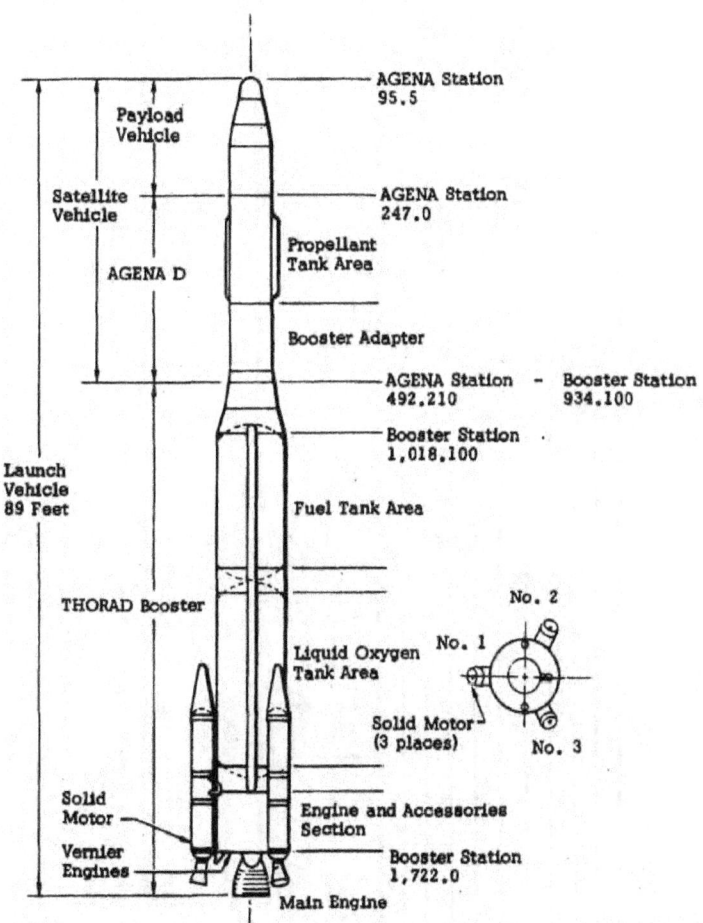

Fig. 1-10 — Major components of the CORONA J-3/CR (KH-4A) Launch Vehicle

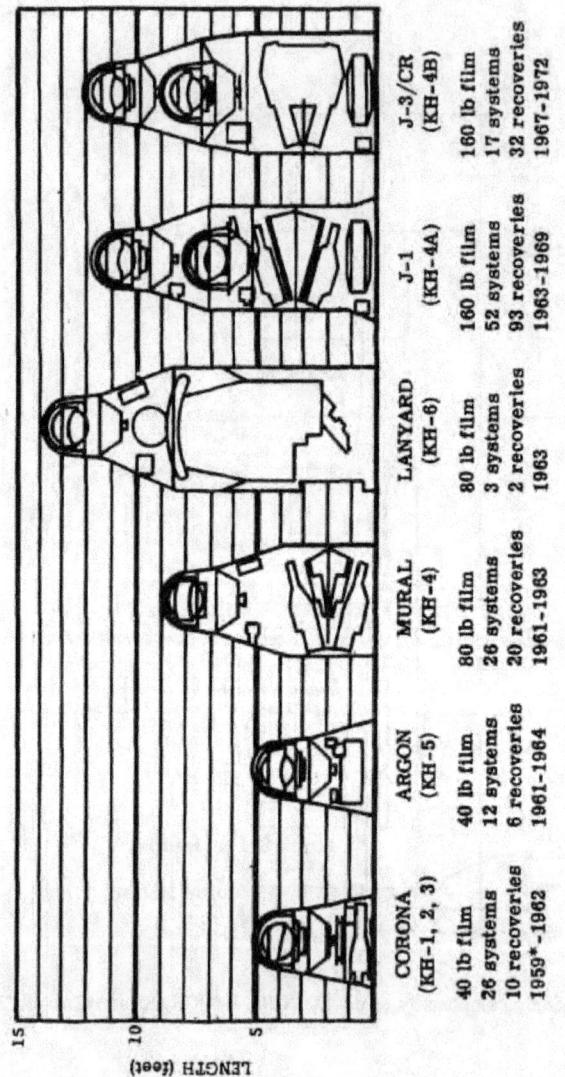

Fig. 1-11 — CORONA Photographic Payload profiles

Table 1-4 — CORONA Payload Characteristics

Designator	C (KH-1)	C' (KH-2)	C''' (KH-3)	ARGON (A) (KH-5)	MURAL (M) (KH-4)	LANYARD (L) (KH-6)	JANUS (J-1) (KH-4A)	J-3/CR (KH-4B)
Camera manufacturer	Fairchild	Fairchild	Itek	Fairchild	Itek	Itek	Itek	Itek
Lens manufacturer	Itek	Itek	Itek	Fairchild	Itek	Itek	Itek	Itek
Design type	Tessar, 24 inch, f/5.0	Tessar, 24 inch, f/5.0	Petzval, 24 inch, f/3.5	Terrain, 3 inch; Stellar, 3 inch	Petzval, 24 inch, f/3.5	Hyac, 66 inch, f/5	Petzval, 24 inch, f/3.5	Petzval, 24 inch, f/3.5
Camera type	70° pan, vertical, reciprocating	70° pan, vertical, reciprocating	70° pan, vertical, reciprocating	Frame	70° pan, 30° stereo, reciprocating (2)	90° pan, 30° stereo, (roll joint)	70° pan, 30° stereo, reciprocating (2)	70° pan, 30° stereo, rotating (2)
Exposure control	Fixed	Fixed	Fixed	Fixed	Fixed	Fixed	Fixed	Slits (4) selectable
Filter control	Fixed	Fixed	Fixed	Fixed	Fixed	Fixed	Fixed	Filters (2) selectable
Primary film (film/base)	1213/ acetate 5.35 mil*	1221/ acetate 2.75 mil	4404/ estar 2.5 mil	3400/ estar 2.5 mil	4404/ estar 2.5 mil	3400/ estar 2.5 mil	3404/ estar 2.5 mil	3404, 3414/ estar 2.5 mil
Recovery vehicles	1	1	1	1	1	1	2	2
Subsystem (stellar/index)	None	None	None	N/A	1 S/I,† 80-mm stellar 38-mm terrain	1 S/I, 80-mm stellar 38-mm terrain	2 S/I's, 80-mm stellar 38-mm terrain	DISIC (Fairchild) 3-inch stellar 3-inch terrain

* Support thickness
† Index only missions 9031-9044

After Bissell and Ritland had worked out the arrangements for the overt cancellation and covert reactivation of the program, they began to address the technical problems associated with the design configuration they had inherited from WS-117L. The subsystem contemplated the use of the Thor IRBM as the first stage booster and, as a second stage, a Lockheed-modified satellite vehicle or spacecraft that had been designed around the Bell Aircraft engine developed for the B-58 Hustler Bomber. It carried the Hustler designation during the development phase of WS-117L but soon came to be known as the "Agena," the name it carried throughout the program.

Several important design decisions were implemented in this organizational period of CORONA. Recognizing the need for good ground resolution to meet the intelligence objectives, it was concluded that the previously developed concept of physical film recovery did indeed offer the most promising approach for a usable photographic return in the interim time period and should be pursued. This resulted in the design of a recovery pod or capsule with General Electric selected as the recovery vehicle contractor. The decision to pursue film recovery proved in retrospect to be one of the most important decisions made in United States reconnaissance activities. History shows that during the crucial decade of the 1960's, intelligence needs could not have been served by the state-of-the-art in readout technology—the alternative concept developed under WS-117L. It should also be noted that both the manned and unmanned United States space recovery programs which followed have leaned very heavily on the re-entry technology developed for CORONA.

Other major decisions for the new CORONA Program resulted from a 3-day conference in San Mateo, California among representatives of CIA, Air Force Ballistic Missile Division, Lockheed, General Electric, and Fairchild. Discussion at the San Mateo meeting got into the need for immediate contractual arrangements with the various suppliers. Bissell remarked that he was "faced with the problem at present of being broke" and would need estimates from all of the suppliers as soon as possible in order to obtain the necessary financing to get the program under way. The suppliers agreed to furnish the required estimates by the following week. However, following that meeting, the project quickly began taking formal shape. Within a span of about 3 weeks, approval of the program and of its financing was obtained, and the design of the payload configuration evolved.* By late March and early April of 1958, lengthy and serious consideration of different camera and spacecraft configurations proposed by Fairchild Camera and Instrument Company (FCIC) and Itek Corporation was culminated. Interest shifted toward the design submitted by the Itek Corporation (primarily formed from resources of Boston University). Itek proposed a long focal length camera which would scan within an earth-center-stabilized pod. This concept promised substantially better ground resolution performance. The Itek design was based on the principles of the Boston University HYAC High Acuity, Panoramic Camera. Bissell recalls that he personally decided in favor of the Itek design, but only after much agonizing evaluation.

---

*With program approval there came a proliferation of payload design ideas, all in a highly compartmentalized environment. As an example, one concept visualized a Ryan Tip Pod Camera in a satellite, spinning much like a well-thrown football forward pass. The camera would trigger on the horizon, advance the film as it progressed across the nadir to the other horizon, and then would be off while the lens completed its arc through the zenith. This "football" was the basis of program manager J. W. "Jim" Plummer's work at the covert Hiller facility in Palo Alto, California. The plan was to adapt the G.E. reentry body instrumentation capsule which was basically an 18-inch sphere covered with a cork-like ablative shield. Progress was well under way on this large "egg" mounted behind the shroud of the WS-117L Agena when Washington powers decided to drop this approach in favor of other proposals which had been introduced.

The decision was a difficult one to make because it involved moving from the previously intended method of space vehicle stabilization to one that was technically more difficult to accomplish. It did, however, standardize on the three-axis stabilization which was being pursued in the WS-117L Agena development and which has been a part of all subsequent photoreconnaissance systems.

The final project proposal was forwarded to: General Goodpaster; Mr. Roy Johnson and Admiral John Clark of ARPA; Mr. Richard Horner, Assistant Secretary of the Air Force for Research and Development; General Ritland; and Dr. James Killian. The proposal was approved, although not in writing. The only original record of the President's approval reportedly was in the form of a hardwritten note on the back of an envelope by General Cabell, then Deputy Director of Central Intelligence.

Although the original intent was that CORONA would be administered in a manner essentially the same as that of the U-2 program it actually began and evolved quite differently. It was a joint CIA-ARPA-Air Force effort, much as the U-2 was a joint CIA-Air Force effort, but it lacked the central direction that characterized the U-2 program. The project proposal described the anticipated administrative arrangements, but fell short of clarifying the delineation of authorities. It noted that CORONA was being carried out under the authority of ARPA and CIA with the support and participation of the Air Force. CIA's role was further explained in terms of participating in supervision of the technical development, especially with regards to the actual reconnaissance equipment, handling all covert procurement, and maintenance of cover and security. The work statement prepared for Lockheed, the prime contractor, on 25 April 1958 noted merely that technical direction of the program was the joint responsibility of several agencies of the Government.

The imprecise statements of "who was to do what" in connection with CORONA allowed for a range of interpretations. The vague assignments of responsibilities caused no appreciable difficulties in the early years of CORONA when the organization was small and the joint concern was primarily with producing as promised, but they later (1963-1964) became a source of friction between CIA and the Air Force. Bissell gave this description on how the program was initially managed:

> "The program was started in a marvelously informal manner. Ritland and I worked out the division of labor between the two organizations as we went along. Decisions were made jointly. There were so few people involved and their relations were so close that decisions could be and were made quickly and cleanly. We did not have the problem of having to make compromises or of endless delays awaiting agreement. After we got fully organized and the contracts had been let, we began a system of management through monthly suppliers' meetings—as we had done with the U-2. Ritland and I sat at the end of the table, and I acted as chairman. The group included two or three people from each of the suppliers. We heard reports of progress and ventilated problems—especially those involving interfaces among contractors. The program was handled in an extraordinarily cooperative manner between the Air Force and CIA. Almost all of the people involved on the Government side were more interested in getting the job done than in claiming credit or gaining control."

The schedule of the program, as it had been presented to the CORONA group at its meeting in San Mateo in late March 1958, called for a "count-down" beginning about the first of July 1958 and extending for a period of 19 weeks. It was anticipated that the equipment would be assembled, tested, and the first vehicle launched during that 19-week period, which meant that the fabrication of the individual components would have had to be completed by 1 July. By the time Bissell submitted his project proposal some 3 weeks later; it had become apparent that the earlier scheduling was unrealistic. Bissell noted in his project proposal that it was not yet possible to establish a firm schedule of delivery dates, but that it appeared probable that the first firing could be attempted no later than June 1959.

It is pertinent to note here that there was no expectation in 1958 that CORONA would still be operating over a decade later. The CORONA Program got under way initially as an interim, short-term, relatively low-risk development to meet the intelligence community's requirements for area search photographic reconnaissance pending successful development of other more sophisticated systems planned for WS-117L. The original CORONA proposal anticipated the acquisition of only 12 vehicles, noting that at a later date it might be desirable to consider whether the program should be extended, with or without further technological improvement.

Having settled on the desired configuration and having received Presidential approval of the program and its financing, the CORONA management team moved forward rapidly with the contractual arrangements. The team of contractors for CORONA differed from the team on the WS-117L as a consequence of selecting Itek's panoramic camera and the film recovery approach. Itek was brought in as one of the two major subcontractors to Lockheed (General Electric being the other). However, to soften the financial blow to Fairchild, Itek was made responsible for the design and development of the camera subsystem with Fairchild producing the camera under subcontract to Itek. Under this arrangement the first 20 cameras were produced. [Following a contract award to Itek in 1959 for a new camera design known as the C''' (C triple prime) (Fig. 1-12), the cameras for the remainder of the program were produced by Itek.] The contractor team continued throughout the CORONA Program, although in 1961 the relationship was changed as a cost savings measure to the Government to that of Associate Contractors. The contractor relationships on the CORONA Program were as friendly and cooperative as any that could have been set up, and this team dedication to this program is one of the primary reasons for the overall success which the program enjoyed. The final contractors were selected on 25 April, and a work statement was issued to Lockheed on that date. The contractors began systems design on 28 April and submitted them for first review on 14 May. The designs were frozen on 26 July 1958.

Thus, by mid-1958, the contractor's responsibilities to the program were moving well toward meeting the goal of a first launch no later than mid-1959. The Government side, however, was running into difficulties. The first had to do with money, the second with cover, and the two were inextricably intertwined. The ▓▓▓▓▓ cost estimate for the 12-vehicle program had assumed that the cost of the Thor boosters would be absorbed by the Air Force by diverting them from the cancelled WS-117L subsystem. That assumption proved to be incorrect. An additional ▓▓▓▓▓ had to be found to pay for the 12 Thors. Further it had been decided that an additional four launch vehicles would be required for testing of launch, orbit, and recovery procedures; and, that an additional three would be required for biomedical launches in support of the CORONA cover story. ARPA could not see its way clear to making DoD funds available merely for testing or for cover support when there were other DoD space programs with pressing needs for money. Consequently, CORONA management had to go back to the President for approval of a revised estimate.

It had also become apparent to the project managers that the original, but as yet unannounced, cover story conceived for the future CORONA launchings (an experimental program within the first phase of WS-117L) was becoming increasingly untenable. WS-117L had by then become the subject of fairly widespread public speculation identifying it as a military reconnaissance program. It was feared that linking DISCOVERER to WS-117L in any way would inevitably place the recon-

MCS HISTORY

Fig. 1-12 — C Triple Prime (C''') Camera (KH-3) in test stand

naissance label on DISCOVERER; and, given the hostility of the international political climate to overflight reconnaissance, there was the risk that the policy level of Government might cancel the program if it should be so identified. Some other story would have to be contrived that would dissociate CORONA from WS-117L and at the same time account for multiple launchings of stabilized vehicles in low polar orbits and with payloads being recovered from orbit.

It was decided, therefore, to separate the WS-117L photoreconnaissance program into two distinct and ostensibly unrelated series: one identified as DISCOVERER (CORONA-Thor boost) and the other as SENTRY or SAMOS (previously discussed). A press release announcing the initiation of the DISCOVERER series was issued in mid-January 1959 identifying the initial launchings as tests of the vehicle itself and later launchings as explorations of environmental conditions in space. Biomedical specimens, including live animals, were to be carried into space and their recovery from orbit attempted.

The new CORONA cover concept, from which the press release stemmed, called for a total of five biomedical vehicles; and three of the five were committed to the schedule under launchings three, four, and seven. The first two were to carry mice and the third a primate. The two uncommitted vehicles were to be held in reserve in event of failure of the heavier primate vehicle. In further support of the cover plan, ARPA was to develop two radiometric payload packages designed specifically to study navigation of space vehicles and to obtain data useful in the development of an early warning system (the planned MIDAS infrared series). It might be noted here that only one (DISCOVERER III) of the three planned animal carrying missions was actually attempted, and it was a failure. ARPA did develop the radiometric payload packages, and they were launched as DISCOVERER's XIX and XXI in late 1960 and early 1961.

The photoreconnaissance mission of CORONA necessitated a near polar orbit, either by launching to the north or to the south. There are few suitable areas in the continental United States where this can be done without danger of debris from an early in-flight failure falling into populated areas. Cooke Air Force Base near California's Point Arguello met the requirement for down-range safety because the trajectory of a southward launch would be over the Santa Barbara channel and the Pacific Ocean beyond. Cooke AFB was a natural choice because it was the site of the first Air Force operational missile training base and also housed the 672nd Strategic Missile Squadron (Thor). Two additional factors favored this as the launch area: (1) manufacturing facilities and skilled personnel required were in the near vicinity, and (2) a southward launch would permit recovery in the Hawaii area by initiating the ejection/recovery sequence as the satellite passed over the Alaskan tracking facility. The name of this base was changed from Cooke to Vandenberg AFB in October 1958.

Unlike the U-2 flights, launchings of satellites from U.S. soil simply could not be concealed from the public. Even a booster as small as the Thor (small relative to present day space boosters) launches with a thunderous roar that can be heard for miles; the space vehicle transmits telemetry that can be intercepted; and the vehicle can be detected in orbit by radar skin-track. Although the fact of a launch having been made could not be concealed, maintenance of the cover story for the DISCOVERER series required that the launchings of the uniquely configured photographic payloads be closed to observation by uncleared personnel. Vandenberg was excellent as a launch site from many standpoints, but there was one feature of it that posed a severe handicap to screening the actual launches from unwanted observation. This handicap was that the heavily traveled Southern Pacific Railroad passes through it. Operational parameters, including the requirement for daylight recovery and for seven denied area passes during daylight with acceptable sun angles, dictated a launch from Vandenberg in the early afternoon. Trains passing through the area broke up this afternoon launch window into a series of successive windows, some of which were of no more than a few minutes' duration. Even today, the space program at Vandenberg is plagued by having to time the launches to occur during one of the intervals between passing trains.

Recovery presented problems in the early development period and throughout the early operational period. The planned recovery sequence involved a series of maneuvers, each of which had to be executed to near-perfection or recovery would fail. Immediately after injection into orbit, the Agena vehicle was yawed 180° such that the recovery vehicle faced to the rear. This maneuver minimized the control gas which would be required for reentry orientation at the end of the mission and protected the heat shield from what at that time was a deep concern for molecular heating. (Later in the J-3 design when these concerns had diminished, the vehicle was flown forward until reentry.) When reentry was to take place, the Agena would then be pitched down through 60 degrees to position the satellite reentry vehicle (SRV) for retrofiring. Then the SRV would be separated from the Agena and spin-stabilized by firing the spin rockets to maintain it in the attitude given it by the Agena. Next, the retrorocket would be fired slowing down the SRV into a descent trajectory, and the spin of the SRV would be cancelled by firing the despin rockets. The retrorocket thrust cone was then separated, followed by the heat shield and the parachute cover. The drogue (or deceleration) chute would then deploy, and finally the main chute would open to lower the capsule gently into the recovery area. The primary recovery technique involved flying an airplane across the top of the descending parachute, catching the chute or its shrouds in a trapeze-like hook suspended beneath the airplane, and then winching the recovery vehicle aboard. Initially, C-119 aircraft were used, but C-130 aircraft replaced them later in the program. If the air catch failed, the recovery vehicle was designed to float long enough for a water recovery by a helicopter launched from a surface ship. The recovery sequence and aerial recovery were similar to those of the HEXAGON program shown in Section 3 of Part II.

While the vehicle was still in the construction stage, tests were conducted of the aerial recovery technique by the 6593rd Test Squadron with poor results. Of 74 drops using personnel type chutes, only 49 were recovered. Using one type of operational drop chute, only four were recovered out of 15 dropped, and an average of 1.5 aircraft passes were required for the hookup. Eleven drops of another type of operational chute resulted in five recoveries and an average of 2 aircraft passes for each snatch. Part of the difficulty lay in weak chutes and rigging and crew inexperience; however, the most serious problem was the fast drop rate of the chutes. Parachutes that were available to support the planned weight of the recovery vehicle had a sink rate of about 33 feet per second. What was required was a sink rate approaching 20 feet per second so that the aircraft would have time to make three or four passes, if necessary, for hookup. Fortunately, by the time space hardware was ready for launching, a parachute had been developed with a sink rate slow enough to offer a reasonable chance of air recovery.

The launch facilities at Vandenberg AFB were complete, and the remote tracking and control facilities which had been developed for WS-117L were ready for the first flight test of a Thor-Agena combination in January 1959. The count-down was started for a launch on 21 January; however, the attempt aborted at launch minus 60 minutes. When power was applied to test the Agena hydraulic system, certain events took place that were supposed to occur only in flight. The explosive bolts connecting the Agena to the Thor detonated, and the ullage rockets fired. The Agena settled into the fairing attaching it to the Thor but did not fall to the ground, however appreciable damage was done. A program review conference was held in Palo Alto two days after the launch failure to examine the possible causes of these events and to assess its impact on the planned CORONA launch schedule. Fortunately, the problem was quickly identified as a timer malfunction. The design was corrected, and the program was ready for resumption of test launches at the rate of about one per month.

At the conference, General Electric surfaced a new problem having to do with the stability of the nose cone during reentry. The cone was designed for a film load of 40 pounds, but the first missions would only be able to carry 20 pounds. GE reported that about 3 pounds of ballast would have to be carried in the forward end of the cone to restore stability. The program officers decided to add an instrument package as ballast for diagnostic purposes and for support of the biomedical cover story, thus converting what could have been dead weight into an extra advantage for this test series.

The test plan contemplated arriving at full operational capability at a relatively early date through sequential testing of the major components of the system, beginning with the Thor-Agena combination alone; then adding the nose cone to test the ejection/reentry/recovery sequence; and finally installing a camera for a full CORONA systems test. Whatever confidence the project planners had in the imminence of success at the start, however, soon must have begun to wane. Beginning in February 1959 and extending through June 1960, an even dozen launches were attempted with eight of these vehicles carrying cameras. All twelve were failures, and no film capsules were recovered from orbit. Of the eight camera-carrying vehicles, four failed to achieve orbit. Of the four vehicles that went into orbit, three experienced camera or film failures, and the fourth was not recovered because of a malfunction of the reentry body spin rockets.

By 1981 state-of-the-art technology, the performance record of early CORONA launches would be totally unacceptable. But it must be remembered that this was the genesis of space exploration, with this ambitious and complex program pioneering in technical fields about which little was known. Even after the program had become operational and was routinely returning photography from space, there were uncleared scientists in the country who were of the opinion that the satellite approach to reconnaissance was not viable.

In the midst of these hectic times in 1959 there were some highlights that are very interesting, in retrospect. For example, the SRV on DISCOVERER II, launched on 13 April 1959 had ejected on the 17th orbit as planned, but a timing malfunction caused by a human programming error resulted in the ejection sequence being initiated too early. The capsule was down, probably somewhere in the near vicinity of Spitsbergen Island north of Norway. In fact, there were later reports that the falling capsule had actually been seen by Spitsbergen residents. The Air Force announced on the 16th that the Norwegian government had authorized a search for the capsule, which would begin the following day. Planes scoured the area, and helicopters joined the search on the 20th. Nothing was found, however, and the search was abandoned on the 23rd. There was speculation at the time and some actual reconnaissance by the Norwegian Air Force which indicated that the capsule may have been recovered by a Soviet rather than an American recovery team. The incident later became the subject of a book by Alistair MacLean, Ice Station Zebra, and of a 1968 movie of the same name. The fictionalized version departed rather substantially from the facts, and it is clear that no one who was involved in the CORONA Program acted as a technical consultant to the film producer.

Another example believed to be of interest is an incident associated with DISCOVERER III. Since part of the CORONA cover story called for some of the early launches to be biomedical vehicles, an experiment involving four live mice was organized for DISCOVERER III. Black mice were chosen in order to ascertain the possible hair-bleaching effects of cosmic rays. The mice were members of the C-57 strain, a particularly rugged breed. They had been "trained", along with 60 other mice, at the Air Force Aeromedical Field Laboratory at Holloman AFB. They were 7 to 10 weeks old and weighed slightly over 1 ounce each. A 3-day food supply was provided which consisted of a special formula containing peanuts, oatmeal, gelatin, orange juice, and water. Each mouse was placed in a small individual cage about twice its size, and each had a miniscule radio strapped to its back to monitor the effects of the space trip on heart action, respiration, and muscular activity. The lift-off on June 1959 was uneventful, but instead of injecting approximately horizontally, the Agena apparently injected downward driving the vehicle into the Pacific Ocean. The second try at launch several days later with a backup mouse "crew" was also a near abort when the capsule life cell humidity sensor suddenly indicated 100 percent relative humidity. The panic button was pushed and troubleshooters were sent up to check. They found that when the vehicle was in a vertical position, the humidity sensor was directly beneath the cages and it did not distinguish between plain water and urine. The cages were dried out and the vehicle launched; however, it again was unsuccessful, falling into the ocean.

Another amusing experiment on an early flight was a means for concealing the payload doors from inquisitive eyes while the vehicle was on the launch pad. The scheme that was hurriedly devised was to cover the doors with fairings made of paper under which were strung two lengths of piano wire with ping-pong balls attached to the forward ends of the wires. The thought was that as the vehicle accelerated during launch, the air flow along the vehicle skin would blow the ping pong balls to the rear, thus tearing off the paper and exposing the payload doors. The strip-away fairing was tested by attaching it to the side of a sports car and driving the car at high speed along the Bayshore Freeway (U.S. Highway 101) late one evening. The test proved two things: (1) that the fairing would tear off as intended, and (2) that the California Highway Patrol could easily overtake a vehicle traveling at 90 miles per hour. Since the test indicated a "go" situation, at 2 a.m. on a foggy, chilly morning under a blaze of floodlights, a few cents worth of paper, piano wire, and ping-pong balls were affixed to a multimillion dollar space vehicle. In parallel with the paper/ping-pong ball fix, a security and environmental shroud was being designed. These shrouds proved extremely valuable to the program in protecting the sensitive thermal surface from salt water spray.

The very first recovery (according to plan) of an object orbited in space was made by the United States on 11 August 1960. The vehicle was DISCOVERER XIII, launched as a repeat of DISCOVERER XII diagnostic flight without camera and film. On 10 August 1960 the vehicle was launched and successfully inserted into orbit. The recovery package was ejected on the 17th orbit, and retrofiring and descent were normal, except that the capsule came down well away from the planned impact point. The nominal impact area was approximately 250 miles south of Honolulu where C-119 and C-130 aircraft circled awaiting the capsule's descent. The splash-down occurred about 330 miles northwest of Hawaii. The airplanes were backed up by surface ships deployed in a recovery zone with dimensions of 250 by 550 miles. Although beyond the range of the airborne recovery aircraft, the capsule descended near enough to the staked out zone to permit an attempt at water recovery. A ship reached the scene before the capsule sank and fished it out of the ocean. (This water recovery technique developed for and perfected by the CORONA Program has been used extensively by the U.S. manned spacecraft programs with the recovery of astronauts after splashdown in the ocean.) Since DISCOVERER XIII was a diagnostic flight, it was given extensive publicity concerning this success in recovering an object from orbit, in large measure to support the cover story of DISCOVERER as an experimental space series. President Eisenhower displayed the capsule to the press, and it was later placed on exhibit in the Smithsonian Institution for public viewing.

The successful recovery of a CORONA SRV, even though it contained no film, was the first assurance of imminent success for a photographic reconnaissance satellite capability.

The next vehicle, DISCOVERER XIV, was launched on 18 August 1960, just one week after the successful water recovery of the DISCOVERER XIII capsule. The vehicle carried a camera and a 20-pound film load. The camera operated satisfactorily, and the full load of film was exposed and transferred to the recovery capsule. The satellite recovery vehicle was ejected on the 17th pass, and the film capsule was recovered by air snatch.

As expected, the resolution was lower than on the U-2 photography; however, this one CORONA mission yielded more photographic area coverage than the total of all U-2 missions that had been flown over the Soviet Union.

The primary purpose of the imagery from the panoramic cameras of the CORONA system was to collect essential intelligence on foreign areas, but the satellite imagery program also included the aspect of providing for the accurate geographic orientation of military and other essentials features on target charts, geodetic positioning for missile system operations, and improving the general accuracy of maps and navigation charts. To cope with these types of orientation and mapping, charting, and geodetic requirements, specialized frame-type cameras were developed to obtain lower resolution but geometrically strong imagery—that is, imagery whose locational (geodetic) characteristics are more accurate because all of the features on a frame are imaged at the same instant of time.

The U.S. Army Corps of Engineers had started geodetic programs utilizing artificial satellites in 1957. The first of these programs was to track the early satellites such as GREB and Echo from islands and, by using the known orbit and time of track, determine the location of the tracking station with respect to the center of mass of the earth and subsequently to the North American datum. By September 1960, it became obvious that our knowledge of orbital perturbations was not sufficient to achieve the precision required. The solution to this problem was the Sequential Correlation of Range (SECOR) program. This program used the principal of trilateration to eliminate the orbital parameters from the solution of the location. This was accomplished by launching a multiple frequency transponder into orbit and measuring the range to the satellite from four stations at the same instant in time. Three of the stations were on the North American datum and the fourth was at a location with undetermined coordinates. The three known stations fixed the satellite in space and the location of the fourth was then determined from three fixed in the satellite. This system became the primary geodetic position system and involved a number of transponder launches. The early launches were designed to use the GREB ball but securing launches for these balls was very difficult, and the number of vehicles that could carry these payloads was very small. This situation was discussed with the DISCOVERER program office and with the integrating contractor, Lockheed Missiles and Space Company (LMSC). Bert Bulkin of LMSC and Major Albert W. Johnson from the SAFSP program office indicated that the Agena aft instrument rack had space for separable packages that were small, and in fact, had launched other spacecraft such as the Amateur Radio Operators Satellite (OSCAR). Bert Bulkin and William Williamson from the Corps of Engineers designed a SECOR spacecraft based on the OSCAR package (same size and weight) which the Air Force program office approved for launch on the DISCOVERER and subsequent CORONA programs. The early launches carried non-separable SECOR transponders on the DISCOVERER launches of 23 October, 5 November, and 12 December 1961. These were all successful and proved the transponder design.

DISCOVERER XX was the first of a dozen launches extending over a period of three years carrying mapping cameras, a program sponsored by the U.S. Army which the President had approved for inclusion within the CORONA project. The purpose of the mapping program, which was known as ARGON (KH-5) (Fig. 1-13), was to obtain precise geodetic imagery to allow an extension of existing datum planes within the Soviet Union. The ARGON, which was operational between May 1962 and August 1964, had a focal length of 76 mm (3 inches) and used 5-inch-wide film. Of the 12 systems launched, (9 attained orbit) only 6 were recovered; however, because of its higher orbit (165 nautical miles) these missions were able to collect virtually complete worldwide coverage at a scale of approximately 1:4,000,000. The imagery from this system permitted a U.S. Army Engineers' contractor to compile the first photographic mosaic of an entire continent—Africa. This mosaic enabled DoD personnel to derive information on and to correct the major hydrographic, vegetation, terrain, and geological patterns shown on contemporary maps.

The CORONA launches, notwithstanding problems, proceeded at an amazing frequency, with 37 launches or launch attempts between April 1959 and 13 January 1962. After DISCOVERER XXXVII, the cover story for DISCOVERER had simply worn out. With the improved record of success and the near certainty of an even better record in the future, it seemed likely that there would be as many as two dozen launches per year for perhaps years to come. The cover story that DISCOVERER was an experimental series had ceased to be tenable, and no other cover story was available to account for the number of launches and their unique mission profiles. So, beginning with the 38th launch, CORONA missions were announced merely as being Air Force satellite launches. On 18 April 1962, the Air Force announced the issuance of a new directive classifying all information pertaining to military satellites and eliminating the DISCOVERER, SAMOS, and MIDAS series designators.

During 1961, Itek developed the MURAL (M) camera system (Fig. 1-14) which provided stereoscopic photography. It is an axiom of aerial reconnaissance that the information content of photography is improved by a factor of two and one-half times with stereo coverage. Thus, the introduction of the M system marked a major step forward in the CORONA Program.

The M system consisted of two C''' cameras on a common mount, one looking 15 degrees aft from vertical and the other 15 degrees forward. Each camera was fed from an individual supply spool (40 pounds of film) mounted on the back of the camera's main plate. The film was panoramically exposed through 70 degrees of lens cell assembly rotation. After exposure, the film from each camera was fed into individual takeup spools in a common cassette. When the forward-looking camera photographed a scene, this same scene would be photographed six frames later by the aft-looking camera, thus providing a 30-degree covergent angle for stereo photography. Simultaneous operation of both cameras was required for stereo photography. The M system configuration further improved CORONA reliability by mounting the two cameras back-to-back. Because the cameras operated (scanned) in opposite directions, they tended to offset any operating imbalances and thereby improved overall system dynamic balance. The M system was capable of a 6 to 7-day mission compared to the 3 to 4-day missions of the C''' and the earlier 1-day missions. The system was designed for nominal altitudes of 110 nautical miles. Dynamic resolution was 80 to 110 lines per millimeter. The first M system, mission 9031, was launched on 27 February 1962, and the stereo photography was excellent.

In 1962, two small frame cameras were introduced into the CORONA system; first an Index camera with 38-mm focal length lens and later a Stellar-Index (SI) camera (Fig. 1-15). The Index cameras were flown starting with the first MURAL mission on 27 February 1962, and the SI cameras were introduced on program flight number 52 on 29 September 1962. Thereafter,

Camera System with Cannister

Camera Showing Terrain and Stellar Lenses

Film Supply

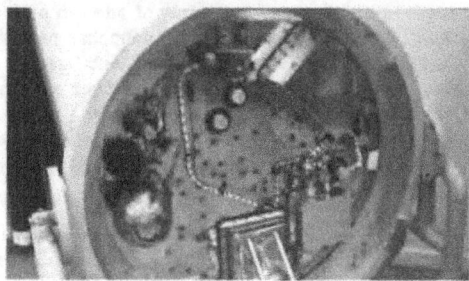
Clock, Pneumatics, and Film Chute

Camera Installed in the Structure

Fully Assembled ARGON System

Fig. 1-13 — ARGON (KH-5) Army Mapping Camera

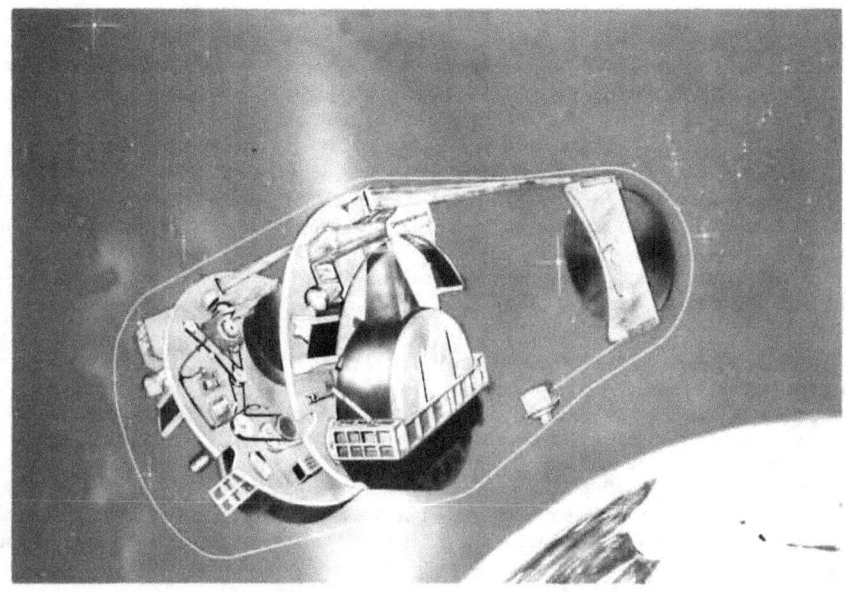

Fig. 1-14 — MURAL (KH-4) Twin Panoramic Camera System

Fig. 1-15 — Index and Stellar-Index Cameras

CORONA missions, with a few exceptions, carried an SI of some vintage. Both the Index and the Stellar Index cameras were manufactured by Itek. The Index had a terrain or down-looking lens only, of 38-mm focal length; the Stellar Index camera also had a 38-mm terrain lens but included an 80-mm focal length stellar lens as well; the terrain camera used 70-mm film, the stellar camera used 35-mm film. Coverage was world-wide, providing supplementary coverage to main camera imagery for more accurate geographic orientation, initially of intelligence targets and later for mapping purposes. A calibration of the knee (90-degree) angle between the Index and Stellar units, as well as the distortion of these two lenses, was established on a precision goniometer. This calibration in conjunction with the mid-exposure time of the shutters (panoramic cameras, Index cameras, and Stellar Index camera) established the position in space for each photographic acquisition. Many relatively small scale maps were made from this combined photography.

As a side benefit, the CORONA Performance Evaluation Team (PET) used the terrain imagery extensively as an aid to main camera photography evaluation. On many occasions throughout a mission, the panoramic imagery would appear slightly degraded without any indication of camera malfunction. By examining the terrain photography from the SI, which covered such a wide area in each frame, cloud patterns and heavy haze not detectable in the pan photography were clearly visible. One of the initial and most challenging tasks of the PET was to educate some of the key management personnel on the subject of image degradation resulting from the atmosphere, "look angle," and sun angle in continuously varying combinations, without the capability to continuously compensate through film/filter/exposure optimization.

In early 1963, the boosting capacity of the first-stage Thor was substantially increased by strapping on three small solid propellant rockets which were jettisoned after firing. This thrust-augmented Thor (TAT), was first used for the launching of the heavier LANYARD camera system (Fig. 1-16), developed by Itek under contract to SAFSP as a follow-on to the SAMOS E-5 camera. It was a panoramic spotting camera with an oscillating lens cell which viewed a large mirror aimed at a 45-degree angle toward the earth. Movement of the mirror enabled the system to produce stereo or mono-photography. The 5-inch film was fed from a supply spool (capacity = 8,000 feet/ 80 pounds of film) to the platen for exposure and then to a takeup cassette in the recovery system. Servo drive rollers controlled the film movement. Because of the limited scan angle of the lens, a roll joint (Z) was incorporated in the structure to increase the scan capability. The effective focal length of the optical system was 66 inches.

The LANYARD camera system had been intended for interim use only until the Air Force's spotting camera system GAMBIT (KH-7) was fully developed and operational. There were three launches of this system; program flight numbers 61, 64, and 68 on the dates 18 March, 18 May, and 30 July 1963, respectively. The first vehicle failed to achieve orbit due to failure of hydraulic steering on the Agena booster. The second achieved orbit but the camera never received the turn-on command—the satellite reentry vehicle (SRV) was recovered after 33 revs. The third achieved orbit and the camera functioned, but only for 23 revs. The system was designed for a 4-day mission but with indication of camera failure, the RV was brought in after 32 revs. The photography had covered 450,000 square nautical miles primarily of communist areas for intelligence purposes. The LANYARD system did not have an active thermal system to control focus and as it turned out on the abbreviated mission, none of the photography had been collected at best focus. However, the using community recorded examples wherein the design specification of 5-foot GRD had been achieved.

By mid-1963, after 68 of the program total of 145 flights, the CORONA program had launched systems consisting of 24-inch single panoramic cameras [C, C' (or C prime), C''' (or C triple prime)], twin 24-inch panoramic cameras (MURAL), an ARGON Mapping Camera, the long focal length panoramic camera LANYARD (L), and the Stellar Index camera.

Program flight number 69, launched on 24 August 1963, introduced the next major upgrading of the CORONA system, the first configuration with two SRV's, commonly referred to as buckets. The new modified system, known as the JANUS or CORONA J-1 (Fig. 1-17), retained the MURAL stereoscopic camera concept, but now with two SRV's in the system, the film capacity was increased to 160 pounds, approximately 32,000 feet of film per mission. Also, on the dual recovery series, two Stellar Index subsystems were carried on each vehicle, one SI feeding its film into SRV-I, the second into SRV-II.

The SI cameras were carried on most CORONA flights between February 1962 and August 1967, at which time a new SI camera came into use. This camera was developed by the Fairchild camera and Instrument Company (FCIC) under contract to SAFSP. In an effort to upgrade the frame camera performance for the MC&G community, the focal length of the terrain camera was increased to 3 inches (as on the ARGON), two stellar cameras with 3-inch focal length lenses were incorporated, and the film width for the terrain or down-looking camera was increased from 70 mm to 5 inches. The new camera had the designation of Dual Improved Stellar Index Camera, commonly referred to by its acronym DISIC. The DISIC resolution was 30 to 60 meters (110 to 200 feet). Coverage was world-wide, initially providing supplementary coverage to main camera imagery for accurate geographic orientation—later versions could be operated independently, exclusively for mapping purposes. Six DISIC cameras were launched on the CORONA, J-3 or CR series (Fig. 1-18), the first on program flight number 120 on 15 September 1967, the last on program flight number 134 on 23 July 1969. The next nine CORONA missions carried a mix of the 38-mm SI's, the last two missions flew without an SI of any description.

The J-3 or Constant Rotating (CR) system was the last in the CORONA evolution. Apart from the significantly improved photographic capability of the hardware, the most significant advance represented by the J-3 was in the flexibility it allowed in command and control of the camera operations.

There were a total of seventeen J-3 systems flown. The recovery of the last J-3 imagery on 31 May 1972 brought to a close the active portion of the CORONA program, though most of the personnel who had gained extensive experience continued in important roles on other space programs. The technological improvements engineered under CORONA advanced the system from a single, vertical pointing panoramic camera having a design goal of 20 to 25-feet ground resolution and an orbital life of 1 day, to a twin camera panoramic system producing stereo-photography. From this point, it became a dual recovery system with an improvement in ground resolution to approximately 7 to 10 feet with twice the film load, to finally the J-3 system with a constant rotator, selectable exposure, and filter controls, planned orbital life of 18 to 20 days, and yielding nadir resolution of 5 to 7 feet. It is important to note that the focal length remained constant at 24 inches throughout this evolution. The dramatic increase in performance was brought about through improvements in design, manufacturing and testing, improvements in thermal design together with a better understanding of orbital temperature effects on focus, new/improved films and processing, and improvements in the orbital vehicle stability.

L System Mockup

View of Large Berylium Mirror

Z-roll Joint

System in Test

Fig. 1-16 — LANYARD (KH-6) Panoramic Camera System

Fig. 1-17 — Artist's view of the J-1 (KH-4A) Camera System with dual SRV's

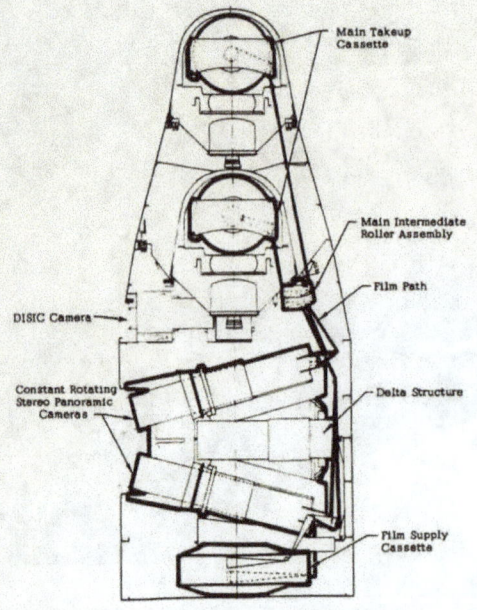

Fig. 1-18 — Major components of the J-3 (KH-4B) Subsystem

The totality of CORONA's contributions to U.S. intelligence holdings on denied areas and to the U.S. space program in general is virtually immeasurable. Its progress was marked by a series of notable firsts: (1) the first to recover objects from orbit; (2) the first to deliver intelligence information from a satellite; (3) the first to produce stereoscopic satellite photography; (4) the first to employ multiple reentry vehicles; and (5) the first satellite reconnaissance program to pass the 100+ mission mark. The CORONA program synopsis is presented in Fig. 1-19.

By March 1964, CORONA had photographed 23 of the 25 Soviet ICBM complexes then in existence; 3 months later it had photographed all of them. The value of the CORONA derived intelligence effort is given dimension by this statement in a 1968 intelligence report: "No new ICBM complexes have been established in the U.S.S.R. during the past year." This statement was made because of the confidence held by the analyst that if an ICBM complex were there, then CORONA photography would have disclosed it. In addition to the program's intelligence contributions, it must not be overlooked in summarizing that the CORONA vehicle had been the carrier for <u>mapping cameras</u> during the initial decade of satellite reconnaissance.

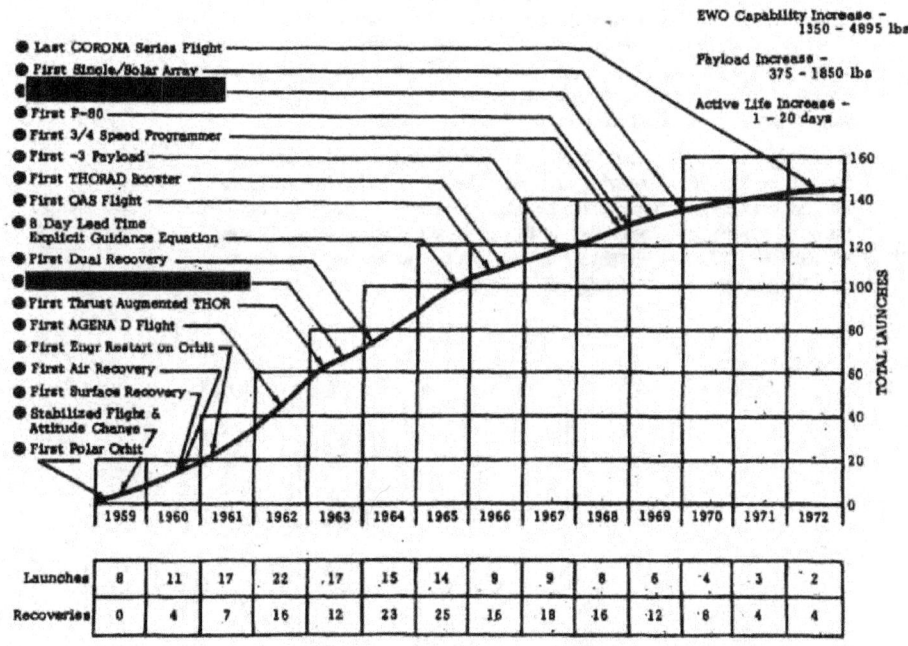

Fig. 1-19 — CORONA Program synopsis

Thus, the program that was so "informally" started as an interim, short-term development had performed superbly for a decade. As recounted recently by Mr. James Plummer, initial Lockheed CORONA Program Manager (now Executive Vice President of Lockheed Missiles and Space Company), "The streamline management techniques and the later incentive contracting methods developed by the CIA and Air Force, in my mind, were as important as the technical problems that were presented in this new field."

Near the close of the program, as previously mentioned, it was suggested by General Lew Allen, Jr., then Director of SAFSP, that a history should be compiled to preserve details of this pioneer in satellite reconnaissance. In response, under CIA sponsorship, a classified history was published and a classified movie was made entitled "A Point in Time." In addition, using recovered hardware from the last flight, development models from the J-3 program, and photographic records from the memorable flights, a classified museum display was set up in the National Photographic Interpretation Center (NPIC), Washington, D.C. (see Fig. 1-20). In his speech dedicating the museum, Mr. Richard Helms, the Director of the Central Intelligence Agency, said:

> "It has been confidence in the intelligence estimates that has allowed President Nixon to enter into the Strategic Arms Limitation Talks and to sign the Arms Limitation Treaty this month. There can be no doubt that the photoreconnaissance satellite represents the primary means of verification for SALT or that CORONA, the program which pioneered the way in satellite reconnaissance, deserves the place in history which we are preserving through this small museum display."

> "A Decade of Glory as the display is entitled, must for the present remain classified. However, as the world grows to accept satellite reconnaissance, we hope it can be transferred to the Smithsonian Institute where the American public can view the work, and the men of CORONA, like the Wright brothers, can be recognized for the role they played in the shaping of history."

Fig. 1-20 — CORONA Program museum display

THE DEFENSE WEATHER SATELLITES

> Note: The Defense Meteorological Satellite Program (DMSP) has continued to provide valuable support to the reconnaissance satellite programs over the years, including of course the HEXAGON Mapping Camera Program. For this reason, following the discussion of early program development, a synopsis is provided covering DMSP advancements in technology and performance. (See Fig. 1-21.)

The justification for development of a military weather reconnaissance satellite lay not solely in the possible delay in NASA NIMBUS* operations, but in two other factors not generally recognized outside the National Reconnaissance Program.[15] First, in 1961, CORONA was starting to return regular packages of photographs of the Soviet Union, and the percentage of cloud-free coverage in the photography seemed likely to improve if timely weather information could be fed into the CORONA operations program. There was an excellent possibility that political objections to NIMBUS operations might limit the quality and quantity of information made available to the U.S. military services. Therefore, CORONA (and other military reconnaissance satellites programmed for later operations) could well be handicapped if they had to depend on data abstracted from a weather satellite program controlled by individuals who honored the "space for peaceful uses" theme NASA continued to proclaim. Second, the timing and quality of weather reconnaissance could not be guaranteed if NIMBUS or some successor civil system were the provider; the operators of various reconnaissance satellite programs were unlikely to have much influence on the operational control of the satellite or on the disposition of its products. If reconnaissance authorities did intrude, there was the danger of public protests that would in effect advise the Soviets that the United States needed weather information (chiefly cloud cover data), and hence presumably was operating reconnaissance satellites. In the early 1960's, those operations had gone underground and there was no immediate prospect of their surfacing in the near future.

As early as 17 November 1960, while the SAMOS program reorganization was in progress, but before the unique structure of General Greer's organization had been clearly established, the Radio Corporation of America (RCA) had proposed to the Air Force the development of a cloud cover reconnaissance satellite system.

The submission, which went to General Greer's deputy, Colonel H. L. Evans, envisioned a 300-pound payload of television components, two readout stations, a satellite control center, and contractor provision of cloud cover analysis services. RCA urged that the system would fill a functional gap in the array of military satellites by exploiting techniques and equipments, many of which had been flight proven in the course of NASA's TIROS weather satellite program. The use of proven or "off-the-shelf" system elements was a particular attraction to both the Directorate of Defense Research and Engineering (DDR&E) and the Undersecretary of the Air Force, Dr. Charyk, who had recently acquired cognizance over the total reconnaissance satellite program Even as early as November 1960 it was clear that both of these authorities would have to approve before a system could be funded for development.

---

*NIMBUS was a NASA "advanced" weather satellite that orbited the earth at an altitude of 500 miles in a north-south direction, taking a band of pictures during each orbit. NIMBUS was equipped with solar panels for electrical power and a horizon scanner to keep the camera pointed toward the earth. Through the system's television cameras, picture signals were sent to an on-board tape recorder where they were stored until "played back" on command from the ground. NIMBUS also carried a radiometer to measure infrared waves coming from the earth.

MCS HISTORY

Appreciating that more was involved than merely evaluating a contractor proposal for a new satellite system, Colonel Evans and his immediate staff devoted some 2 months to its examination and to considering how they might overcome some obvious obstacles to securing system approval. The chief technical difficulty was selecting a launch vehicle; the proposed payload was too heavy for any of the available probe rockets and too light to warrant use of the Thor-Agena. Piggy-back modes were considered in detail and then discarded because of their possible degrading effect on the basic Thor-boosted satellites—which were mostly carrying CORONA payloads.

Having weighed all the available evidence, Colonel Evans in February 1961 suggested to the chief of BMD's Scout booster office, Lieutenant Colonel D. A. Stine, that it might be feasible to develop a variant of RCA's proposed cloud cover satellite for launch via the Scout. A successful combination would be relatively inexpensive, both in payload and booster elements, and it would serve a highly useful function in supporting a variety of Air Force missions—including SAMOS. At Colonel Evans' direction, Stine and his program office people, with the continued assistance of RCA consultants, put together and formally submitted a preliminary development plan. In endorsing it, Colonel Evans directed that it be expanded to include provisions for testing direct readout techniques during system operation.

After surviving a massive overhaul of the Air Research and Development Command in April 1961, and reassignment of some programs, the cloud cover satellite attracted the personal attention of Undersecretary Charyk. On 21 June in the aftermath of a presentation involving several of the "Five Year Plan" systems, he quietly abstracted the main elements and told General Greer's group to put together a "minimum" proposal involving a four-vehicle program. Five days later, Greer approved the "minimum plan" drawn up by his people and sent it forward to Charyk. On 11 July, Charyk submitted it to the DDR&E with a request for approval and funding.

By 15 July 1961, Dr. Harold Brown, DDR&E Chief, had advised Charyk that he would support the "minimum program" if it could be clearly demonstrated that the system had advantages over an expanded TIROS development. The Undersecretary convinced Brown during a conversation on 19 July and later that day telephoned General Greer that DDR&E had approved the proposal— subject to a set of special conditions. Those conditions, though unusual, were nonetheless implied by elements of the development plan and by the 12 July instructions on program security. Essentially, the program was to be based on fixed price contracts, was to be continued only so long as flight schedules remained valid, had to be entrusted to contractors aware of and willing to accept the schedule requirements, and was to be given a new home and conducted under special security provisions. The purpose and effect of the qualifications were clear: in no manner was the "normal" Air Force to become aware of the program's objectives, schedules, or techniques. Use of the widely known term "MISS" (Meteorological Information Satellite System) was to cease.*

A special security policy statement that achieved the desired ends was prepared on the day the instructions arrived in Los Angeles. Like the basic SAMOS under which it now sheltered, the program was exempted from customary review and approval channels, from routine reporting requirements, and from the halo of publicity routinely erected over any military space program that received significant funding support.

---

* "MISS" was earlier used as the acronym for the abortive "Man In Space Soonest" proposals of 1958-1959.

BIF-059W-23422/82
Handle Via
BYEMAN/TALENT KEYHOLE
CONTROL SYSTEMS JOINTLY

Fig. 1-21a — Early weather reconnaissance system

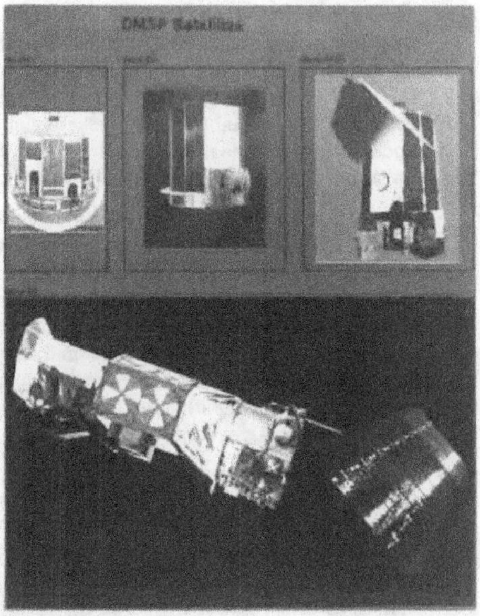

Fig. 1-21b — DMSP System evolution

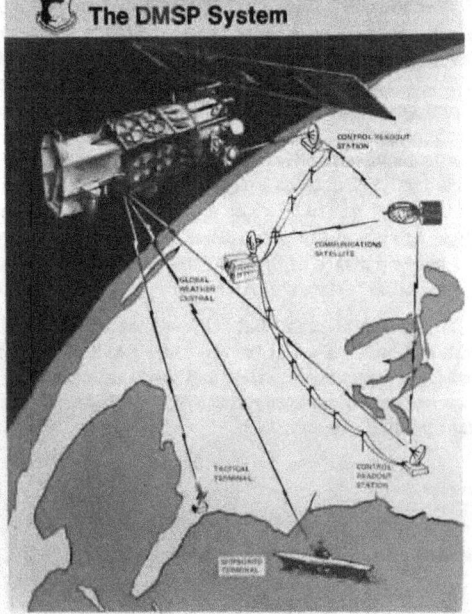

Fig. 1-21c — DMSP System network

In the next two weeks, the program acquired: a director, Lieutenant Colonel T. O. Haig; a name, Program 35;* a cost structure derived from payload (RCA) and vehicle contractor (Chance Vought) data; and a set of accepted procedures. In point of fact, full and final approval from DDR&E still had not been received, but there seemed little doubt that it would come.

Dr. Brown on 5 August approved the program Charyk had proposed. Later that day the Air Force Undersecretary telephoned General Greer the authorization to proceed with contractual commitments as appropriate.

By 31 August, Colonel Haig's embryonic program office had prepared a highly detailed development plan—in three copies circulated only to those with a "must know" status. The document defined the purpose and approach of Program 35, as then understood, and identified the chief technical and operational characteristics of the satellite development.

The goal of Program 35 was a weather observation satellite system that would enhance the effectiveness of SAMOS operations and improve the accuracy of cloud cover predictions for other military satellites. The development was necessary, in large part, because the extant NASA programs had deficiencies in program development time scales, security, and program management controls.

The 120-pound satellite, a 10-sided polyhedron 23.5 inches across and 21 inches high, was to be spin stabilized on an axis perpendicular to the plane of the orbit. The camera, fixed at 90 degrees from the spin axis, would point directly toward the earth once each time the satellite rotated. At programmed intervals, the television camera would take pictures of an 800-mile-square area on the earth, the exposures being made when horizon sensors indicated that the lens was vertical to the earth. The images could either be recorded on tape or read out by any suitably equipped ground station within range. Spin axis orientation was to be controlled by a magnetic torque system developed by RCA and proposed for NASA's TIROS I. NASA had rejected the technique as impossible.

With the satellite in a sun-synchronous 400-mile polar orbit, the system would provide 100 percent daily coverage at latitudes above 60 degrees and 55 percent coverage at the equator. When readout was undertaken during the western hemisphere portion of a pass that included photography, video data on eastern hemisphere cloud cover could be transmitted to the Global Weather Center at Offutt Air Force Base within one hour of its being observed. In a less favorable pass sequence, readout would always be possible within three orbits of sensor activity. Should it prove desirable, either for test purposes or for an actual operation, a direct real-time readout mode could be employed to feed cloud pictures to any ground station within range.

None of those who designed the program expected to encounter any serious technical problems during development. The TIROS type cloud cover sensor system had been flight proven. The Scout vehicle had been little tested but seemed potentially reliable enough (five successes in seven flights). The satellite would rely on proven satellite control and readout systems and stations, with standard airborne command, control, and readout components from such programs as ADVENT and MIDAS completing the technical equipment. Although much of the equipment was experimental in character, it did exist and it was available from contractors who had experience in its fabrication.

---

*At various times, the Air Force-DoD weather satellite program was known as Program 35, Program 698 BH, and Program 417.

Funds were to come from SAFSP resources through an SSD channel, with program management entirely concentrated in the program office. The contractor structure included RCA (spacecraft), Chance Vought (prime booster contractor), Minneapolis Honeywell (guidance and control), and (as solid fuel rocket fabricators) Aerojet General, Thiokol, and Allegheny Ballistic Laboratories of Hercules Powder Company. Assembly, checkout, and launch were responsibilities of the 6565th Test Wing (later the 6595th Aerospace Test Wing), under SSD control. Satellite control functions were to be exercised by the ▮▮▮▮▮▮▮▮▮▮▮▮▮▮▮▮▮▮ The Air Weather Service would do weather analysis in the Global Weather Center located in the SAC underground headquarters at Offutt Air Force Base, Nebraska.

The actual payload, weighing 120 pounds, would consist of a vidicon-camera recorder system similar to the wide angle sensor used in TIROS I and II, plus stabilization and control devices. RCA estimated the sensor system would have an orbital life of 90 days.

The ingenious attitude control system depended on torqueing the axis perpendicular to the orbital plane through an electric-current loop around the perimeter of the satellite. The torque was generated by a command that caused current to flow around the loop in the desired direction.

Spin-up during injection of the fourth stage and payload into final orbit was followed by a decrease to 12 revolutions per minute under the impulse provided by a system of "yo-yo" weights. After rotation speed decreased to nine per minute, spin-up rockets could be used to re-energize the satellite.

Pulses generated by the horizon sensor drove a specialized computer which triggered the camera. The recorder component was capable of storing 32 frames of information. Seven frames were required to cover the area of interest for each pass, but the probability of exposing the first frame was only 50 percent, therefore fixed sequences of eight frames were planned.

Changes from the TIROS transmitters were mostly in the direction of transistorizing. Performance was little affected, although both weight and power requirements decreased. The transmitter broadcast a 62.5-kilocycle bandwidth signal at 250 megacycles per second. Frequency modulation would permit a total output bandwidth of 290 kilocycles with 2 watts of output power. The video data were to be collected by existing 60-foot antennas at Vandenberg and New Boston.

For practical purposes, the program office had not come into existence until August 1961 although the first flight was firmly scheduled for May 1962. In the months that followed there were serious development problems with the fourth stage rocket motor, and interorganizational matters with NASA had to be worked out.

Notwithstanding the many difficulties, the program continued its pace toward a May launch. On 25 April the first West Coast Scout launching failed from improper functioning of the third stage. By 1 May the initial Program 35 Scout had been mated with the first "operational" payload on the pad at Vandenberg. There was a dress rehearsal on 7 May and a launch attempt on 13 May, scrubbed because of various technical problems. Ten days later, after minor holds, the first Program 35 vehicle lifted off the launch pad. A catastrophic failure during second stage burn ended the test.

Following a strained relationship between the Air Force and NASA, the issue of Scout responsibility was settled on 21 June 1962 by means of formal agreements which conceded control of virtually all Scout matters to NASA, including general configuration, modification, launch stand procedures, and most related topics.

Following modifications, revisions, and overhauls, and 11 launch attempt cancellations for several reasons, on 23 August the second Scout with payload was successfully launched and the payload went into a near optimum orbit, in spite of numerous failures or anomalies as discovered during post-launch booster performance analysis. Direct readout was successfully attempted over the New Hampshire station on 26 August and on the following day the first remote readout data on cloud cover over the Soviet Union was taken from the same point in orbit.

With the potential of the satellite thoroughly proven, it was becoming increasingly clear that the greatest danger to continued program success lay in the ineffectiveness of the Scout booster. Studies of alternatives included initially the Thor-Delta (Thor with a Vanguard upper stage) and later were expanded to include Thor-Able Star (Thor with an Aerobee upper stage) and Thor-Agena (the combination used in Program 162, still known as DISCOVERER).

The great talking point still was the first satellite, vehicle 3502, which after 137 days and 1968 orbits (as of 7 January 1963) was returning better cloud cover information than immediately after its launch.

SAC's interest in the 417 satellite (as it was now called) had increased considerably after exposure to early results. The 417 represented a near perfect training device for space operations, being stable, dependable, and relatively uncomplicated. Operation by a field command was not particularly dangerous for the vehicle on orbit. (Haig called the 417 satellite "idiot proof".) Perhaps most important, the system was inexpensive to operate and maintain. Finally, there was increasing pressure from the Navy and the Strike command for a tactical readout system. The Air Force had a small but significant investment in ground stations (and one that would not become rapidly obsolete, either) which underlined the importance of continuing an active space program. There was a substantial opportunity for inexpensive experimentation. Taken together, these represented solid arguments for continuing, indeed, for expanding, 417. So long as it retained its basic characteristics of simplicity, reliability, and economy of operation, it remained attractive.

Colonel Haig seemed convinced, even this early, that the eventual salvation for 417 lay in adoption of a new booster. The studies begun the previous month had identified the Block II Thor (with Bell Telephone Laboratories guidance) and the Agena D as the most economical booster combination. The two provided the greatest injection accuracy, highest theoretical reliability, and heaviest payload potential of all the systems examined.

On 19 February the second 417 satellite went into orbit. It was not placed as accurately as the first, being 50 miles low in apogee and 140 in perigee, as well as having a 2-degree inclination error. Analyses indicated that each of the first three stages had performed badly—although the AB6X-259 third stage engine was the chief offender. The satellite's functioning was generally acceptable, even under such handicaps, until late April, when the primary tape control circuit went bad, eliminating the bird's ability to store primary data. The direct readout mode remained operational and 80 percent effective. By September 1963, however, the satellite would have gotten so thoroughly out of phase with the sun that a 180-degree reorientation would be necessary.

The February 1963 launch (operation 0240) proved the feasibility of still another innovation in cloud cover reconnaissance. Nearly a year earlier, the program office had undertaken a study of radiation measuring subsystem devices which by registering background radiation from the earth's surface could identify night cloud cover. The small investigation had been stimulated by General Greer's observation, in March 1962, that it was unfortunate that a cheap infrared system was not feasible.

In the course of their inquiry, program office people discovered that Dr. V. E. Sumoni of the University of Wisconsin had produced a prototype and one flight model infrared sensor for the Explorer program. Because of changes in that NASA activity, the sensor had never been orbited. It had every indication of meeting the requirements for a secondary payload for the 417 spacecraft.

On 3 September 1962, the program office let a low cost contract covering rehabilitation of Sumoni's prototype and tentatively scheduled it to be flown on the fourth 417 satellite (0240). By December 1962, test results were so promising that contracts were let covering the design, development, and test of five additional subsystems (one prototype and four flight versions) for use with the second group of four 417 spacecraft, then on order.

All went well and the first Sumoni radiation measuring subsystem was aboard 0240 for its February 1963 launch. The device functioned perfectly; by May 1963 infrared data were being routinely extracted and the system still was 95 percent operational. It continued to function until January 1964.

From the data obtained through the Sumoni system, the Third Weather Wing, using computer programs written entirely by Air Weather Service personnel, produced daily operational maps of cloud cover throughout the entire period from October 1962 through January 1963—the span of the Cuban missile crisis and the immediate aftermath. The technique was so successful that extension of the infrared measurement program was subsequently approved and it acquired the role of a semi-permanent element in the total 417 systems.

The general notion, by early March 1963, was that use of Thor-Agena as a booster would provide enough additional lifting capacity to permit launching a new satellite based generally on 417 technology but also incorporating the best proven features of TIROS, NIMBUS, RELAY, and other satellites into a system with minimum requirements for a long and costly development process. The general proposal, known as 417-1, had all the operational potential of NIMBUS plus circuit redundancy which promised enhanced reliability.

The proposal to develop 417-1 also had attractions other than technical. General B. A. Schriever, Air Force Systems Command Chief, heard the 417-1 presentation on 15 March, indicated that he was "intensely interested," and asked that he be provided with a formal development plan at once.

The appearance of two viewpoints on how 417-1 should be conducted reopened all the past arguments about the need for a secure system for military weather reconnaissance. Nevertheless, there were sound indications that Ruebel of DDR&E seriously planned to substitute 417-1 for NIMBUS. The problem was how.

Dr. Charyk was thinking less in terms of a substitute for NIMBUS than a basic 417, improved, modernized, and mated to an improved Scout booster. The details of 417-1 were far from certain, Charyk having expressed marked reservations about the cost of using Thor-Agena, the redundancy of a stabilized Agena, and the prospect of high command and control costs if Agena were employed.

On 26 April, the fourth attempt to orbit a 417 satellite ended in loss of third stage thrust 8 seconds before the scheduled burnout followed by violent tumbling and total destruction of the upper stage and payload.

By early May 1963, the generally unsatisfactory characteristics of the Scout booster had received widespread recognition. Quite apart from the impulse to go to a more powerful booster as a means of putting a heavier and more sophisticated payload into orbit, there was a general determination that either Scout would be significantly improved or the Air Force would go to a more reliable vehicle.

MCS HISTORY

On 11 July 1963, after nearly 10 months of circling the earth, the original 417 satellite (vehicle 3502) responded to ground commands for the last time. Attempts to recapture control by reorienting the bird were unsuccessful later that month forcing the ground controllers to postpone further efforts until the satellite assumed a more receptive attitude as a result of its gradual change of position in orbit. Concurrently, the program office suspended operations involving the second satellite (vehicle 0240) for intervals of 2 or 3 days at a time—and by that time 0240 was functioning only in the direct readout mode.

Almost unnoticed in the mounting clamor against NASA's handling of the Scout problem, the Strategic Air Command on 12 July 1963 began operating the first two military readout stations, one at Fairchild AFB, Washington, one at Loring AFB, Maine. The transition from contractor to SAC operation of the readout stations marked a major turning point in the evolution of an Air Force space capability.

After months of controversy and rework of the Scout booster scheduled for the next launch, Scout 132 (the launch crew had come to refer to it as the X-132), the fifth 417 launch, was attempted on 27 September 1963. This ended in failure due to a malfunction of the third stage control system arising from a premature loss of hydrogen peroxide.

On 3 October, 6 days after the launch disaster, Colonel Haig briefed Undersecretary Dr. Brockway McMillan on the launch and on program status. McMillan seemed to favor a booster other than Scout. His instructions to Colonel Haig were that he should continue with plans for at least one more Scout-boosted program launch, to complete study and planning for a single trial launch using the Thor-Agena combination, and to complete the study of Minuteman potential. McMillan had been rather specific in another direction, Haig had orders to work out an estimate of the money that could be recovered by a complete cancellation of Scout procurement.

The prospect of continuing with Scouts in their current configuration apparently was closed. On 7 October, acting on instructions from Colonel Haig, the Scout directorate at SSD formally cancelled the last two vehicles on the original delivery order and all six of the follow-on order. A stop work order was issued to cover all vehicles subsequent to number 134—later extended to include 134.

On 23 October 1963 McMillan ordered immediate cancellation of all activities connected with the Scout booster, immediate effort to recover every possible dollar from NASA, and assignment of a Thor-Agena from "available resources" to support a December or January 417 launch. The launch was to be in a dual payload configuration originally described by Haig during his 3 October presentation. Development of the "optimum payload capability" was also authorized in the Undersecretary's 23 October instructions.

"Complete and immediate" termination orders went to NASA early on 25 October 1963. Thus ended the Scout phase of Program 417. In five attempts, the program office had one unqualified success, one partial success, and three catastrophic launch failures. The satellite had operated marvelously well, considering the difficulties in launch phases.

To the credit of the Scout phase, the program office had a remarkable record of cost effectiveness, had functioned with a combination of fewer inhabitants and larger responsibilities than any other space vehicle development in Air Force history, and had progressed from concept to satellite in orbit more rapidly than any other earlier organization. Few of the people who operated the program office and developed the satellite were briefed on the existence of such programs as GAMBIT and CORONA. It must be recalled, however, that until 1963 the original

BIF-059W-23422/82
Handle Via
BYEMAN/TALENT KEYHOLE
CONTROL SYSTEMS JOINTLY

SAMOS program maintained an official existence, and most Air Force people, unaware of the GAMBIT and CORONA programs, assumed that SAMOS, in one of its several incarnations, was the intended beneficiary of satellite-based weather reconnaissance. The P-35 system, as it was originally called, had operated under ordinary but rigidly enforced security controls and was not incorporated in the BYEMAN control system although access to information about the program was closely controlled.

x x x x x x

The program transitioned to the Thor booster in 1964 and continued to use it until 1980. Program 417 had two launches on the Thor-Agena with two satellites each—F-6 and F-7 launched on 19 January 1964, F-8 and F-9 launched on 18 June 1964. Both launches were successful and the satellites exceeded their expected lifetime.[16]

A new solid fuel upper stage called "Burner I" was used for the launch of F-10. The satellite failed to separate from the upper stage resulting in mission failure. Burner I was used successfully through the remainder of Block II, with the last launch in 1966.

The Block I and Block II Program 417 Satellites (1962-1966) were limited by on-board recorder capacity to 7 or 8 pictures of approximately 750 x 750 nm each. Lens characteristics and earth curvature resulted in imagery which was badly distorted geometrically, requiring skilled personnel to manually interpret each picture. The limited area covered by each picture did not provide contiguous coverage on successive orbits except at high latitudes. Infrared instruments introduced late in the series provided only a crude measure of the temperatures of the cloud tops with area coverage at extremely poor resolution (115 nm x 115 nm) and a single strip through each visible picture at a resolution of 2 nm. These early attempts at IR measurements clearly illustrated the value of IR data and provided the basis for later sensor technology and data interpretation methods.

The Block 4 series (1966 to 1970) provided much improved area coverage through the use of two vidicon camera systems offset left and right of the orbit track. Tape recorder capacity was increased to provide on-board storage for 54 picture pairs. Block 4 therefore provided contiguous coverage on successive orbits and storage capacity sufficient to provide coverage during the 3 to 4-orbit periods when ground station contacts were not available. IR technology remained essentially the same as the early configuration except that the system was upgraded to provide four 2-nm resolution stripes instead of one.

Other improvements in the Block 4 series included magnetic momentum or spin rate controls eliminating the expendable spin rockets, and a Boeing-developed upper stage for the Thor launch vehicle called Burner II. Orbit weight for Block 4 was 160 pounds.

During the Block 4 series, the program office, under the direction of (then) Major John E. Kulpa, Jr. since January 1965, was working on a revolutionary new concept for the DMSP system. The concept involved replacement of the vidicon camera system with a mechanically scanned radiometer to provide high-quality, distortion-free imagery in both the IR and visible spectral bands. The visible imagery would also be provided on the night side of the orbit under moonlight illumination conditions. The spacecraft would be three-axis stabilized in order to support this new sensor system. Following studies, and a SPO/Aerospace demonstration of the concept aboard an Air Force T-39, contracts were awarded to RCA for the new spacecraft and to Westinghouse for the sensor and ground display equipment. This new Block 5 spacecraft provided a continuous strip image 1,600 nm wide in both IR and visible at a resolution of 2 nm, night and day. Tape recorder storage was sufficient to provide global coverage. In addition, the Eurasian land mass could be covered at a resolution of 0.3 nm in the visible and, later in the program, both IR and visible.

BIF-059W-23422/82
Handle Via
BYEMAN/TALENT KEYHOLE
CONTROL SYSTEMS JOINTLY

The Block 5 system had few limitations. By the time the system had evolved fully there was no limitation on the local time at which the spacecraft could operate. In fact, excellent imagery could be obtained with the spacecraft over the terminator where one side of the picture was illuminated by sunlight and the other side by moonlight. The high quality imagery and operational flexibility provided the ability to operate a spacecraft at any local ascending node time to provide optimum support to the HEXAGON program. Normal operational configuration was one spacecraft in the early morning for forecast and one near noon for forecast verification.

The first three Block 5 spacecraft were launched on the Thor Burner II developed for Block 4. As additions and improvements were made, the spacecraft weight increased from 250 pounds (Block 5A) to 450 pounds (Block 5B and C) requiring an additional stage to be added to Burner II (Burner IIA). Operational experience with the Block 5A, B, C series suggested further development of meteorological satellites and the methods for application of the data products. The imagery for Block 5A, B, and C was of excellent quality but not perfect. Due to the geometry of the orbit and sensor characteristics, the resolution at end of scan was lower than the center. This resolution variation and analog methods for on-board stage and data transmission frustrated the Air Weather Service attempts at computer automation of the data analysis task. A new sensor concept was developed in the early 1970's that compensated for the edge-to-center resolution variation. The utilization of advanced digital technology and digital tape recorders could provide the data in a computer compatible digital format directly from the spacecraft. All sensor function, i.e., optical system, data formatting and storage, data acquisition control functions, etc., were combined into a self-contained system called the Operational Linescan System (OLS).

The attitude control accuracy of the Block 5A, B, and C satellites provided geographic location of the picture with a maximum error of 5 nm at the center of the picture swath degrading to 25 nm at the edge. This error was not a major problem for manual interpretation methods since landmark recognition could be used as a reference. However, for automated analysis methods more accurate control of spacecraft attitude was considered necessary in order to eliminate the reference to landmarks.

The combination of a spacecraft designed to provide a high precision attitude control and the new constant resolution (all digital, OLS sensor) were the drivers for an entirely new system called Block 5D. The Block 5D spacecraft includes three major departures from previous DMSP designs: (1) the attitude control system is referenced to the stars rather than earth horizon in order to provide the required pointing accuracy (0.01° 3 axis), (2) the two upper stages of the launch are integrated into the spacecraft along with an inertial reference package that is used both during ascent and on-orbit (the 5D spacecraft is essentially an orbital stage), (3) the spacecraft has redundant systems where all previous designs were single string. In order to tie all these features together and provide the computations necessary to translate for celestial coordinate to earth coordinates, all spacecraft functions (orbit and boost) are controlled by on-board computers.

The first of the Block 5D series was launched in 1976. There have been five launches, the fifth, launched in May 1980, failing to achieve orbit due to an upper stage problem. The four successful spacecraft provided excellent data with all but one spacecraft exceeding or equaling their expected operational life.

The Air Weather Service has never fully utilized the capabilities of Block 5D. The fully automated processing of high resolution imagery exploiting the precision geographic accuracy of Block 5D has not been demonstrated. However the constant resolution of the OLS sensor has significantly improved some of the automated products over the products derived from the earlier Block 5A, B, and C satellites.

All DMSP satellites since Block 4 have provided direct readout capability to tactical users throughout the world. All of the Block 5 series have provided this data in an encrypted digital format.

There has been an increasing number of other sensors aboard DMSP satellites to support other Air Weather Service requirements. These sensors provide information on atmospheric parameters such as vertical temperature and moisture profiles and ionospheric parameters such as electron density.

DMSP plans to continue with upgraded versions of the Block 5D spacecraft indefinitely. The program will transition to the Altas booster in 1982 for the remainder of the Block 5D Program. Changes are planned only to improve operational life and incorporate improvements in encryption and survivability. No major change in the cloud cover imager is planned. Changes or additions to the other mission sensors are planned or are in study. This includes the use of microwave techniques for measurement of rainfall, soil moisture, and sea surface conditions, and incorporation of improved ionospheric sensors.

THE GAMBIT PROGRAM

Note: GAMBIT, still operational after approximately 18 years, was the first operational American satellite system to return high resolution photography. It is the only remaining operational system from the early satellite reconnaissance programs. During its developmental and operational life, GAMBIT was identified by several identifiers other than its code name. "Cue Ball" and Program 206 were, respectively, the classified non-Byeman cover names and the white program designators. The successor surveillance satellite in the National Reconnaissance Program was GAMBIT-3, also informally referred to as Program 207, ADVANCED GAMBIT, and G^3 or "G Cubed." Since GAMBIT is still an active program, only a brief account of its development phase through first flight will be related here. The GAMBIT (KH-7) payload configuration is shown on the following page (Fig. 1-22).

On 24 March 1960, more than a month before the U-2 affair, Eastman Kodak had informally submitted to the Reconnaissance Laboratory at Wright Air Development Division (WADD) a proposal to develop a high-acuity 77-inch focal length camera for satellite reconnaissance purposes. On 17 June, Eastman followed up the original submission with a relatively detailed proposal for yet another recoverable reconnaissance system, this embodying a 36-inch camera to provide convergent stereo coverage of the Soviet territories (this was to become the E-6, previously discussed). Eastman called the system "Blanket." Still later, on 20 July, Eastman disclosed to WADD a second volume of the technical proposal, this covering the 77-inch camera mentioned originally in March. Suggesting the same technical approach and many of the components defined in "Blanket," Kodak proposed a system capable of providing 2 to 3-foot resolution for spot coverage of selected ground targets. Alluding to the 77-inch focal length strip camera and a currently popular television program, Eastman called the proposed system "Sunset Strip." Procurement of Sunset Strip work was to be undertaken through BMD channels and was to be managed as part of the total SAMOS program rather than as a separate camera development project. The shift of responsibility to BMD meant, in practice, that the existent SAMOS program office became the Air Force focal point for Sunset Strip activity.[17]

In November 1960, a new and highly significant innovation of early November meetings was the proposal to use the E-6 program as a cover for development of the "Sunset Strip" system. Dr. Charyk agreed with General Greer's suggestion that Eastman develop the 77-inch camera under the name Project GAMBIT—a term that Colonel Heran chose, and which was considerably more meaningful than most code designations—while General Electric developed a suitable reentry vehicle. By keeping the physical and environmental limitations of E-6 and GAMBIT compatible with one another, it seemed possible to develop and test GAMBIT without any outward indication that such a program existed.

By this time the proposed "Sunset Strip" development program was so widely known that it would be necessary to invent and circulate a palatable reason for canceling an essentially reasonable approach to satellite reconnaissance. Project personnel achieved that end by having BMD terminate the Eastman study contract for "Sunset Strip" with the excuse that "review of recent proposals for the E-6 camera reveals that future study in this area (77-inch camera) is not required." Simultaneously, the SAMOS office drew up the first of its "black" contracts, authorizing Eastman to continue the development as a covert effort.

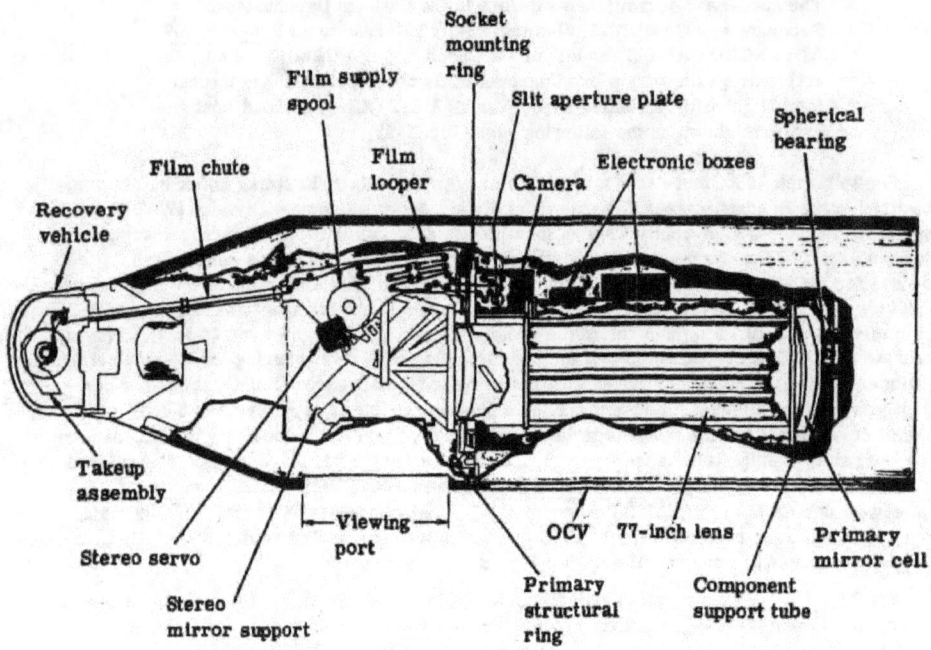

Fig. 1-22— GAMBIT (KH-7) configuration

With GAMBIT, there came into being a tightly contained procurement and program management capability that had no real precedent in the Air Force. Security requirements originating in the President's desire to avoid any implication of military operations in space became so tight that the transition from "extremely secure" to "clandestine" and thence to "covert" was in some sense inevitable. The political vulnerability of the widely publicized "E" programs made GAMBIT even more important than would normally have been true, though the potentially very remarkable performance of the GAMBIT system was in some respects a sufficient justification for emphasizing that program.

As discussed in the E-6 section, the use of E-6 as a cover for GAMBIT had certain disadvantages that were recognized early. The desirable solution, suggested in Greer's notes of December 1960, was total disassociation from the original SAMOS program. In November 1960 he had begun "black" contracting under the philosophy that since "everybody" knew it was impossible for the Air Force to buy anything expensive without going through established review and approval channels, one might do quite a lot of unsuspected buying and contracting by merely obtaining a direct authorization. It occurred to him that the solution to the GAMBIT quandary might be found in the same thesis. He thereby invented the concept of the "null program," a development with no known origin and no specified goal. If such a program were conducted under the aegis of a highly classified payload, it should be entirely possible to purchase boosters, upper stages, and launch services through normal channels. Because "everybody" knew that the entire reconnaissance satellite program was in Greer's keeping, the assignment of "null program" responsibility to the regular Space Systems Division (SSD) organization would serve to convince most observers that it had to have some objective other than reconnaissance. Vague references to precise land recovery (a real but secondary objective of GAMBIT at the time) might serve to induce suspicion that the "null program" actually had a "bombs in orbit" goal.

In July, the first moves toward establishing an activity called "Program 206" were taken. Through the Air Staff, SSD received authorization to buy four "NASA type" Agena B's for launches starting in January 1963—the Agenas to be assigned to no particular space program "for the present." In August, Charyk sent a memorandum to the Air Force Chief of Staff which emphasized the need to protect the USAF's "capability to do future space projects" and which affirmed the desirability of ordering six Atlas boosters (configured to accept Agena B's) to be used starting in February 1963. Again there occurred the phrase about "not assigned to a particular space project."

Having gotten a small batch of Atlas and Agena vehicles on order, SAFSP moved to the next business—formal creation of a "null project." On 25 September 1961 the Air Force Vice Chief of Staff directed General B. A. Schriever, AFSC commander, to establish "Project EXEMPLAR." That code phrase, which was classified confidential, was defined as covering four launches from the Pacific Missile Range starting in February 1963. The authorizing message noted that the Secretary of the Air Force had separately ordered the necessary Agena and Atlases "on an unassigned basis." "They are hereby assigned EXEMPLAR," the teletype read. The "white" correpondence that was to be associated with Exemplar stated requirements for the usual sort of elaborate documentation (development plans, cost projections, and the like) that had become customary for new programs. All of the "white" elements were gathered under EXEMPLAR which for reasons of administrative convenience had the additional and unclassified nickname "Cue Ball."* Colonel Q. A. Riepe was named the Cue Ball program director. A complex network of

---

* "Cue Ball" was chosen to add spice to conjectures about orbiting bombs and means of returning them to precise sites on the earth.

nominal and actual reporting channels linked Cue Ball to General Greer. (Greer then had an additional duty assignment as Vice Commander of the Space Systems Division, although his primary responsibility was still for the reconnaissance mission.)

As part of the cover plan, General Greer had decided to have Cue Ball broken up into two elements, Program A and Program B. "A" would include the first four Atlas Agena vehicles and "B" the remaining six needed for the approved 10-launch effort.

Although the Cue Ball plan was successful at getting the program under way, constant maneuvering was required to keep the program funded under this disguise.

By mid-1961, the concept of GAMBIT development, and its technical details, had been worked out in detail. Essentially, GAMBIT differed from E-6 (to which it still maintained a technical likeness) in having substantially higher ground resolution, in possessing a capability for photographing specific targets which were off the immediate orbital track, and in being intended for land recovery.

Because of its need for higher resolution, GAMBIT would fly somewhat lower than E-6. A photographic altitude of 90 nautical miles was generally considered desirable. The resolution requirement, over a period of several days, required more precise altitude control than in E-6 and an ability to rotate the camera section about the vehicle's roll axis. Land recovery implied extremely precise deboost velocities and reentry programming.

GAMBIT mirrors were larger, made to closer tolerances, and lighter than in any previous system. Thermal gradients between the reflecting surfaces and the rear supporting surfaces had forced consideration of metal rather than glass backing, further complicating the problem.

The performance of the GAMBIT camera depended as much on vehicle stability as on any inherent photographic quality. Pointing had to be extremely precise, requiring extreme accuracy in the horizon sensors, and a stable platform gyrosystem that would allow the sensors to stay locked on the horizon while the vehicle rolled to point toward targets on either side of the orbital track. Because the ground swath width of GAMBIT cameras was only approximately 10 miles, more photographs would be taken from a canted than from a vertical position.

The complexity of orbital operations derived from the inability of the launch system to put the orbital vehicle on a predetermined orbit with the precision required by the narrow swath width. Command programming had to be changeable in flight, and further complexity derived from the need to set highly accurate roll positions for photography on either side of the vehicle's track.

By 1962 it was determined that the GAMBIT system was 500 pounds over design weight, and most of the overweight derived from complications introduced by the land recovery requirement. Moreover, the reasons for distrusting air-sea recovery modes had become much less valid since 1960. Successful CORONA recoveries were proving to be less difficult as time passed. Overwater recovery, as developed in the CORONA program, seemed a very simple process when compared to the planned land recovery scheme. In its descent toward the ocean, a CORONA reentry vehicle could safely shed all sorts of accessories—hatch covers, and the ablative cone being the most obvious. Such jetsam fell into the ocean without danger to anything below, and then sank into the secure obscurity of a cluttered sea bottom. A land recovery vehicle could shed nothing that might come to earth as a lethal projectile which, if discovered, might breach the security of the satellite reconnaissance effort.

On 24 August following additional briefings by Captain Frank B. Gorman (USN), General Greer's plans chief, and Colonel Riepe, Charyk authorized Greer to begin immediate development of a CORONA type recovery system for GAMBIT, planning on a June 1963 first flight date.

On 30 October General Greer announced to members of the Program 206 office that Colonel William G. King was assuming management responsibility for their project and that Colonel Riepe had been detailed to a new and demanding SSD program.

One of King's first moves after moving into GAMBIT management was to advise General Greer that he thought the design of the adapted capsule represented much more of a change than Greer had intended. After evaluating Greer's recommendation that the entire 206 program be transferred to SAFSP, Charyk concurred in the "desirability" of this move. Greer and King then set about changing the technical character of GAMBIT.

There was more to "Hitchup," as the notion of keeping the orbital control vehicle attached to the Agena was called, than met the unwitting eye. An elaboration of the scheme involved use of the "Roll Joint" coupling invented for LANYARD. Should the orbital control vehicle prove generally unreliable, it might be possible to introduce the LANYARD Roll Joint between the Agena and the payload end of GAMBIT, eliminating reliance on the stability and control elements of General Electric's orbital control vehicle.

The chief difficulty in this idea was devising a non-compromising means of bringing the Roll Joint part of the technique into the GAMBIT program. As was the case with the CORONA reentry capsule, the Roll Joint was quite unknown to most GAMBIT people, and because of the security compartmentalization that existed within the reconnaissance program structure, it seemed highly unwise to disclose the existence of LANYARD to large numbers of GAMBIT workers. So "Charyk's" message of 30 November, actually written by General Greer, contained the "suggestion" that Greer contact Lockheed about the Roll Joint as ". . . he (Charyk) believes a similar idea was once proposed and possibly designed in connection with another space program." The resulting cover story was that Lockheed would be empowered to "develop" the earlier "idea," delivering finished Roll Joints to GAMBIT as though they were new items with no relationship to any other reconnaissance program.

In a full scale program review on 14 December, additional measures for ensuring the success of the first GAMBIT were proposed, one technical innovation being a Lifeboat provision. Lifeboat was another provision technique originated in the CORONA program; it involved the provision of independent reentry command circuitry (including a receiver), a separate magnetometer, and its own stabilization gas supply. All were independent of the main systems. If the primary reentry systems became inoperative for any reason, Lifeboat could be separately actuated. The magnetometer used lines of magnetic force around the earth as a longitudinal stabilization reference, permitting the device to place the Agena (or any other suitably equipped orbital vehicle) in a proper attitude for the start of deboost, relying entirely on its own gas supply for attitude control and a taped command sequence for the recovery process. In several experiences with CORONA vehicles, Lifeboat had proved highly reliable.

On 19 December, the Undersecretary formally authorized the Lifeboat, Hitchup, and Roll Joint expedients for GAMBIT. Lifeboat was to be a permanent part of the total system, Hitchup was to be incorporated in the first four vehicles [but a determination on use would be made on a flight-by-flight basis, while Roll Joint was to be developed as a bona fide operational substitute for the OCV (orbital control vehicle) roll system].

As vehicle and funding problems were being worked, the camera seemed to be coming along nicely, demonstrating in tests an equivalent 2.7-foot ground resolution at better than 115 lines per millimeter. The only problem that appeared to offer any particular difficulty was the motor speed drive, and it was far from insurmountable. A mirror mounting problem, that had earlier given trouble (and which was similar to a problem then holding up LANYARD), had been essentially solved by November 1962.

One additional change in the basic configuration of GAMBIT was recommended in January 1963 and approved for adoption on 28 February. This was a stellar-index camera, earlier treated as "purely an auxiliary" but now considered quite important. The National Photographic Interpretation Center made the original recommendation, CIA's Herbert Scoville endorsed it, and Charyk approved its inclusion. The camera itself—developed for the CORONA MURAL—was to be procured through a black CIA contract with Itek. Because of procurement and installation delays arising from the advanced stage of completion of the first lot of GAMBIT payloads, the fourth GAMBIT was the first which could be scheduled to incorporate a stellar-index system.

By virtue of circumstances, the fourth GAMBIT vehicle became the first in what was essentially a remodified configuration. Hitchup capability was provided in all of the first six, but Lifeboat was an Agena installation in the first three, being shifted to the GE vehicle thereafter, and roll-joint capability was scheduled to be incorporated starting with the fourth system, as was the stellar-index camera. (As it turned out, the SI was not incorporated until the 7th flight.)

By early May, study of the problems of supplementary launch, standby, and quick reaction had been sufficient to show that a high launch rate could be maintained by keeping at least three pads in a GAMBIT configuration and by building up a modest stockpile of boosters and GAMBIT systems.

The possibility of tandem recovery vehicles for GAMBIT was examined but it was General Greer's judgment that nothing serious should be attempted in the matter of tandem configuration GAMBIT's until the original system had been well proven.

By the time such matters were resolved, attention was turning toward the impending first launch of GAMBIT. Booster payload assembly had begun in February, after some delay because of the late arrival of prime components and the need to incorporate hitchup provisions. In order to protect schedules, Colonel King had agreed that it would be permissible to put the missing components into the total system during functional testing.

Then, during the late afternoon of 11 May, a faulty valve in combination with a deficient fuel loading sequence caused a loss of internal pressure in Atlas 1900 being used in checking out procedures for the first GAMBIT flight. The booster collapsed on its stand, dumping both the GE orbital vehicle and the Agena on the concrete hardstand. The GE vehicle was severely damaged, the Agena to a lesser degree. Surprisingly, there was neither explosion nor fire, although 13,000 gallons of liquid oxygen and a full load of fuel sloshed over the stand and the nearby terrain. Equally fortunate, the payload did not split open, so there was no compromise of GAMBIT security. But the camera system was rendered permanently useless, a large part of the optics system being demolished, and the recovery vehicle was so battered that further use seemed imprudent. Neither the camera nor the orbital vehicle was that scheduled for the first GAMBIT flight; the Agena, however, was supposed to be used in that launch.

One other set of developments had been continuing parallel to the technical aspects of launch preparations. These involved security and deception. In addition to the concern by the CIA over preserving CORONA security, the missile assembly building would have to be cordoned off to separate GAMBIT from other programs, such as the E-6, that shared the facility. One of the problems peculiar to pretending that GAMBIT was a non-camera project was that a certain number of Eastman Kodak people had to be at the launch stand during final checkout. The problem decreased appreciably when Lieutenant Colonel John Pietz and Colonel J. W. Ruebel ran a careful study of needs and discovered that no more than four or five camera specialists were actually needed. In dress rehearsals for the first launch, they were literally smuggled into the launch area in the back of an unmarked van. The practice was dropped, however, when the driver wrecked the empty truck while returning from one delivery run. Thereafter the needed specialists entered the launch zone as inconspicuously as possible, but using more conventional means of transportation.

Following final launch preparations, which included an elaborate deception scheme worked out by Colonels Ruebel and Pietz, Major David Bradburn, and Lieutenant Colonel Ralph J. Ford, the first GAMBIT was launched at 1344 hours Pacific Daylight time, on 13 July 1963, just 22 months and 17 days after the National Security Council decision to proceed with development of a "covert" alternative to SAMOS.

Climb-out, separation, and orbital injection occurred as planned. Both Atlas and Agena operated normally, apogee being 116 nautical miles and perigee 107.

On the fifth orbital revolution, command controllers turned on the camera for eight strip exposures of 20 seconds each, commanding an identical maneuver on each of the next two orbits. On orbits eight and nine, two stereo pairs and five 20-second strips were exposed—after which the premature exhaustion of Agena stabilization gas forced discontinuance of camera operations.

With the depletion of Agena control gas, the Lifeboat became the only means of recovering the film capsule. The GAMBIT-Agena coasted through eight uncontrolled orbits after stabilization gas was exhausted during orbit nine, ground control activated the "Lifeboat" circuitry during the 17th pass, and on orbit 18 an "execute" signal from the ground station went to "Lifeboat." Routine separation and recovery followed. There was no drama—and nobody minded.

Evaluation of the recovered film indicated an out-of-focus condition apparently caused by uncompensated temperature changes that affected the surface of the primary mirror and by faulty image motion compensation settings. Nevertheless, best resolution on the 74 exposed frames (and nine stereo pairs) was on the order of 3.5 feet; 5-foot ground resolution occurred on several stretches of the 198 feet of exposed film, and average resolution was about 10 feet.

Operational use of the original GAMBIT system continued from 12 July 1963 until 4 June 1967. During this time 38 vehicles were launched, notwithstanding a period of serious orbital control problems.

In its successor configurations, with longer focal length optics and dual recovery capability, $\underline{G^3}$ has produced exceptionally high resolution photography. But these details will be presented in a GAMBIT history supplement at some future time.

THE KH-11 SYSTEM

The KH-11, the most recent and the most sophisticated of the satellite reconnaissance systems, became operational in 1976, adding still another exceptional capability to the inventory of intelligence collecting devices.

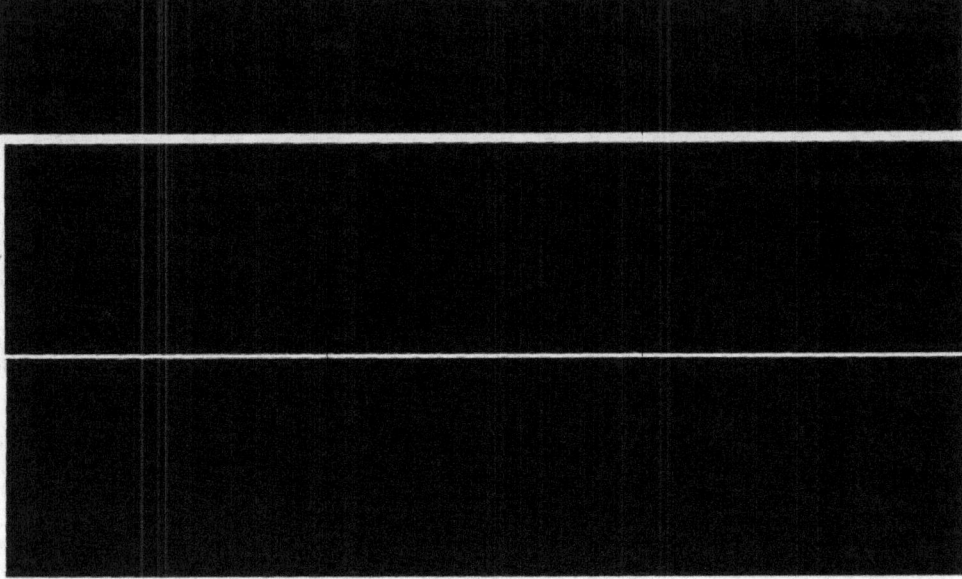

MCS HISTORY

EXHIBITS OF TYPICAL IMAGERY FROM PHOTINT PROGRAMS

The following examples of imagery from the programs described in Part I reflect evolutionary progress. Negatives from which these prints were made were furnished by NPIC from one of their briefing manuals. Although the examples are considered to be typical of early systems' performance, they are not necessarily representative of the exceptionally high resolution capability of the currently configured/programmed KH-8 and KH-9 systems with the improved films now available.

Table 1-5 presents a summary of the panoramic, mapping, and strip camera characteristics for the systems that were used in photographing the exhibits. Table 1-6 presents the vertical ground coverage footprints for the various camera systems. It is intended to depict that the length of a "burst" with strip cameras will vary depending upon target coverage requirements.

Table 1-5 — Camera System Exhibit Data

Panoramic Camera Systems

Figure	Program/System Designator	KH Number	Optics	Film	
1-23	CORONA (C)	KH-1	24-in., f/5.0 Tessar	1213 acetate	5.25-mil
1-24	CORONA (C')	KH-2	24-in., f/5.0 Tessar	1221 acetate	2.75-mil
1-25	CORONA (C''')	KH-3	24-in., f/3.5 Petzval	4404 estar	2.50-mil
1-26	CORONA (MURAL)	KH-4	24-in., f/3.5 Petzval	3404 estar	2.50-mil
1-28	CORONA (JANUS-J1)	KH-4A	24-in., f/3.5 Petzval	3404 estar	2.50-mil
1-30	CORONA (JANUS-J3-CR)	KH-4B	24-in., f/3.5 Petzval	3404/3414 estar	2.50-mil
1-32	CORONA (LANYARD)	KH-6	66-in., f/5.0 Hyac	3400 estar	2.50-mil
1-36	HEXAGON	KH-9	60-in., f/3.0 Mod. Schmidt	SO-315 estar	1.20-mil

Mapping Camera Systems

Figure	Program/System Designator	KH Number	Optics	Film	
1-27	Stellar Index (SI)	Flown on KH-3, 4, 4A	1.5-in., f/4.5 Biogon	3400 estar	2.50-mil
1-29	DISIC	Flown on KH-4B	3.0-in., f/4.5 Ikogon	3400 estar	2.50-mil
1-31	ARGON (DAFF)*	KH-5	3.0-in., f/2.5 Geocon	3400 estar	2.50-mil
1-35	HEXAGON (MCS)	KH-9	12.0-in., f/6 Biogon	SO-315 estar	1.20-mil

*Army Mapping Program

Strip Type Camera Systems

Figure	Program/System Designator	KH Number	Optics	Film	
1-33	GAMBIT (Program 206; Cue Ball)	KH-7	77-in., f/5 Maksutov	3404 estar	2.5-mil
1-34	Advanced GAMBIT (G³)	KH-8	175-in., f/4 Newtonion prime focus with Ross corrector	SO-312 (9-in.) SO-209 (5-in.)	1.2-mil 1.5-mil
1-38	(Intentionally left blank)	KH-11 ▮▮▮	(Intentionally left blank)		
	(Intentionally left blank)	KH-11 ▮▮▮	(Intentionally left blank)		

Table 1-6 — Vertical Ground Coverage Footprints

Panoramic Camera Systems—95-n.mi. altitude

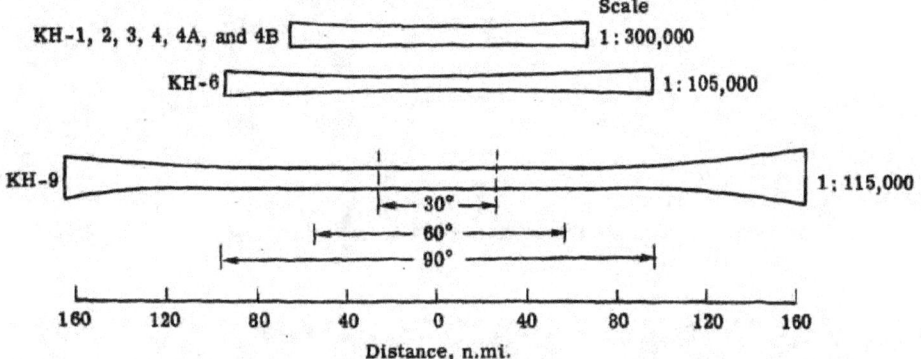

Mapping Camera Systems—95-n.mi. altitude

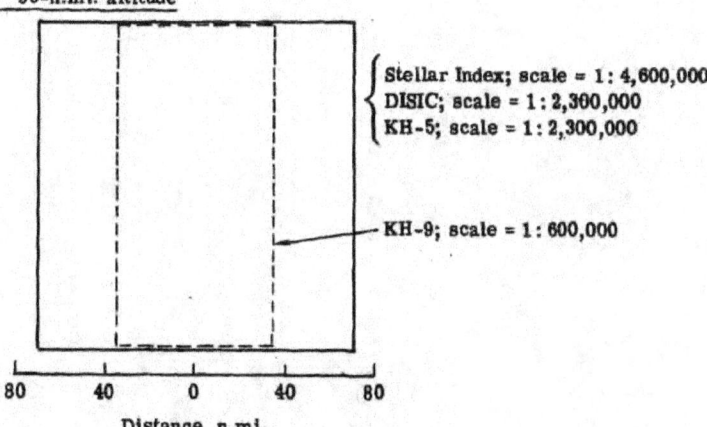

Strip Type Camera Systems—80-n.mi. altitude

KH-7 Variable strip | Scale 1:76,000

KH-8 1:33,000

KH-11

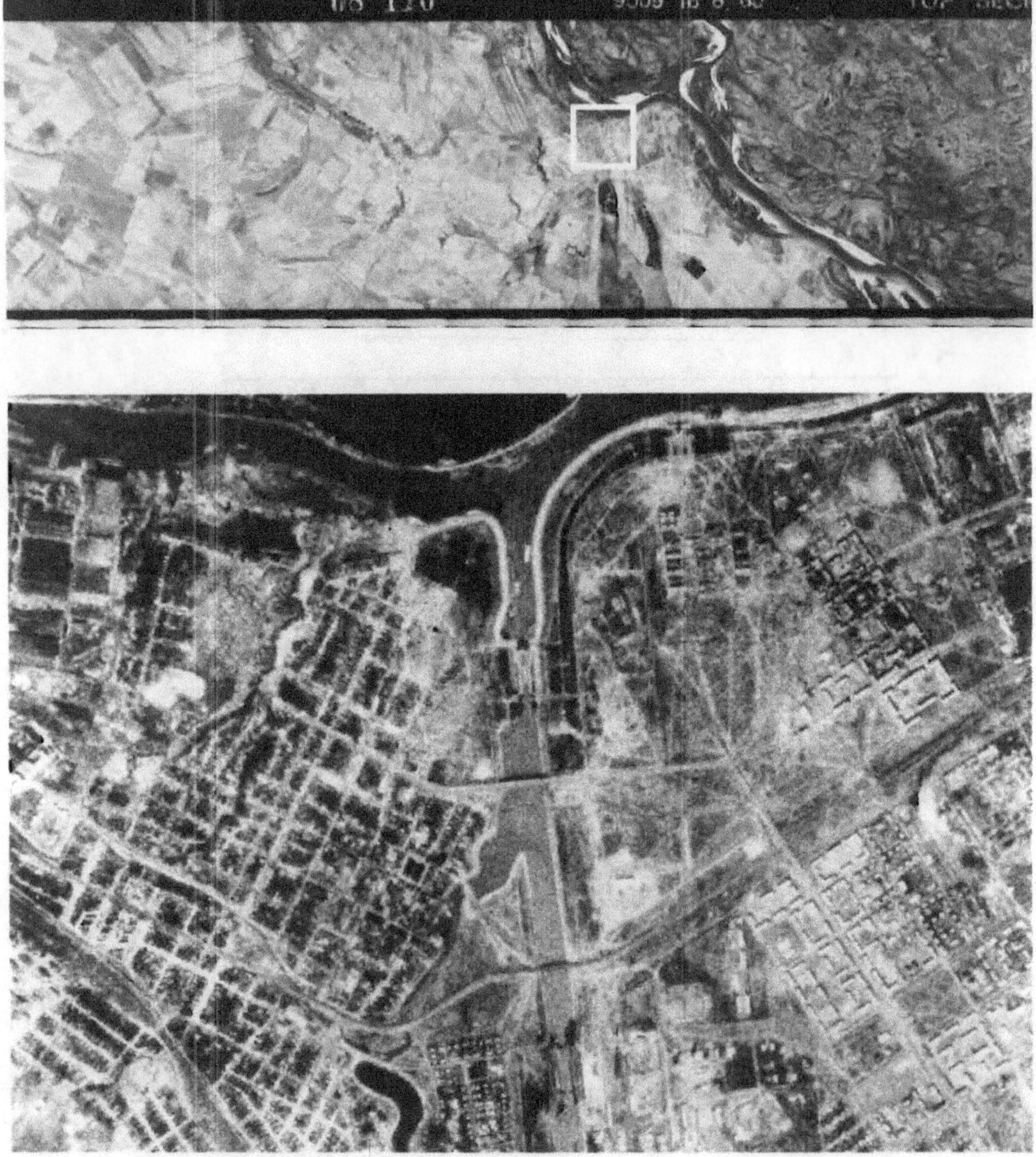

Fig. 1-23 — KH-1 Camera System imagery (CORONA C); contact (partial frame) and 20× enlargement

MCS HISTORY

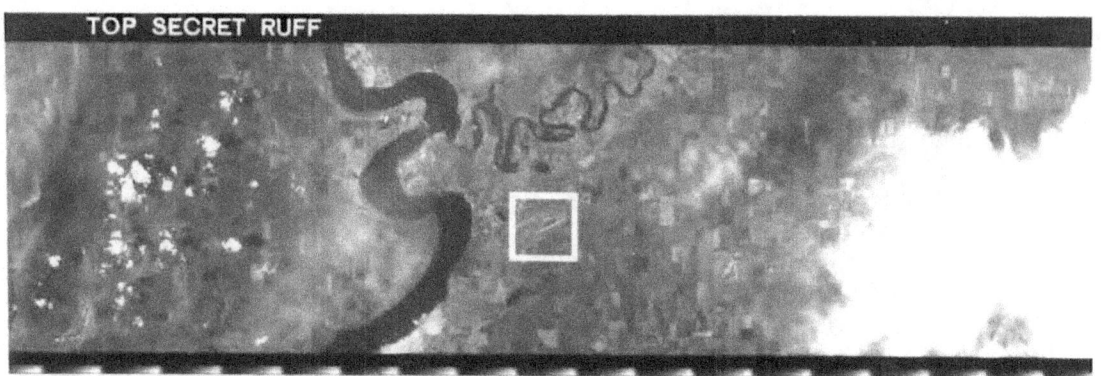

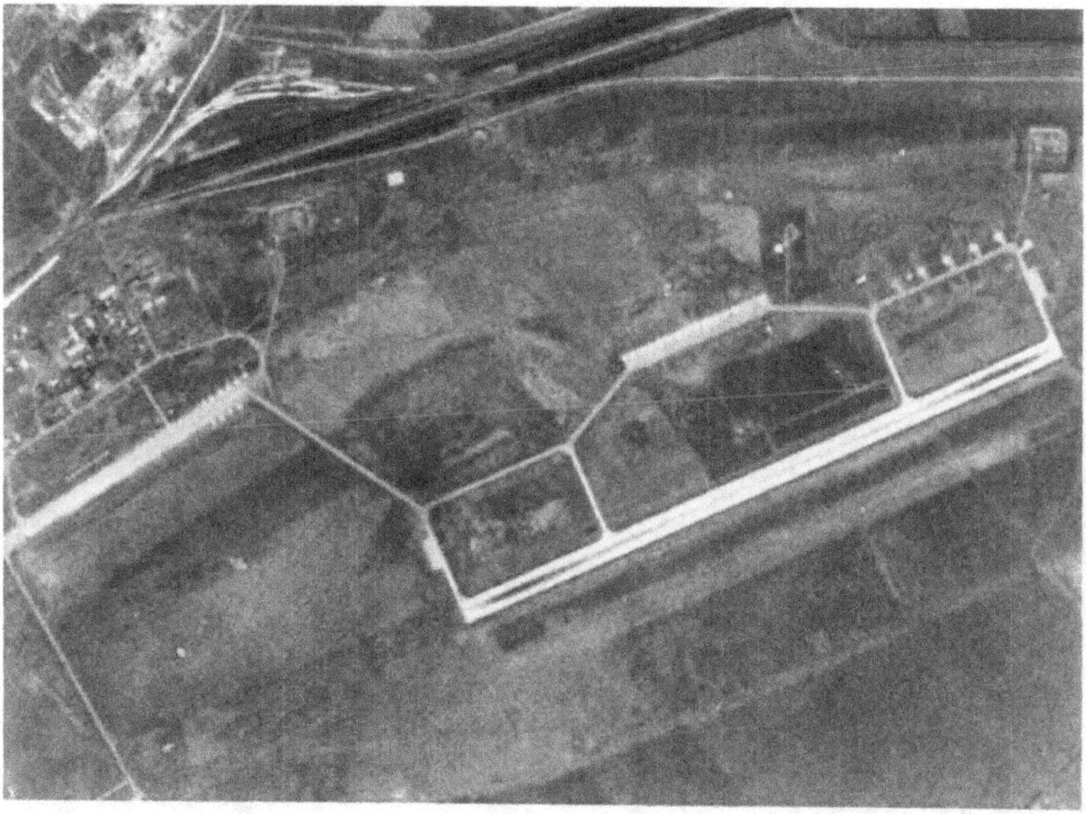

Fig. 1-24 — KH-2 Camera System imagery (CORONA C′); contact (partial frame) and 20× enlargement

BIF-059W-23422/82
Handle Via
BYEMAN/TALENT KEYHOLE
CONTROL SYSTEMS JOINTLY

Fig. 1-25 — KH-3 Camera System imagery (CORONA C'''); contact (partial frame) and 20× enlargement

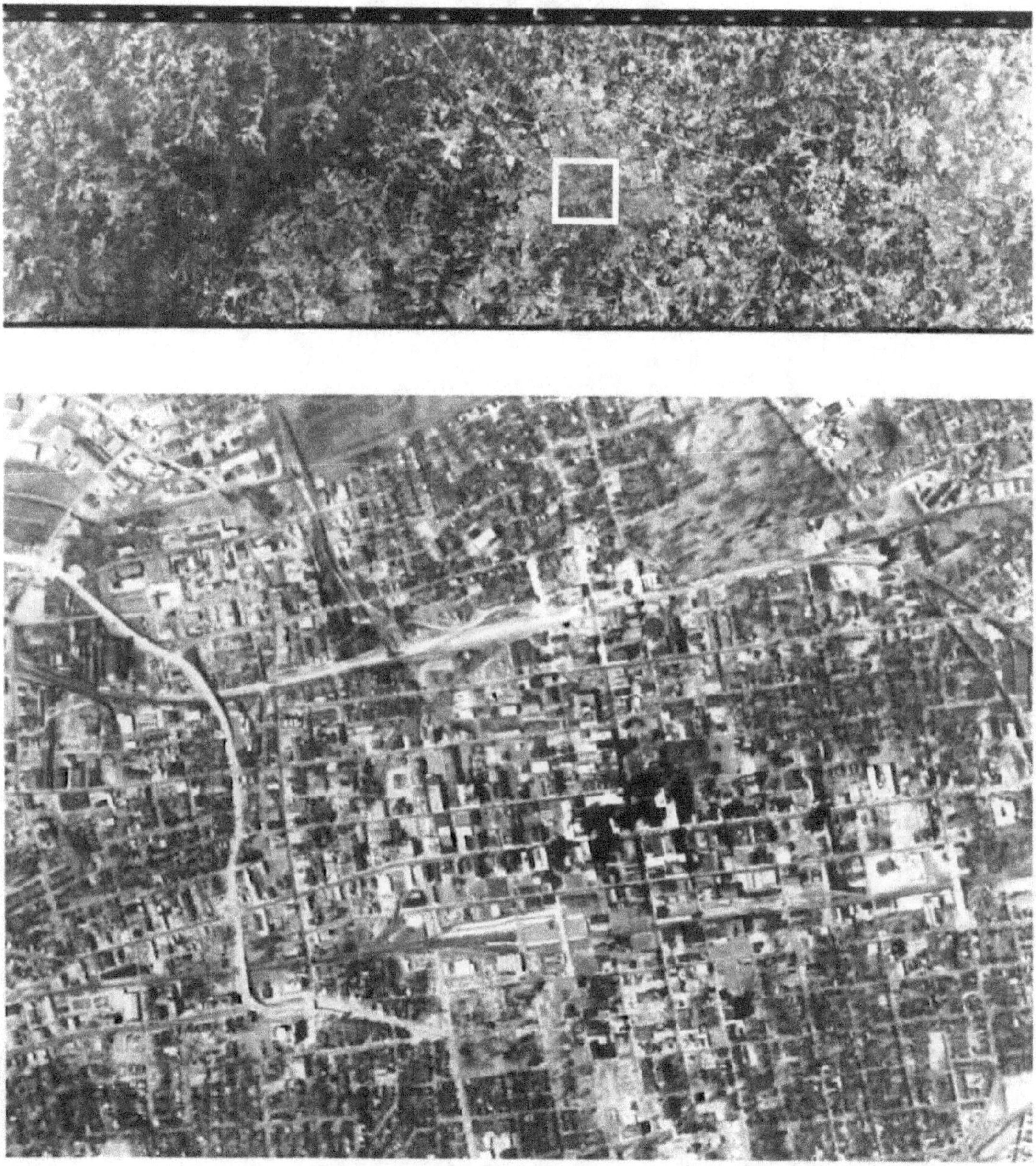

Fig. 1-26 — KH-4 Camera System imagery from panoramic camera (CORONA-MURAL); contact (partial frame) and 20× enlargement

Fig. 1-27 — KH-4 Mapping Camera System imagery from stellar/index (S/I) camera (CORONA-MURAL); contact and 5× enlargement

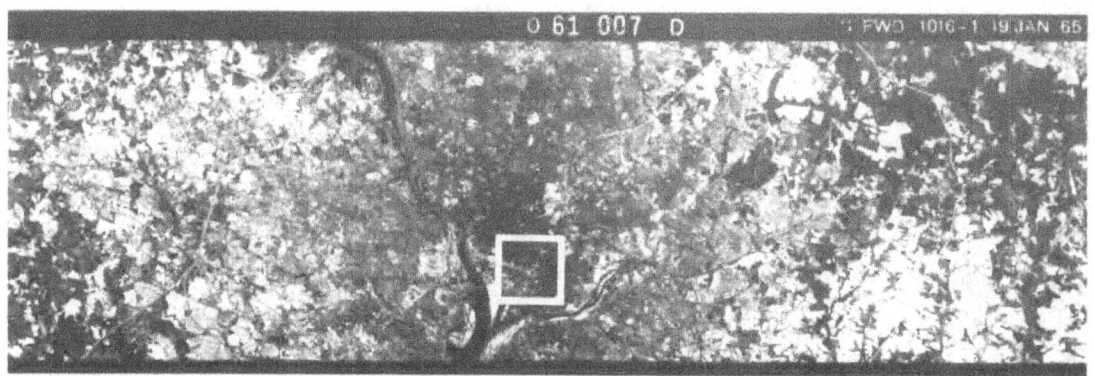

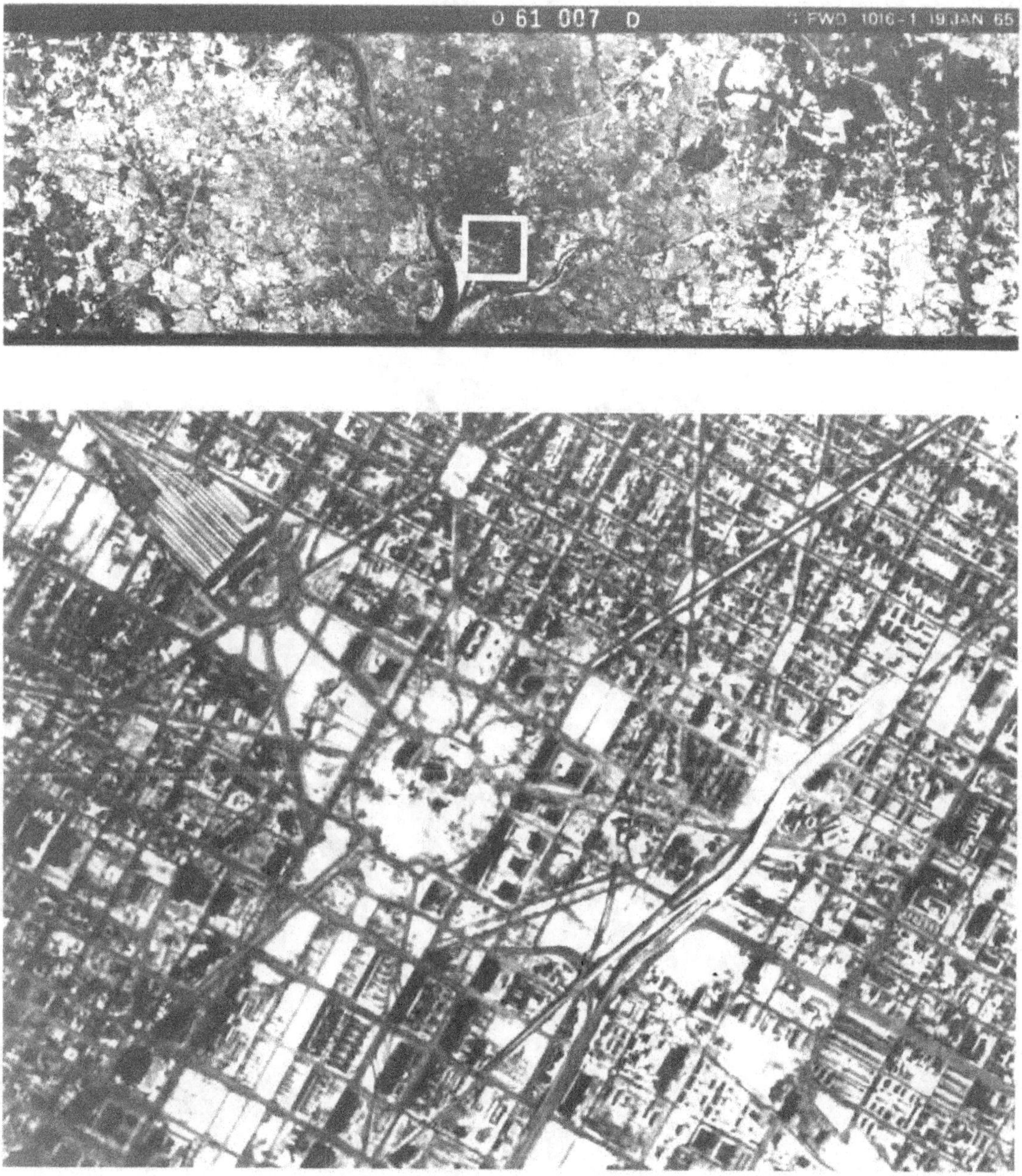

Fig. 1-28 — KH-4A Camera System imagery (CORONA-JANUS J-1); contact (partial frame) and 20× enlargement

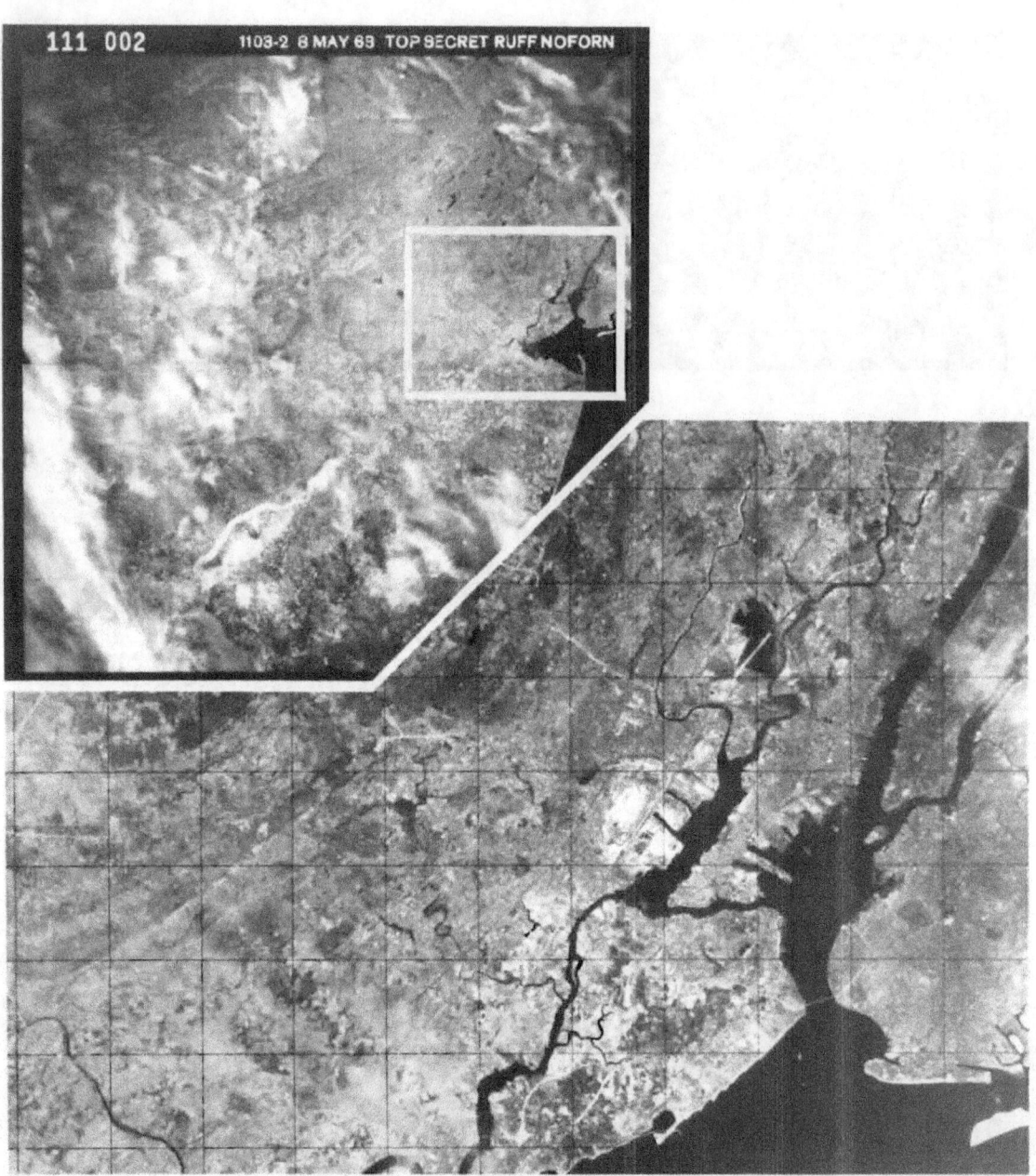

Fig. 1-29 — KH-4B Mapping Camera System imagery (DISIC); contact (partial frame) and 7× enlargement

MCS HISTORY

Fig. 1-30 — KH-4B Pan Camera System imagery (CORONA-JANUS J-3); contact (partial frame) and 20× enlargement

BIF-059W-23422/82
Handle Via
BYEMAN/TALENT KEYHOLE
CONTROL SYSTEMS JOINTLY

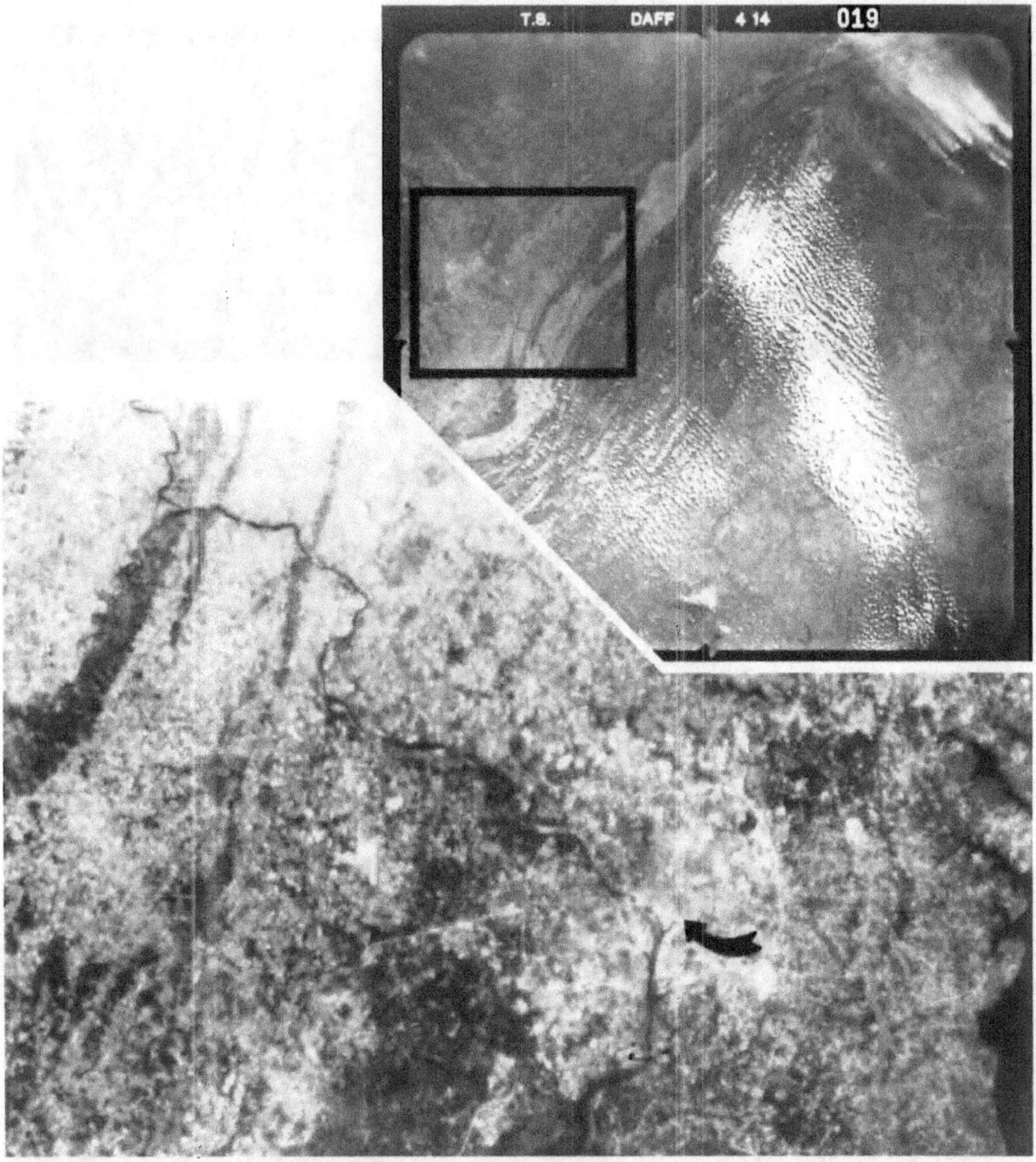

Fig. 1-31 — KH-5 Army Mapping Camera System imagery (CORONA-ARGON); contact (partial frame) and 7× enlargement

Fig. 1-32 — KH-6 Camera System imagery (CORONA-LANYARD); contact (partial frame) and 20× enlargement

Fig. 1-33 — KH-7 Camera System imagery (GAMBIT-PROGRAM 206, CUEBALL); contact (partial frame) and 25× enlargement

Fig. 1-34 — KH-8 Camera System imagery (GAMBIT-G-CUBED); contact (partial frame) and 20× enlargement

Fig. 1-35 — KH-9 Mapping Camera System imagery (HEXAGON-MCS); contact (partial frame) and 20× enlargement

MCS HISTORY

Fig. 1-36 — KH-9 Camera System Imagery (HEXAGON-PAN); contact (partial frame) and 70x enlargement

BIF-059W-23422/82
Handle Via
BYEMAN/TALENT KEYHOLE
CONTROL SYSTEMS JOINTLY

REFERENCES

1. "The U.S. Military in Space, Its Inheritance and Bequest," the United States Air Force Academy Military Space Doctrine Working Group, Mar 1982 (UNCLASSIFIED).
2. Project RAND, "Project Feed Back Summary Report," R-262, vols. I and II, edited by J. E. Lipp and R. M. Salter, 1 Mar 1954 (SECRET).
3. Author.
4. Barnett, Clyde H. Jr., "History of Project JACK POT," Aeronautical Chart and Information Center, 25 Sept 1956 (SECRET).
5. Interview with Brig. General William G. King, Jr., USAF (Ret) (UNCLASSIFIED).
6. Physics and Medicine of the Upper Atmosphere. A report of a symposium sponsored by the USAF School of Medicine and the Lovelace Foundation for Medical Education and Research, edited by Clayton S. White, M.D. and Brig. General Otis Benson, Jr. The University of New Mexico Press, Albuquerque, 1952 (UNCLASSIFIED).
7. Augenstein, Bruno W., "U.S. Military Space Programs: A Brief Analytical History and Interaction with Operational Forces, 1945-1975," August 1975 (UNCLASSIFIED).
8. From author's review of records including "TRW's Space Log," winter 1968-69, vol. 8, no. 4 (UNCLASSIFIED).
9. Perry, Robert, "A History of Satellite Reconnaissance," vol. I, BYE-17017-74, prepared under direction of the NRO (TOP SECRET/BYE).
10. Perry, Robert, "A History of Satellite Reconnaissance," vol. I, BYE-17017-74, prepared under direction of the NRO (TOP SECRET/BYE).
11. Perry, Robert, "A History of Satellite Reconnaissance," vol. I, BYE-17017-74, prepared under direction of the NRO (TOP SECRET/BYE).
12. Interview with Dana Jones, Itek Corporation (SECRET).
13. Perry, Robert, "A History of Satellite Reconnaissance," vol. I, BYE-17017-74, prepared under direction of the NRO (TOP SECRET/BYE).
14. "CORONA Program History, Vol. I Program Overview," produced by the Directorate of Science and Technology, Central Intelligence Agency, 19 May 1976 (TOP SECRET/BYE).
15. Perry, Robert, "A History of Satellite Reconnaissance," vol. IIA, BYE-17017-74, prepared under the direction of the NRO (TOP SECRET/BYE).
16. DMSP data from 1964 forward provided by David Nelson, Aerospace Corporation, El Segundo, Calif. (SECRET).
17. Perry, Robert, "A History of Satellite Reconnaissance," vol. IIIA, BYE-17017-74, prepared under the direction of the NRO (TOP SECRET/BYE).
18. KH-11 System Technical Manual (TOP SECRET/RUFF).

PART II
The HEXAGON (KH-9) Mapping Camera System

Mission 1216 (SV-16) launch on 17 June 1980

MCS HISTORY

INTRODUCTION

Preface to Introduction

Mr. Robert Perry was identified in Part I as a principal author of draft histories of early satellite reconnaissance systems. This work was accomplished under special arrangements between the Rand Corporation and SAFSP.

In 1973, following updating (in 1972) of the G and G^3 flight histories, another volume was published in which the HEXAGON program was brought up to an agreed terminal point of July 1973. Mr. Perry's research for this volume was supported by Robert A. Butler, at the time of writing a consultant with ▆▆▆▆▆▆▆▆▆▆▆▆▆▆▆ of Santa Monica, California. The history was prepared under terms of a contract between the Director NRO Program A (Director of Special Projects, Office of the Secretary of the Air Force), and the Technology Service Corporation.

This report,"A History of Satellite Reconnaissance, Volume IIIB,"[1]* in which the HEXAGON Program is addressed, has been used as the primary reference source for the introduction to Part II, pages I-1 to I-10. From this interesting and detailed report, highlights have been selected to provide an overview of HEXAGON development, which of course had to precede incorporation of the Mapping Camera, and without which there would have been no KH-9 Mapping Camera capability in this time period.

The program highlights selected, which hopefully provide sufficient detail for the scope of this report, were taken essentially verbatim from the referenced report, "A History of Satellite Reconnaissance, Volume IIIB."

HEXAGON was the outgrowth of effort undertaken in two earlier pseudo-program enterprises known as <u>FULCRUM</u> and <u>S-2</u>. The problems that beset HEXAGON development from 1966 to 1971 were unmistakably derived, in considerable part, from the assumptions, plans, schedules, and concepts that characterized those predecessor activities. As several officials of the sponsoring development agencies later conceded, HEXAGON was prematurely advanced from engineering development to system development. Unwittingly, it became at once the most costly and the most lengthy of the several ambitious developments undertaken by the National Reconnaissance Program. But it became one of the most successful, largely offsetting whatever criticisms might have been leveled at its preoperational phases.

---

* References are listed in Section 10.

BIF-059W-23422/82
Handle Via
BYEMAN/TALENT KEYHOLE
CONTROL SYSTEMS JOINTLY

FULCRUM had begun as an Itek study initially funded by the Central Intelligence Agency in 1964. But FULCRUM was preceded by an extended period of technological rummaging about in the requirements for a new search system—a replacement for CORONA and for the cancelled E-6.

The conduct of FULCRUM and the subsequent emergence of a HEXAGON program were marked by two years of variously intense controversies about requirements, schedules, technology, and organizational prerogatives.

Once the dual-camera, stereo-capable CORONA-MURAL system had been proven technically feasible, it was inevitable that a still better system based on CORONA concepts and hardware would be proposed. In March 1962, the CIA endorsed an Itek proposal to develop what came to be called the M-2 search system (for CORONA-MURAL-2). It involved the substitution of a single 40-inch, f/3.5 lens and a dual-platen film system for the dual-camera CORONA-MURAL then in use.

The M-2 proposal was formally presented for NRO review on 24 July 1962. The E-6, with its two 36-inch focal length cameras (vs. CORONA 24-inch cameras) was expected to provide better search coverage capability and thus would be the "successor system". But as discussed in the E-6 section of Part I, although showing great promise in test, the operational phase of the E-6 program had not been successful—and at the same time CORONA was returning film images with resolution on the order of 13 feet, with a dual-RV version under design and scheduled for launch in May 1963. Research undertaken after cancellation of the original E-6 SAMOS Program together with the search phase studies led toward Eastman's S-2 designs of 1964.

In the spring and early summer of 1963, CIA reconnaissance specialists had proposed two alternatives to M-2 as candidates for the "next generation" reconnaissance satellite. One was a vehicle that could be flown covertly, that could be represented to be something other than a reconnaissance vehicle.

The second concept suggested the need for a system that could perform wide-area coverage at very high resolutions, the proposed requirement emerging from a series of studies conducted by CIA system analysts in early 1963. Needless to say, both concepts were objects of controversy. Such requirement uncertainties were passed on to the "Purcell Panel", a special reconnaissance study group established by John A. McCone, Director of Central Intelligence, in the spring of 1963.*

The Purcell Panel concluded that "the natural incompatibility of wide coverage and high resolution within a given payload, is becoming more acute. . .as the art advances." An effort to combine the two functions in a single system "with only a modest improvement in resolution. . . would not be a wise investment of resources", the committee decided. Rather than focus immediately on development of a new system, the NRO was urged to concentrate on improving the average quality of returns from CORONA. The Purcell Panel made a number of specific suggestions for lines of research that promised to lead in that direction. But the Panel suggested that a new system, though ultimately needed, was for the moment a lesser requirement.

The Panel did not accept the findings of an earlier study group organized by General Greer, at Undersecretary Charyk's direction in April 1963. Concerned with the broad issue of what should be developed in the way of a new search system, the West Coast group (headed by Colonel Paul Heran) had decided that an "improved" E-6 CORONA-style recovery capsule should be developed

---

*The "Purcell Panel", headed by E.M. Purcell, included A.F. Donovan, E.G. Fubini, R.G. Garwin, E.H. Land, D.P. Ling, A.C. Lundahl, J.G. Baker, and H.C. Yutzy. Many of the "Purcell Panel" members subsequently became members of the "Land Panel", which between 1965 and 1972 operated as the principal advisor for reconnaissance matters to the President's Scientific Advisory Group and the President's Science Advisor.

in parallel with the proposed Itek M-2 system, the more promising of the two being produced, once its superiority had been verified.

Following cancellation of all the SAMOS E-Series, with LANYARD in some difficulties, and GAMBIT still unproved, high-risk technology was in disfavor in the summer of 1963. A new NRO director, Undersecretary Dr. Brockway McMillan, ordered cancellation of M-2 work at Itek in July 1963.* Itek's efforts were to be principally focused on improving CORONA product quality. To that end, General Greer's directorate made a number of specific suggestions for detail changes. CIA technical specialists in reconnaissance, then concentrated under Dr. Albert D. Wheelon, concluded that the proposals were inadequate, so in October 1963 Wheelon called into being a new special study group (the Drell-Chapman Committee) "to explore the whole range of engineering and physical limitations on satellite photography...". The group was to be concerned not merely with CORONA improvements, but also with standards and needs for new systems.

The most attractive prospect for new program creation still was in the search area. True, an ultra-high-resolution camera was also on the general requirements list, but it seemed several years in the future and, in any case, in 1963-1964 the surveillance concept that seemed most promising was embodied in the Manned Orbiting Laboratory(MOL)-DORIAN System, still embryonic but certain to be an Air Force undertaking.

Two events followed in close order. On 18 November 1963, the NRO's West Coast Directorate contracted with Itek for general feasibility studies of a new broad-area search system and for the preliminary parametric design of such a system. Not quite two months later, the CIA separately authorized Itek to study a remarkably similar set of problems, but specified a somewhat more ambitious design goal based on the findings of in-house CIA analyses. The CIA action was a delayed response to the Drell-Chapman Committee findings of late 1963, but it indirectly represented a continuation of the search system research approach embodied in the M-2 studies undertaken by the CIA in an effort to find a feasible improvement mode for CORONA-MURAL. The "West Coast Itek Study" led to S-2; the "CIA-funded Itek study" was the genesis of FULCRUM.

As described by Itek in June 1964, FULCRUM was to be a Titan II-boosted system built around a pair of rotating 60-inch focal length cameras and a transport system for 7-inch film, the general arrangement somewhat resembling what later became CORONA J-3. The increase in focal length was intended to provide resolution in the order of 2 to 4 feet across a ground swath 360 miles wide. Carrying about 65,000 linear feet of film, the system would nominally be able to photograph more than 10 million square miles of the Earth on each mission.

S-2, as first conceived, was in some respects a simpler system than FULCRUM. Intended to have both panoramic and pointing capability, it would have good resolution in a pointing mode (3 to 4 feet), and would cover a swath about 150 miles wide during search operations. The "early S-2" embodied new optics and camera mechanism, but would rely on the Atlas-Agena booster combination and an enlarged GAMBIT-style recovery vehicle. Interestingly, the first "engineering models" undertaken in the two programs were the optics of the S-2 and the film transport of the FULCRUM. Itek remained the principal FULCRUM system contractor; Greer's organization brought Kodak and Fairchild into the camera study program in September 1964 and subsequently funded space vehicle studies by both Lockheed and General Electric. Perkin-Elmer declined an invitation to bid for participation in the embryonic S-2 camera studies, but undertook some work in support of FULCRUM.

---

*Nonetheless, the elements of M-2 reappeared, in proposal form, at frequent intervals in later years, not finally disappearing until the availability of an operational HEXAGON became reasonably certain in 1971. In subsequent incarnations the basic M-2 was given several transitory names, CORONA-J-4 being the best known.

The various studies of 1963-1964 and the generous investment in pre-design research to that time encouraged the July 1964 statement of a new and formal search system requirement. Issued under the imprimatur of the United States Intelligence Board on 29 July, it called for a single-capability search-surveillance system with the area coverage equivalence of CORONA at resolutions equal to those provided by GAMBIT.

While the CIA-managed effort continued, chiefly under contract to Itek but also with Philco and Perkin-Elmer, the West Coast group was devoting equivalent attention to camera system studies being prepared by Itek, Eastman Kodak, and Fairchild. General Electric and Lockheed were performing space vehicle and reentry system research for both CIA and NRO sponsors. It seemed inevitable that some version of the solid-rocket Titan III would serve as the boost vehicle, whatever the final system configuration.

Of the several contractors involved in some aspect of camera system design, Eastman seemed, to the S-2 program office, to have the most promising concept. The CIA clearly favored Itek's approach (which incorporated an optical bar system sponsored by the CIA's in-house lens specialists).

The relatively even tenor of development in parallel was readily disturbed in February 1965; Itek abruptly renounced any intention of continuing FULCRUM development work on observation satellites rather than pursue the FULCRUM task as then defined. The decision was motivated by Itek's continuing disagreements with the CIA's technical monitors and the Agency's insistence that Itek defer to Agency specialists in technical matters.

Itek's action resulted in Perkin-Elmer becoming the principal FULCRUM camera system contractor. Then, in May 1965, the S-2 project office designated Eastman for the S-2 development.

Among significant personnel changes in the satellite reconnaissance program in 1965, Undersecretary Dr. Brockway McMillan was replaced by Assistant Secretary Dr. Alexander H. Flax as Director, National Reconnaissance office (DNRO), Major General Robert E. Greer retired in July, and Dr. Albert Wheelon resigned in October.

The NRO-preferred configuration of S-2 in early August provided for a four-bucket recovery system (with potential growth to a six-bucket design) associated with a camera capable of providing 3-foot resolution (at nadir) from an altitude of 120 miles. The payload would satisfy both search and surveillance coverage requirements if launched at a rate of six to nine systems per year. Carrying 1,000 pounds of primary film (and 63 pounds of film for a stellar-index camera), S-2 would have a length of 50 feet, a diameter of 7.5 feet, and an on-orbit weight of 12,000 pounds for a 25-day mission.

At this time, there was a special concern being experienced at Eastman Kodak. That concern was over-extension; Eastman was then producing GAMBIT-1 payloads, developing and building initial lots of GAMBIT-3 payloads, building a Lunar Survey payload for NASA under NRO cognizance and developing the S-2 payload. Added to that formidable set of tasks was DORIAN, the high-resolution camera payload scheduled to be secretly incorporated in the Manned Orbiting Laboratory vehicles being built for the Air Force. Dr. McMillan's solution to this was to propose transfer of the Eastman S-2 design to Itek, with Itek also continuing development of the second-preference S-2 camera already in process. Although complex, the transfer was not unprecedented, Itek's original FULCRUM camera design having been shifted to Perkin-Elmer in the aftermath of the February 1965 dispute between Itek and the CIA.

By mid-November 1965, owing to various delays in the search system program, it appeared that CORONA operations would have to be extended for at least a year past the point at which the new system had been earlier scheduled to enter service. One of the interactive complications was the necessity of diverting to the procurement of additional CORONA systems some of the funds earlier planned for allocation to search system development.

It was April 1966 before the several search-surveillance system proposals were eligible for the transition to a formal competition stage.

Dr. Flax had designed the management mode for HEXAGON to comply with the provisions of the 11 August 1965 NRO charter and related agreements between the CIA and the Department of Defense. That essentially meant that the CIA would retain responsibility for sensor development and sensor-related activities (for HEXAGON), and the NRO's Special Projects Directorate (in Los Angeles) for all else in the total program. The two agencies would, for each segment of their responsibilities, provide system engineering, system integration, and management.*

Given those fundamentals, Flax proposed to distribute a system operational requirements, an RFP (request for proposal) covering the sensor system, a management plan, and a schedule of planned NRO actions.

Sensor source selection, the first order of business, was assigned to a board headed by Leslie C. Dirks of the CIA and composed of four additional members, two from the CIA and two from the Directorate of Special Projects. They were scheduled to receive formal inputs from Itek and Perkin-Elmer by 22 July. Booster source selection was entrusted to a similarly constituted board chaired by Colonel W.R. Talioferro of the Titan III System Program Office. Booster proposals were due by 1 September; Flax expected contract negotiations to be completed by early November 1966.

On 30 April 1966, both the Special Projects Directorate and the CIA officially established HEXAGON project offices in their respective organizations. Flax confirmed the nomination of Donald Patterson of the CIA to direct sensor development and named Colonel Frank S. Buzard to head the HEXAGON Systems Program Office on the West Coast.

With the approval of a HEXAGON program and assignment of sensor subsystem responsibility to the CIA, existing S-2 contracts with Itek had to be terminated. Colonel Buzard negotiated the essential contract agreements with Itek between 6 May and 23 May 1966, and on the latter date Itek formally began work preliminary to a proposal for HEXAGON camera system development. With issuance of the request for proposals on 23 May, both Itek and Perkin-Elmer became contractors to the CIA's newly created Sensor Subsystem Project Office.

On 25 May, Flax authorized the creation of a source selection board for the Satellite Basic Assembly (SBA) under Buzard's direction. The board included four NRO and two CIA members. The formal requests for proposals went to Lockheed, General Electric, McDonnel-Douglas, North American, and Hughes. (Hughes subsequently decided against participating in the competition.) Proposals were due by 22 August, one month after the scheduled receipt of the sensor system proposals.

Continuing problems with the stellar-indexing (SI) camera specifications were to delay the selection of a system to provide useful mapping data, principally to the Army. During the spring of 1967, Perkin-Elmer proposed a system (dubbed SIMEC) based on the concept of printing calibrated reseau lines on normal HEXAGON panoramic photography for mapping reference. Doubts about the quality of SIMEC induced Dr. Flax to convene a joint technical evaluation committee to examine the Perkin-Elmer proposal. The committee members (from Program A, the CIA, and

---

*This two-agency management mode continued until 1974, at which time the CIA HEXAGON sensor subsystem functions were transferred to SAFSP.

such other groups as the Army Mapping Service and the National Photographic Interpretation Center) were not impressed. They concluded that SIMEC could not meet the Army's requirements for 1:50,000 scale maps, that it promised to be excessively costly, and that the reseau pattern would obscure the underlying HEXAGON imagery to an unacceptable degree. The committee's recommendation was to abandon efforts to incorporate mapping capability in the HEXAGON panoramic cameras.

Although the System Program Office had earlier concluded that a 12-inch SI camera was needed to satisfy Army mapping requirements, and the Directorate of Defense Research and Engineering later formally urged that a 12-inch (focal length) SI camera be used in HEXAGON, action to that end was not immediately feasible. The cost implications were alarming, given that HEXAGON was edging toward substantial price increases in several areas, the added weight of a 12-inch camera and associated hardware would be substantial, and there was the CIA argument that the HEXAGON should not carry mapping equipment at all. Further, the addition of any film-using camera to HEXAGON presented the problem of recovering the film; should it be transported into the existing buckets, or have its own satellite recovery vehicle?

Although a firm decision had not been reached on the number of recovery vehicles, on 6 July Dr. Flax agreed to the commencement of reviews of recovery vehicle proposals and agreed to issuance of requests for proposals by 19 July. The issuance of a request for proposal for the Stellar Terrain (Mapping Camera) in late August completed the formal actions needed to get HEXAGON development underway, but hopes that the development itself could proceed as expeditiously were to prove unduly optimistic. Almost two years were to pass before the recovery vehicles were at last put on contract, although initial estimates of first launch data for the new system postulated availability of all subsystems within 18 months of program start.

On 30 August (precisely as scheduled) the sensor source selection board reported its findings to Dr. Flax. The evaluations unanimously concluded that Perkin-Elmer had the better proposal and recommended that sensor development be assigned to that contractor.

Flax received notice of the findings of the source selection board for the Satellite Basic Assembly on 26 September 1966 and during November received initial reports of the source selection boards for the recovery vehicles and the stellar-index camera. He accepted the recommendation that Lockheed develop the satellite but withheld approval of the start of satellite vehicle work until mid-July 1967.

Proposals for both recovery vehicles and stellar-index cameras were returned for further work. McDonnell Douglas eventually won the recovery vehicle competition, and Itek the stellar-index camera competition which had been narrowed down to just Itek and Fairchild. Formal contracts appeared on 30 September and 15 November 1968, respectively. McDonnell-Douglas would provide four Satellite Reentry Vehicles (SRV's) to retrieve the Perkin-Elmer panoramic camera film, while film from the 12-inch Itek Mapping Camera would be brought back in a modified General Electric Mark V bucket as SRV No. 5.

In May, Dr. Flax settled the who-does-what argument over camera-vehicle integration responsibilities by accepting the CIA's contention that Perkin-Elmer would do the job of installing the camera system in the vehicle assembly more effectively than could Lockheed, thus permitting disposition of several lesser questions still hinging on that fundamental issue.

FULCRUM, the 1963 proposal that eventually led to HEXAGON, had initially been conceived as a search system to replace CORONA. Eventual approval of HEXAGON development expanded that concept to include surveillance by incorporating the 1964 "CORONA coverage at GAMBIT resolutions" statement. Between 1964 and 1968, considerable advances in reconnaissance technology had affected CORONA and GAMBIT; the former had become a highly cost-effective search system with remarkably good reliability, and the latter a surveillance system with a demonstrated ▮▮▮▮ resolution capability and evident growth capability to about ▮▮▮▮ "best resolution". The clandestine aspect of the Air Force Manned Orbiting Laboratory (MOL) program that also began in 1964 included a DORIAN camera with nominal ▮▮▮▮ capability. Several proposed unmanned camera systems with at least that resolution potential were beginning to demand attention by 1968. Further, some of the more optimistic participants in the satellite reconnaissance effort had by that time concluded that it was now feasible to undertake development of a high-resolution readout system with near-real-time capability. In the growing national uproar over the costly Indochina War, defense budgets were becoming tighter; one consequence was that the development of expensive new satellite reconnaissance systems was becoming increasingly dependent on finding the necessary money within ceiling-limited NRP budgets. HEXAGON was the single most expensive item of the 1968-1970 National Reconnaissance Program.

Starting in mid-1968, therefore, and continuing for a full year, proposals for reorientation, cutback, or cancellation of HEXAGON were frequent, serious, and loud. They began routinely enough in budget bureau suggestions that HEXAGON program costs were excessive and that the mission HEXAGON had been designed to perform could be as well performed by other, less costly systems. That entirely legitimate issue tended to get submerged in the subsequent advocacy of particular "other" systems, partly because the McNamara tradition of proposing "alternatives" had become a fixture of the system evaluation process, partly because various groups within the satellite reconnaissance community had taken to sponsoring one particular system, and partly because any decision to cancel or reduce expenditures on HEXAGON could not but enhance the prospects of some other proposal for reconnaissance satellite development and operations.

Such issues as the types of systems required to verify compliance with the Strategic Arms Limitations Talks (SALT) began to concern the NRP Executive Committee during the summer of 1968. Late in that summer, Deputy Secretary of Defense Paul Nitze, alert to the increasing costs of the HEXAGON program, the remarkable new capabilities being demonstrated by other reconnaissance satellites, and the potential value of HEXAGON in a SALT-agreement verification setting, instructed Dr. John Foster, Director of Defense Research and Engineering, to undertake a comprehensive evaluation of HEXAGON.

Cost was in no wise a new issue. But during the summer and fall of 1968 it became apparent that substantial reductions in prospective NRP budgets for fiscal years 1969 through 1973 were inevitable and that one way of offsetting them would be to cancel HEXAGON. The objection, of course, was that HEXAGON returns seemed essential to satisfaction of approved NRP objectives for the post-1972 period. At that point in the discussions, the Bureau of the Budget revised an earlier suggestion that the combination of GAMBIT-3 and an improved CORONA (presumably some variant of what was generally known as the CORONA J-4 proposal) would satisfy the requirement at a cost perhaps ▮▮▮▮ below that anticipated for HEXAGON. The CIA, DIA, NPIC, and NRO responded in concert that without a complete redesign (with costs then estimated to be equal to those of completing HEXAGON development), CORONA could never provide search resolutions much better than about 4.5 feet—and all those agencies were agreed that search resolutions better than 3.0 feet were essential to verification of arms limitations agreements. The Bureau of the Budget rejoinder that a 1.5-foot difference in resolution could not possibly be worth the ▮▮▮▮ it would surely cost by 1973 had no evident effect.

In November 1968 the American electorate chose Richard M. Nixon to succeed Lyndon B. Johnson as President. Nixon appointees took office in January 1969. John Foster, Director of Defense Research and Engineering, and Richard Helms, Director of Central Intelligence, were among the few senior officials to carry over from one administration to the other. Nitze was succeeded by David Packard as Deputy Secretary of Defense, and Clark Clifford, President Johnson's last Secretary of Defense, by Melvin Laird. Clifford had delegated responsibility for virtually all matters concerned with the National Reconnaissance Program to Nitze; Laird did the same for Packard, but kept closer tabs on NRP policy decisions than had Clifford. Laird's instructions from President Nixon were to reduce defense expenditures below the levels proposed by the Johnson Administration, and he did not propose to exempt the NRP from funding cutbacks. The new Director of the Bureau of the Budget, Robert P. Mayo, had received similar instructions: he found a ready advocacy of NRP funding cuts embedded in the permanent staff of the bureau.

Very shortly after taking over the budget bureau, Mayo proposed cancelling HEXAGON and substituting a CORONA-GAMBIT capability. Packard saw little merit in the idea (he had concluded that if any major reconnaissance program were to be cancelled it should be MOL-DORIAN, a measure that would have about the same financial effect as a HEXAGON cancellation), and for the moment Mayo received no support from the White House.

Late in March, Mayo again marshalled budget bureau arguments against HEXAGON and carried them to the President. On 9 April 1969, President Nixon ordered HEXAGON to be cancelled and approved carrying MOL-DORIAN to completion.

Whatever the reasoning behind the 9 April decision, reconsideration was immediate. At Helm's urging, the President delayed action on HEXAGON cancellation for two weeks. In that interval Helms and Packard made their objections known to the President, and on 21 April, Mayo reversed his original stand. The three brought Laird to their way of thinking by late April. The fundamental argument they settled on (eventually presented by Mayo) was that HEXAGON would provide a much better capability for validating any arms limitation agreement than MOL-DORIAN.

What may have been a clinching argument against MOL appeared as an independent recommendation of the Land Panel which reached the President on 6 May 1969. Dr. Land and his group favored cancelling the manned part of the MOL program, developing an unmanned high resolution satellite using DORIAN-derived optics, and diverting funds saved by the MOL cancellation to a "real-time-readout" system. President Nixon was sufficiently intrigued by the potential of the readout system Dr. Land advocated to make that capability the principal reconnaissance satellite objective of his administration. To implement that decision he reversed his earlier verdict on HEXAGON and ordered cancellation of MOL-DORIAN. Laird publicly announced that aspect of the decision on the morning of 10 June 1969; the endorsement of HEXAGON received no public notice, of course.

The June 1969 decision was conclusive, and before long was irreversible. To have cancelled HEXAGON after the summer of 1969 would have decimated the national capability for search-satellite operations. Proposals for extending CORONA production and even for stockpiling CORONA's against some future need (which presumably could have included the failure of the HEXAGON development program) gained an occasional hearing thereafter, but never again did they have high-level support. Enough CORONA systems had been ordered to protect against a serious gap in coverage should HEXAGON be delayed in development—which proved notably wise—and the development of a reasonably effective and not too costly GAMBIT modification (Higherboy) represented another hedge against delayed HEXAGON availability.

During the first 2 years after HEXAGON program approval, delays had largely arisen due to uncertainties of program definition and design. Some delays in implementing elements of HEXAGON other than the camera subsystem were deliberate, based on judgment that the camera subsystem required more time for development than most other components due to the fact that the contractor had less experience and facilities for his task. Their effect had been to cause a significant slippage in program schedules. Although their advocates had represented both S-2 and FULCRUM to be fit for full system development by late 1965, not until the spring of 1966 had a development start been approved, and not until 1968 were all of the essential elements of the HEXAGON system under contract. Decisions on booster configuration, recovery vehicle configuration, the selection of a stellar index and mapping camera, and accommodation of the orbital vehicle to the changing design of the camera system had been delayed for longer than could reasonably have been anticipated.

After system definition had finally been completed, an event that was difficult to date but could most accurately be assigned to mid-1968, HEXAGON began to encounter the sorts of engineering and test problems that had marked the development of all earlier photographic satellites. Although the HEXAGON program schedule made some allowances for slippages caused by unforeseen engineering difficulties, in the end they proved to be insufficient.

The first unrecoverable slippage of any kind was acknowledged early in 1969 (while the scheduled first launch date still was 1 October 1970); Perkin-Elmer spent an unprogrammed $2\frac{1}{2}$ months of additional work in completing and testing the final qualification model of the camera-vehicle midsection assembly. The disclosure of that misadventure had been preceded by a rather unsettling special review of HEXAGON engineering work undertaken through the end of 1968; the review highlighted 14 major and 28 lesser system faults that required prompt attention.

Although to that time only about 2 months of unrecoverable slippages in the total HEXAGON program had been positively identified, and schedules had been designed to accommodate at least that much slack, in June 1969 Dr. McLucas* assigned to his principal deputy, Dr. F. Robert Naka, the task of determining the viability of the HEXAGON launch schedule (which then called for first flight no later than December 1970). Naka's evaluation† forwarded to members of the NRP Executive Committee on 20 June, contained carefully qualified expressions of caution.

In addition to evaluating the probability that HEXAGON would be launched as scheduled, Naka estimated the degree of confidence the NRO should have that the first HEXAGON mission would be successful, and looked at various ways of optimizing search mission products at least possible costs. An unavoidable parallel issue was whether CORONA vehicles additional to those then on order should be purchased as a safeguard against a lapse in search coverage that might occur if HEXAGON operations began appreciably later than December 1970.

Naka calculated a 95 percent probability for a first HEXAGON launch no later than June 1971, and rated at 75 percent the probability to launch no later than January 1971. He concluded that about 75 percent confidence should be assigned to the possibility of mission success on the initial flight and foresaw a 95 percent probability that at least one of the first three missions would be successful. Given those odds, he suggested that the 12 CORONAS programmed for launch at about 2-month

---

*Dr. J.L. McLucas succeeded Flax as Director, National Reconnaissance Office, in April 1969.
†Dr. Naka signed and reported the findings as spokesman for a committee that included ▮▮▮▮▮▮▮▮ of the CIA's sensor project office and Colonel L.S. Norman of the NRO's Directorate of Special Projects. Although preliminary findings were forwarded to the Executive Committee in June, formal reports seem not to have been prepared until September 1969.

intervals between June 1970 and July 1971 should be rescheduled to allow for at least two missions after July 1971, thus ensuring a minimum overlap of CORONA with HEXAGON and providing some search coverage in the event of either a HEXAGON slippage past June 1971 or mission failure. Given the existing uncertainties of HEXAGON scheduling, Naka also cautioned that the need for more CORONA's should be reassessed in December 1969.

Both Perkin-Elmer and Lockheed had advised program managers of potentially massive HEXAGON cost growth. Costs were not unrelated to schedules, of course, and in the late months of 1969 schedules were becoming almost as worrisome as costs. To maintain the required pace of progress, several contractors had resorted to double and triple shifts and the extensive use of overtime. HEXAGON overtime and multiple-shift work was necessary to meet schedules that were based on the planned expenditure of existing stocks of reconnaissance satellites, chiefly CORONA systems.

In October 1969, Dr. Naka again reviewed HEXAGON status, and although an indicated additional slippage of at least 1 month had appeared since August, he recommended that the decision on an additional CORONA buy be postponed until January 1970. By January there had been no significant change, so the Committee somewhat reluctantly decided to forgo the option of ordering more CORONA vehicles.

In December 1969, Brigadier General W.G. King (who in August 1969 had succeeded Major General John L. Martin, Jr., as NRO head of Program A, the Directorate of Special Projects) convened a special meeting of HEXAGON principals from the program office, the sensor office, and the major contractors to reevaluate the prospect of meeting the scheduled December 1970 launch date. All agreed that although the schedule was getting tighter with the gradual disappearance of slack time that had earlier been provided to accommodate inevitable engineering and test difficulties, the December 1970 deadline was reasonable—but staying on schedule would require "vigorous action" by all concerned.

On 7 July, at the Perkin-Elmer plant, the first flight-article twin camera assembly (P-1) suffered a catastrophic failure during low temperature chamber testing. It had been scheduled for 31 July shipment to Lockheed. The extent of damage was so great that no possibility of timely repair and recalibration could realistically be entertained. On 10 July, therefore, the sensor program office confirmed the contractor's judgment that the second sensor system (P-2), originally scheduled for 5 December shipment, had to be substituted in first-flight schedules. It was conceivable that P-2 could be qualified and shipped by 26 August, but given the earlier disappearance of virtually all remaining slack time in the flight readiness schedule, there was slight prospect of meeting the 17 December 1970 first flight target date.

Following arrival of the second camera payload (P-2) at Lockheed's Sunnyvale facility, major problems with the film transport mechanism again stalled the test program. Formal acknowledgement of the inevitable launch date slippage came from General King on 15 September.

By January 1971 it had become apparent that "March 1971" (which had widely been interpreted to mean "about 1 March") had better be restated as April, and 9 April became the new official target date—although in private session the Executive Committee received advice from Dr. Naka that "about May 10, 1971" was a better estimate. Somewhat less inclined than in the past to accept schedule assurances at their face value, the NRP Executive Committee endorsed Dr. McLucas' action in providing additional insurance against extended HEXAGON trouble by authorizing work on a GAMBIT modification (Higherboy) that would permit GAMBIT to operate as a makeshift search system by flying at altitudes of about 525 nautical miles. At that distance, GAMBIT swath widths would approximate those of CORONA, and resolution would be about the same. The first of three

Higherboy kits ordered for insurance would be ready by November 1971 but would not be needed before April 1972, in the worst possible case.

Dr. Naka's cautious appraisal of the worth of "official" HEXAGON launch schedules proved sound almost immediately. By the end of March, problems encountered in acoustic and thermal tests of the first payload-vehicle assembly caused program managers to reschedule the initial launch for "not earlier than 3 May 1971", and by April it had become apparent that the 4-month allowance for payload integration and checkout should have been 7 months. Late in April new delays intervened, and 20 May became the target date. Then on 26 April the program office learned that extended testing of the shutter on the second and third camera payload sections had disclosed that failure was liable to occur after only 28,000 cycles of shutter operation. Colonel Buzard sadly advised Brigadier General Lew Allen, new Program A Director, that because the shutters in the payload then being prepared for launch had already experienced 20,000 and 28,000 cycles operation respectively, there was a high probability of shutter failure on orbit. He therefore proposed to delay the first launch until at least June.

Diagnosis and shutter modification (and retesting) had chewed up so much time that "about 14 June" had to become the new launch target date. (Because HEXAGON payload vehicles would not be trucked over California highways on weekends, when traffic was heaviest, and because the payload would not be ready for trucking before 28 May, four additional days delay were imposed by the unfortunate coincidence of the Memorial Day weekend and the completion of payload testing at Sunnyvale.)

But that was the last. Payload delivery was on schedule, pre-launch checkout was almost uneventful, and on 15 June 1971 the first HEXAGON satellite went into orbit. Carrying HEXAGON from program approval to first launch had taken 5 years rather than 2 and had cost rather more than twice as much as initially estimated, mostly for camera development, which cost three times as much as the CIA had anticipated, but a launch had been brought off. And in the end the critical scheduling estimates provided by Dr. Robert Naka and his associates in 1969 had proved remarkably accurate: HEXAGON did indeed fly in June 1971 (the "95-percent probability" date), and it did indeed function successfully (the "75-percent confidence evaluation").

SECTION 1

THE HEXAGON SATELLITE

GOVERNMENT ORGANIZATIONS

The government organizations participating in HEXAGON operations: Committee on Imagery Requirements and Exploitation (COMIREX); the Global Weather Center (GWC) of the Air Force Weather Service; the Imagery Collection Requirements Subcommittee (ICRS) of COMIREX; the National Photographic Interpretation Center (NPIC); the Satellite Operations Center (SOC)* of the National Reconnaissance office (NRO); the Satellite Test Center (STC) of the Air Force Satellite Control Facility (SCF); the 6595th Satellite Test Group, the 6594th Aerospace Test Group, Hickam AF Base, Hawaii; the Sensor Subsystem Project Office (SSPO) of the CIA; the HEXAGON Program Office at the NRO's Directorate of Special Projects (SAFSP); the Air Force Special Projects Production Facility (SPPF)†; and the Defense Mapping Agency (DMA) (and its predecessors).

Both Program A and B Directors, General Lew Allen at the time of first launch, and the CIA's Director of Reconnaissance, John Crowley, reported to Dr. Alexander H. Flax (DNRO) for purposes of managing the operational aspects of HEXAGON. The System Program Office (Los Angeles) and the Sensor Subsystem Project Office (Langley, VA)‡ were respectively responsible for mission operational software (computer programs) and participation in the development and analysis of the software.

It would be virtually impossible to list all of the government employees by name who have made significant contributions to the HEXAGON Program. However, a list has been compiled to identify the heads of the various government organizations having substantial control over the shaping of HEXAGON from its origin to the present time.§

Government Organizations That Control the HEXAGON Program

DNRO—Director, National Reconnaissance office

USAF—Director, Secretary of the Air Force Special Projects (SAFSP)

- Maj. General John E. Kulpa, Jr.   Aug 1975 to Present
- Maj. General David D. Bradburn   June 1973 to Aug 1975
- Maj. General Lew Allen, Jr.   Apr 1971 to June 1973
- Brig. General William G. King, Jr.   May 1969 to Apr 1971
- Maj. General John L. Martin, Jr.   July 1965 to May 1969
- Maj. General Robert E. Greer   1960 to July 1965

---

*Initially and until September 1978 when the responsibility was assigned to ▮▮▮▮▮
†Initially and until September 1975 when this capability was transferred to SAFSP/NPIC.
‡Initially and until 1974 when these functions were assigned to SAFSP.
§Due to the sensitivity of some satellite reconnaissance programs, this list of organizations and names is not complete.

MCS HISTORY

DMA—Defense Mapping Agency

- Maj. General Richard M. Wells — 1 July 1981 to Present
- Maj. General William L. Nicholson, III (USAF) — June 1979 to 30 June 1981
- Lt. General Abner B. Martin (USAF) — Aug 1977 to June 1979
- Vice Admiral Shannon D. Cramer (USN) — Aug 1974 to Aug 1977
- Lt. General Howard W. Penny (USA) — July 1972 to Aug 1974
- Colonel ▆▆▆▆ (USA)* — Dec 1969 to Apr 1972
- Colonel ▆▆▆▆ (USAF)* — Jan 1964 to Nov 1969

CIA—Deputy Director for Science and Technology (DDS&T—or equivalent)

Aerospace Corporation—Advanced Orbital Systems Division

- James R. Henry — June 1979 to Present
- C. James Crickmay — Apr 1973 to June 1979
- Bruce L. Adams — Dec 1969 to Apr 1973
- Leonard C. Lidstrom — Jan 1969 to Aug 1969
- John W. Luecht — Aug 1967 to Dec 1968
- John D. Sorrels — Dec 1966 to July 1967
- George M. Kelsey — To Dec 1966

SAFSP HEXAGON Program Office

- Colonel Lester S. McChristian — July 1978 to Present
- Colonel Raymond A. Anderson — Aug 1973 to July 1978
- Colonel Robert H. Krumpe — June 1971 to Aug 1973
- Colonel Frank S. Buzard — July 1966 to June 1971

SAFSP MCS Payload Division

- Lt. Colonel Guy F. Welch — July 1976 to July 1982
- Lt. Colonel William G. Powell — Aug 1973 to July 1976
- Lt. Colonel Albert W. Johnson — May 1971 to Aug 1973
- Captain Guy F. Welch — July 1970 to May 1971
- Captain David F. Berganini — Feb 1966 to June 1970

---

*Director, MC&G Directorate, Defense Intelligence Agency, predecessor of DMA.

ASSOCIATE CONTRACTORS

Project HEXAGON is a team effort consisting of the governmental organizations listed and several major contractors throughout the United States (Fig. 1-1). These contractors provide a coordinated effort by using Interface Control Documents as binding technical agreements for responsibilities and performance of their respective equipments.

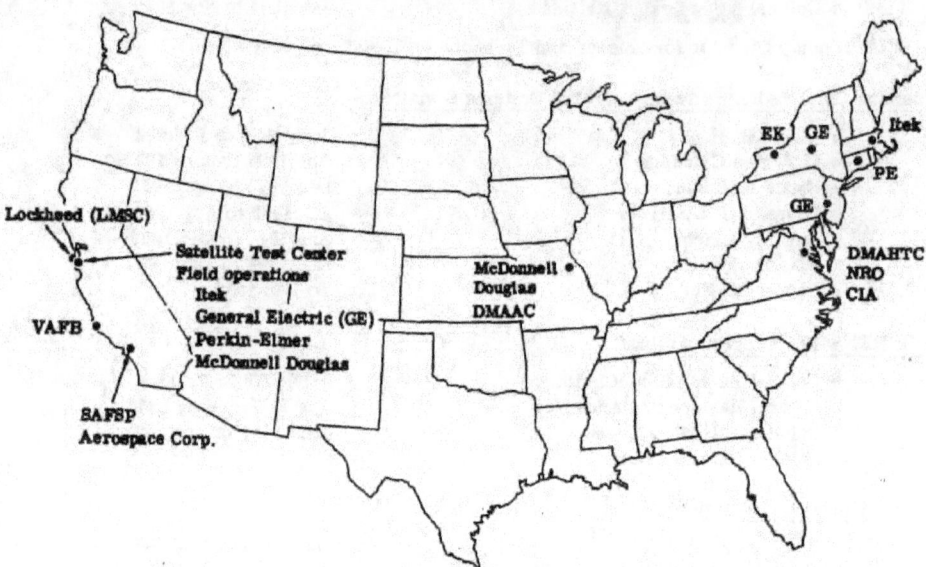

Fig. 1-1 — Locations of government and contractor facilities supporting HEXAGON Program

HEXAGON Program Contractors

The contractor team program managers who served the HEXAGON Program over the past years are identified below. An asterisk by the contractor's name indicates direct support to the Mapping Camera System.

*Lockheed Missiles and Space Company (LMSC)

- Steve P. Treat   Feb 1980 to Present
- Paul J. Heran   Feb 1970 to Feb 1980
- Dr. Stanley I. Weiss   July 1967 to Feb 1970

HEXAGON Program Contractors (Cont.)

*Itek Corporation Optical Systems Division (OSD)

- Maurice G. Burnett — Oct 1976 to program completion 1981
- Jean R. Manent — Apr 1974 to Oct 1976
- William J. Reusch — Aug 1973 to Apr 1974
- D. David Cook — Feb 1972 to Aug 1973
- Paul J. Mailhot — Jan 1970 to Feb 1972
- John F. Doyle — Nov 1968 to Jan 1970
- John T. Watson — June 1968 to Nov 1968

*General Electric Company, Reentry Systems Division (GE-RSD)

- John S. Kleban — Feb 1975 to program completion 1981
- Stephen F. Csencsitz — Mar 1970 to Feb 1975

*General Electric Company, Aerospace Electronic Systems Department (GE-AESD)

- Francis Smith — 1980 to Present
- Elmer B. Tamanini — 1978 to 1980
- James O. Moore — 1969 to 1978
- Robert M. Larkin — 1967 to 1969
- Norman N. Feldman — 1966 to 1967
- John H. Griswald — 1964 to 1966

*TRW

- David M. Yaksick — Aug 1979 to Present
- Clair D. Calvin — Apr 1975 to Aug 1979
- David M. Yaksick — Nov 1973 to Apr 1975
- Gerald K. Lambert — Sept 1972 to Nov 1973
- William V. Buck — Mar 1972 to Sept 1972
- Winston W. Royce — Nov 1971 to Mar 1972
- Thomas A. Magness — Oct 1969 to Nov 1971

Perkin-Elmer Corporation, Optical Technology Division (OTD)

- Victor Abraham — Oct 1980 to Present
- Kent H. Meserve — July 1979 to Oct 1980
- Michael A. Mazaika — Oct 1977 to July 1979
- Bernard Malin — May 1975 to Oct 1977
- Paul E. Petty — Dec 1972 to Oct 1975
- Harry W. Robertson — Dec 1970 to Dec 1972
- Michael F. Maguire — Oct 1967 to Dec 1970
- Kennett W. Patrick — Jan 1967 to Oct 1967
- Richard W. Werner — Oct 1966 to Jan 1967

McDonnell Douglas Astronautics Company (MDAC)

- Forrest D. Blanton — 1974 to Present (1981)
- Logan T. McMillian — 1968 to 1974

The project HEXAGON system and contractor team consists of:

Search/Surveillance (Stereo Panoramic)

- Two-camera assemblies—Perkin-Elmer, Danbury, Conn.
- Film supply and takeup units—Perkin-Elmer, Danbury, Conn.
- Shroud, mid- and forward-section structure—Lockheed, Sunnyvale, Calif.
- Reentry vehicles (Mark 8)—McDonnell Douglas, St. Louis, Mo.
- Film—Eastman Kodak, Rochester, N.Y.

Mapping, Charting, and Geodesy System (Missions 1205-1216)

- Stellar and terrain cameras—Itek, Lexington, Mass.
- Reentry vehicle (Mark V)—General Electric, Philadelphia, Pa.
- Structure—Lockheed, Sunnyvale, Calif.
- Film—Eastman Kodak, Rochester, N.Y.

Satellite Control Section

- Telemetry, power, and pyros—Lockheed, Sunnyvale, Calif.
- Command system—General Electric, Utica, N.Y.
- Attitude control and orbit adjust—Lockheed, Sunnyvale, Calif.
- Structure and booster adapter—Lockheed, Sunnyvale, Calif.

Booster Vehicle—Titan IIID

- Stage 0 solid propellant—United Technology's Chemical System Division, Sunnyvale, Calif.
- Stage I and II liquid propellant—Martin Marietta Corporation, Denver, Colo.

Mission Support Software

- TUNITY—TRW, Sunnyvale, Calif.
  - Vehicle command and control
  - Camera operations
  - Requirements management
- Satellite Control Facility Software—System Development Corporation, Santa Monica, Calif.

Technical Support Organization

- Technical support for the Air Force System Program Office (responsible for the HEXAGON Program technical interface coordination) Aerospace Corporation, El Segundo, Calif.

THE AEROSPACE VEHICLE

The HEXAGON Satellite Vehicle is launched by the Titan IIID Booster Vehicle from Space Launch Complex - 4 East, Vandenberg Air Force Base, California. When mated together, the entire assembly is termed the "Aerospace Vehicle" (see Fig. 1-2). The complete SV including the shroud is mated to the booster vehicle 14 days prior to launch. The Aerospace Vehicle is then functionally checked and all propellants and gases are loaded.

The booster vehicle can place 24,000 pounds into an 82 x 144-nautical mile (perigee x apogee) orbit with an inclination (~97 degrees) that provides the nearly sun synchronous condition needed for long-life missions.

The Titan IIID booster vehicle is a three-stage booster consisting of the standard liquid core for stages I and II plus two solid rocket motors (SRM's) as stage O.

The flight control system stabilizes the vehicle from launch to SV separation in response to: (1) attitude data, (2) rate data, and (3) command data issued by the flight control computer and/or the radio guidance system via ground tracking station.

Electrical power for the flight control system instrumentation, flight safety, and electrical sequence system is provided via silver-zinc primary batteries and solar panels which are deployed after achieving SV stability on rev 1.

The HEXAGON vehicle performs two major functions: (1) world-wide search and surveillance missions with two cameras that provide stereo panoramic photography; (2) mapping and geodesy missions with stellar and terrain frame cameras (missions 1205 through 1216). The film from the search and surveillance missions is recovered as each of four large reentry vehicles (Mark 8) is filled; the Mapping Camera System (MCS) film is retrieved via a single reentry (Mark 5) vehicle mounted on the HEXAGON vehicle nose (Fig. 1-3). Accurate vehicle location for the mapping mission is determined with the Doppler Beacon System (DBS) and by the Navy Navigational System (NAVPAC). In addition to the stereo panoramic cameras and the Mapping Camera System, the HEXAGON vehicle can also carry:

The SV configuration (Fig. 1-4) incorporates overall mission success considerations as well as weight minimization and structural efficiency. The panoramic camera system film supply, cameras, and RV's are arranged in line for film path simplicity; the two-camera assembly is relatively close to the attitude control system in the aft section to enhance pointing accuracy. The Mapping Camera System (MCS), the reentry vehicle (RV), and the Doppler Beacon System (DBS) are contained completely in the Auxiliary Payload Structure Assembly (APSA) which is mounted to the front of the SV (Figs. 1-3, 1-4). Aft section electronic/electric equipment, mounted on trays in a modular fashion, is accessible through removable panels during factory and pad repairs. Access is provided to the RV's, two-camera assembly, and film supply for necessary servicing. Propulsion/control force elements are grouped in a module for testing efficiency, and brazed plumbing is used to ensure the integrity of the propellant system through handling, launch, and flight.

In the factory, the SV is brought to flight readiness by acoustic and thermal vacuum testing of the assembled vehicle; vehicle instrumentation is designed for such system level testing with RF command and data links.

The SV is shipped flight-ready to the launch base, with validation prior to launch. When required, equipment is replaced on a module/box basis to preserve factory verifications.

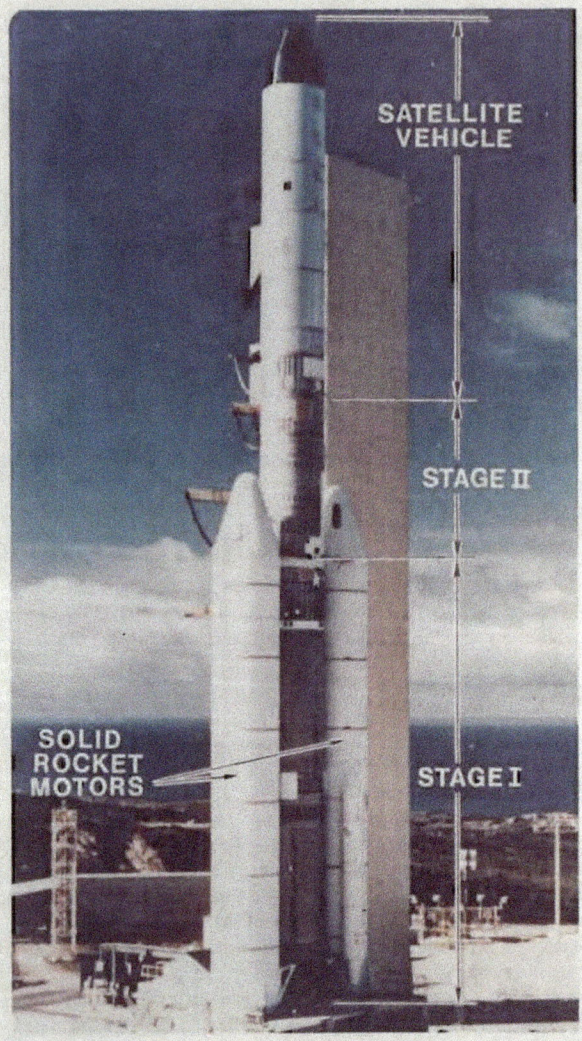

Fig. 1-2 — The aerospace vehicle

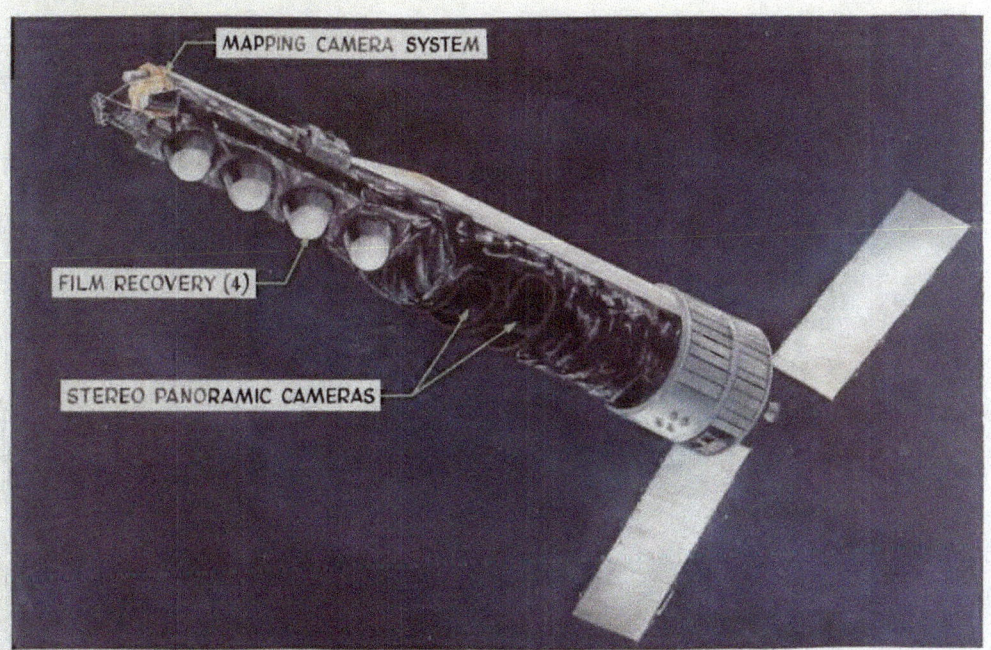

Fig. 1-3 — HEXAGON vehicle on orbit

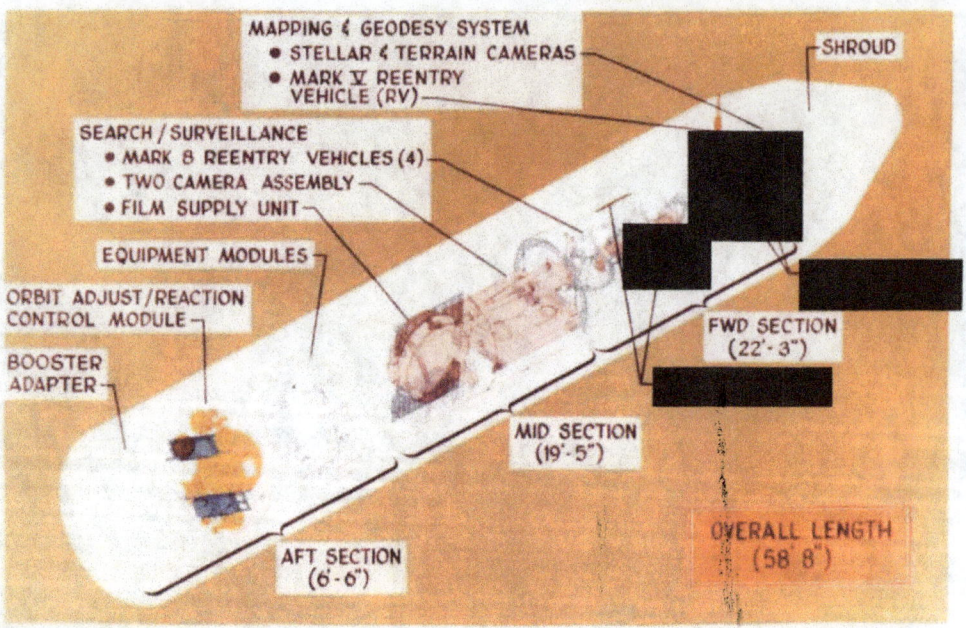

Fig. 1-4 — Satellite vehicle configuration

Provision has been made for alignment of critical elements during assembly and for verifying the alignment of the Attitude Reference Module with the two-camera assembly at the launch pad.

The overall length in orbit of the SV illustrated is 52 feet. At launch, with shroud and booster adapter, the length is 58.75 feet. The shroud, which protects all but the aft section, is 52 feet long. The solar arrays, when deployed, extend to 17 feet outboard on each side of the vehicle. Injection weight for the SV illustrated is approximately 24,000 pounds.

Satellite Basic Assembly Structure

The SBA structure, shown in the cut-away drawing (Fig. 1-5), is of semimonocoque construction. The booster adapter section has aluminum skin, rings, and stringers. This section contains the booster separation joint, which uses $2\frac{1}{2}$ grain/ft of mild detonating fuse to break a circumferential beryllium strip.

The OAM/RCM section has corrugation-reinforced aluminum skin with aluminum and magnesium internal structure. This section contains the propulsion elements and the solar array modules.

The equipment section has twelve removable corrugation-reinforced aluminum skin panels bolted to an aluminum tubular internal structure which supports honeycomb equipment panels. Guidance, communication, command, and power components are mounted on these panels as subsystem modules.

The mid-section has a short titanium conical section and a cylindrical section of magnesium skin, with magnesium hat-section longitudinal stiffeners. A magnesium and titanium internal structure supports the primary payload.

The forward section has aluminum and magnesium skin with magnesium hat-section longitudinal stiffeners. The internal magnesium and aluminum structure with titanium fittings supports the four reentry vehicles. The Mapping Camera System ███████████████ are supported on the external surfaces of the forward section.

The Mapping Camera System is supported in the Auxiliary Payload Structure Assembly (APSA).

Satellite Basic Assembly-Aft Section

The aft section (Fig. 1-6) consists of an equipment module, a booster adapter section, and an orbit adjust module/reaction control module (OAM/RCM). It is 10 feet in diameter and 5 feet long. This section is a semimonocoque structure with a corrugated aluminum external skin. It weighs approximately 3,500 pounds including all equipment, less expendables. The aft section provides environmental protection and thermal control during ground, ascent, and orbital operations. The structure is capable of withstanding the dynamic and static conditions imposed during all phases of ground handling, launch, ascent, and orbit. The aft section interfaces with the booster, mid-section, aerospace ground equipment, main electrical umbilical, pressurization and propellant loading lines, and the battery cooling lines.

The booster adapter section mates the Satellite Vehicle to the Titan IIID booster. The adapter is equipped with 70 square inches of vent area. The separation joint, with a redundant pyrotechnic system, is a part of this section.

The OAM/RCM section houses and supports the OAS/RCS hydrazine systems which provide orbit and attitude control, the independent lifeboat freon gas system which provides emergency attitude control, and the solar array modules which generate power. This section interfaces with ground pressurization and propellant loading lines. The solar array modules which mount on the aft bulkhead adjacent to the OA engine nozzle are not shown in the photograph.

The equipment section consists of 12 equally spaced, equally sized bays, each capable of supporting up to 500 pounds of equipment on individual trays. Each equipment bay provides sufficient access to allow complete module installation and removal at the factory and pad as shown in the lower completely open bay. The other bays as shown have non-flight panels with ground access doors used in factory assembly and test. This section interfaces with the main electrical umbilical and the mid-section.

Attitude Control

The Attitude Control System (ACS) (Fig. 1-7) provides earth-oriented attitude reference and rate sensing. It develops Reaction Control System (RCS) thruster firing signals to bring the vehicle to a commanded attitude and to maintain attitude and rate with the accuracies shown in Table 1-1. The ACS also provides measurements of vehicle attitude and rate during search/surveillance operations to the accuracy shown.

The ACS is a three-axis rate gyro-integrator system with updating in pitch and roll by horizon sensor and in yaw by gyrocompassing. Error signals generated by the gyros and horizon sensor are combined in the flight control electronics, and modulated by pseudo-rate circuits in each axis to provide thruster firing commands with the impulse bit control necessary to meet the tight rate control and short settling-time requirements.

All elements are redundant for malfunction correction. Cross-strapping between redundant and primary ACS components (horizon sensors, gyros, flight control electronics assembly) is possible to permit selection of non-failed components to drive the RCS thrusters.

Table 1-1 — Vehicle Attitude and Rate Accuracies

	Control Requirements			Measurement Requirements		
	Pitch	Roll	Yaw	Pitch	Roll	Yaw
Search/Surveillance Operations						
Attitude accuracy (deg)	0.7	0.7	0.64	0.4	0.4	0.5
Rate accuracy (deg/sec)	0.014	0.021	0.014	0.001	0.001	0.001
Non-Horizontal Operations						
Attitude accuracy (deg)	3	1	1			
Rate accuracy (deg/sec)	0.15	0.15	0.15			

Settling time from search/surveillance disturbances: Stereo = 0.2 sec; Mono = 6 sec

Orbit Adjust and Reaction Control

An Orbit Adjust System (OAS) and Reaction Control System (RCS) (see Fig. 1-8) provide the forces necessary to control the vehicle orbit and the vehicle attitude in orbit, respectively. The OAS provides injection error correction (if required), drag and perigee rotation makeup, and deorbit of the Satellite Vehicle at the end of the mission. The RCS provides pitch, yaw, and roll control via eight thrusters.

OAS and RCS both use catalytic decomposition of monopropellant hydrazine to generate thrust. For reliability, the systems are pressure-fed, with the pressurizing gas enclosed in the propellant tank with the hydrazine. This results in declining or blowdown pressure characteristics; the thrust level of the OAS engine declines from 250 to 100 pounds and that of the RCS engines from 6 to 2 pounds. A quad-redundant valve operated by the command system controls flow to the OAS engine. The ACS generates signals that control the firing of the RCS engines.

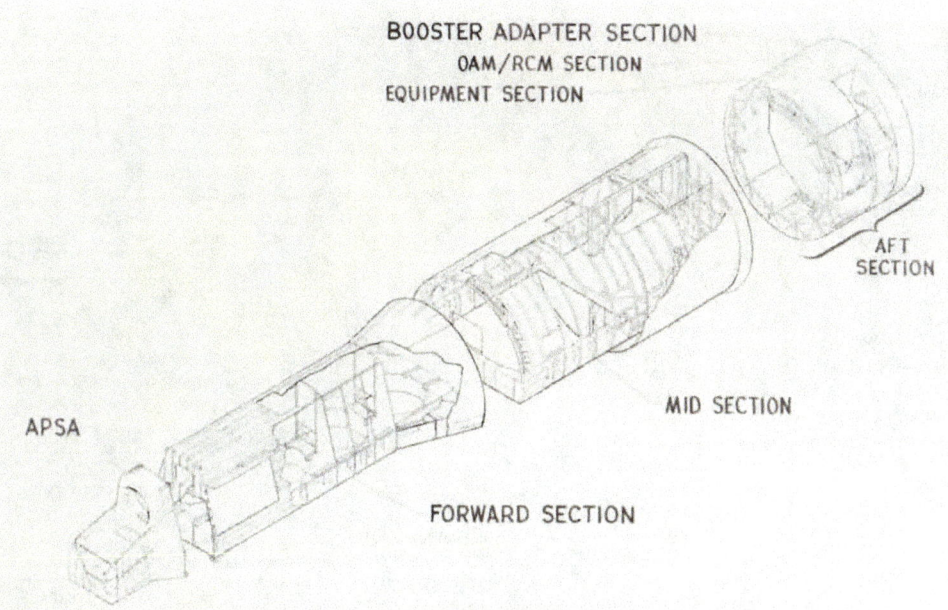

Fig. 1-5 — Satellite Basic Assembly structure

Fig. 1-6 — Satellite Basic Assembly—aft section

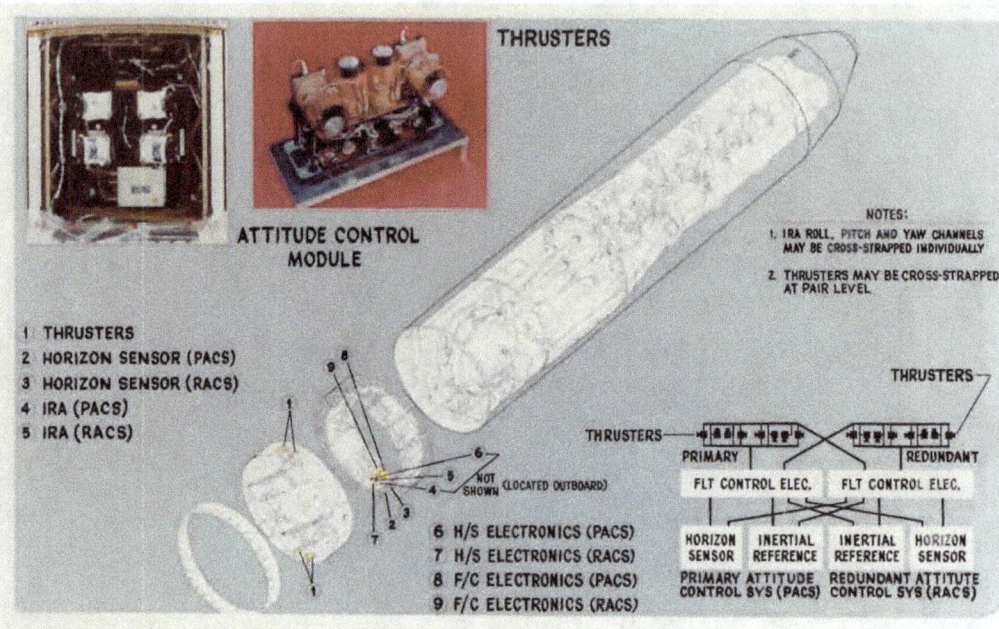

Fig. 1-7 — Attitude control

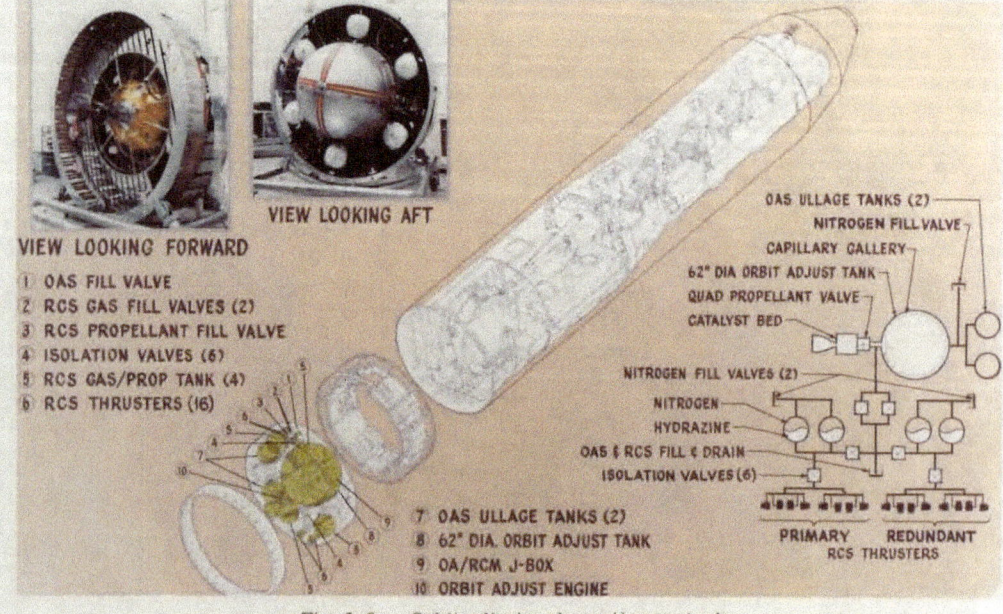

Fig. 1-8 — Orbit adjust and reaction control

On SV-15, the 62-inch diameter OAS tank could be loaded with up to 4,000 pounds of propellant with two spheres containing high pressure nitrogen (isolated by pyro valves and admitted into the OA tank at times selected during the mission) to maintain the pressure within the desired operating range.* This propellant could be utilized in OA burns to provide velocity increments of 2 to 400 ft/sec. A passive (surface tension) propellant management device maintained propellant at the tank outlet at all times, permitting engine firings in any attitude.

The four 22-inch-diameter RCS tanks provide capacity for 450 to 540 pounds of propellant. Propellant orientation is maintained by diaphragm. The thruster impulse bit (0.15 lb-sec or less, depending on blowdown status) is compatible with the tight rate-control requirements. A complete redundant set of thrusters is provided for malfunction protection; either set can be supplied by the four tanks and each pair of thrusters can be driven by the primary or redundant ACS valve drivers.

A transfer line is provided between the OAS and RCS tanks to permit propellant exchange to optimize the use of on-board propellant for each mission.

Electrical Distribution and Power

Power to operate the Satellite Vehicle is provided by solar arrays deployed from the aft section following separation from the booster (Fig. 1-9). Rechargeable NiCd batteries (type-40) provide energy storage to meet dark-side-earth and peak power requirements. Unregulated power is distributed throughout the vehicle to using equipment within a 24 to 33 Vdc range.

The power generation and storage system comprises four parallel segments, with an array section, charge controller, and battery in each to reduce the effect of a failure; a single malfunction will not terminate the mission. Fusing of equipment, limiting minimum wire size, and isolating voltage-critical circuits add to the reliability.

The power system is capable of providing approximately 11,000 watt-hours/day of usable power over a beta angle range of −8 to +60 degrees by adjusting the array angle about the vehicle roll axis. This will support at least 52 minutes per day of search/surveillance and mapping camera system operation.

Power for the lifeboat system is provided by one type-40 battery from the main power system. Equipment necessary for recovery vehicle and Satellite Vehicle deorbit can be switched to this battery for emergency operations. Depletion of the batteries below 55 percent or an excessive load on the main power system will automatically isolate the lifeboat system and its battery. This assures adequate power for the emergency operations. The lifeboat system can be reconnected to the main system by command if the anomaly can be corrected.

Pyro power is provided by either of two type-40 batteries from the main power system and distributed by redundant circuits.

Telemetry and Tracking

The telemetry subsystem provides real-time data (ascent at 48 kbps, engineering analysis at 128 kbps, and orbit at 64 kbps), and tape recorded data (48 kbps played back at 256 kbps). The telemetry provides status data for normal mission operation, test operations and evaluation, command acceptance confirmation, and post-flight evaluation. Tape recorder storage allows the monitoring of the SV temperature profile by periodic sampling. Over 1,500 data sources are monitored—some at up to 500 samples per second.

---

*On vehicles SV-13 and SV-14, the two nitrogen tanks were manifolded directly with the OA tank and provided enough ullage space to permit 3,700 pounds of propellant to be loaded with the operating pressure range.

The SGLS-compatible tracking subsystem provides range measurement information, including slant range (50-foot maximum 1σ bias error and 60-foot rms maximum noise error), range rate (0.2 foot/second maximum 1σ error), and angle-of-arrival (1.0 milliradian maximum 1σ bias error and 1.0 milliradian rms maximum noise error). Schematics of the telemetry and tracking systems are shown in Fig. 1-10.

Command and Timing

The Extended Command System (ECS) (Fig. 1-11) provides real-time and stored-program command capability. The SGLS compatible ECS system with complete redundancy provides 64 real-time and 626 stored-program commands with a memory capability of 1,152 commands. Ninety-six secure command operations are possible. On SV-15 and up, the number of secure command operations were increased to 192. The ECS provides operational commands to perform primary and secondary missions, the capability to configure the vehicle into various operational modes, a pre-flight test and checkout capability, security for critical functions, and a time signal to the PCM and the payload.

The Minimal Command System provides 28 real-time and 66 stored-program commands with a memory capability of 53 commands. Ten secure command operations are available. The Minimal Command System provides lifeboat commands for an independent capability of recovery RV's and initiating SV deboost and the capability to obtain real-time and recorded telemetry data.

The Data Interface Unit (DIU) provides for the generation, storage, and transfer of time information to the search/surveillance camera, mapping camera, telemetry, and pallet. The DIU also provides the Mapping Camera System and pan camera time-request-pulse to the NAVPAC.

Lifeboat II

The lifeboat system provides emergency capability to initiate separation of two Reentry Vehicles (RV) and to deorbit the Satellite Vehicle in the event of a complete failure of the main power system, the attitude control system, or the extended command system.

Emergency operational control is provided by the 375-MHz receiver and Minimal Command System, with capability for real-time, stored-program, and secure commands.

Attitude control for RV releases and SV deorbit is provided by earth-field sensing magnetometers, rate gyros, and a cold gas (Freon-14) control force system. Lifeboat is capable of RV releases and SV deorbit operations on both south-to-north and north-to-south passes.

Power to keep the system ready for use, and for the emergency operations, is provided by a type-40 battery and 1/4 of the solar arrays from the main power system. The OAS engine and the redundant SGLS, PCM, tape recorder, and other equipment necessary for RV release, SV deorbit, and recovery of vehicle diagnostic data are switched from the main power system to the lifeboat bus for the emergency operations. In a nominal tumbling mode, enough power is generated to keep this emergency mode operating until the vehicle reenters. Please see Fig. 1-12 for schematics of Lifeboat II.

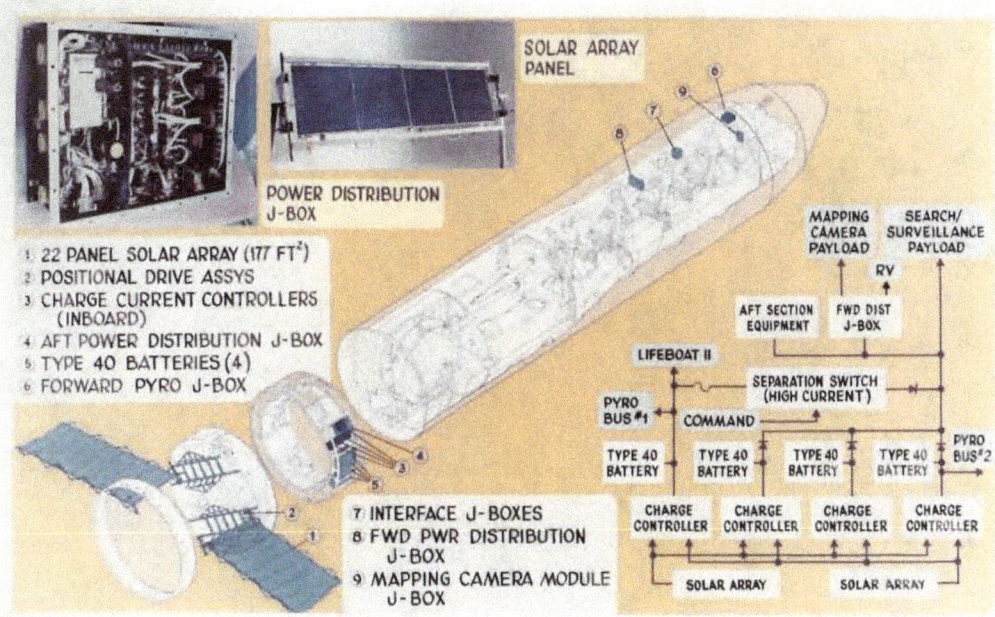

Fig. 1-9 — Electrical distribution and power

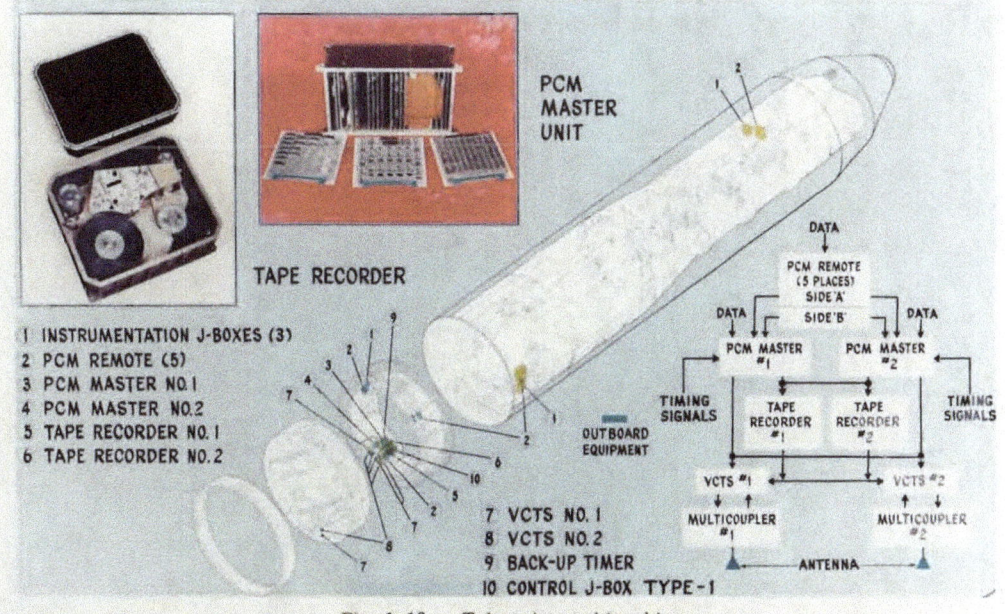

Fig. 1-10 — Telemetry and tracking

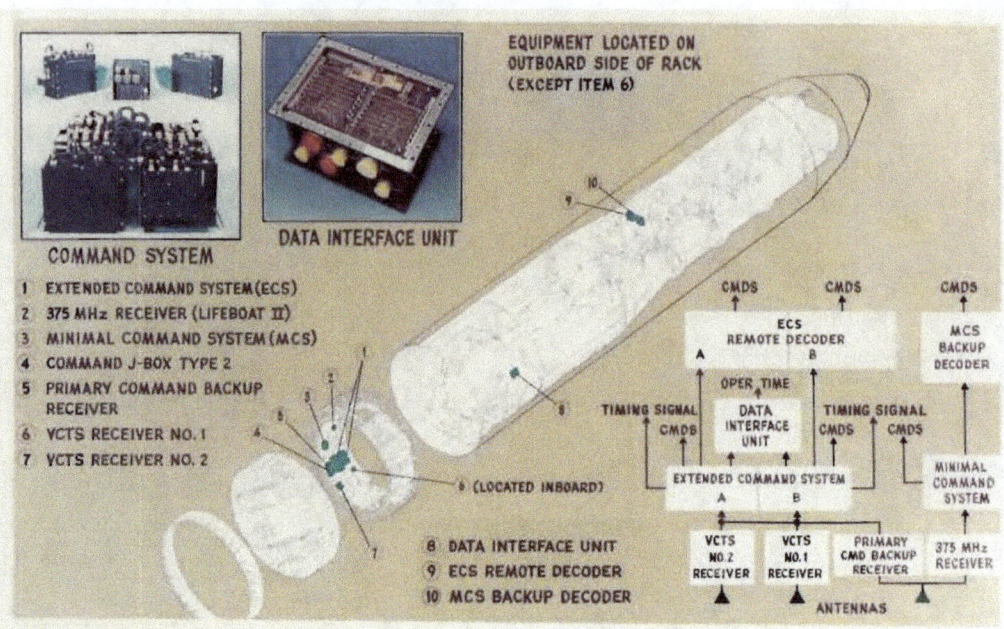

Fig. 1-11 — Command and timing

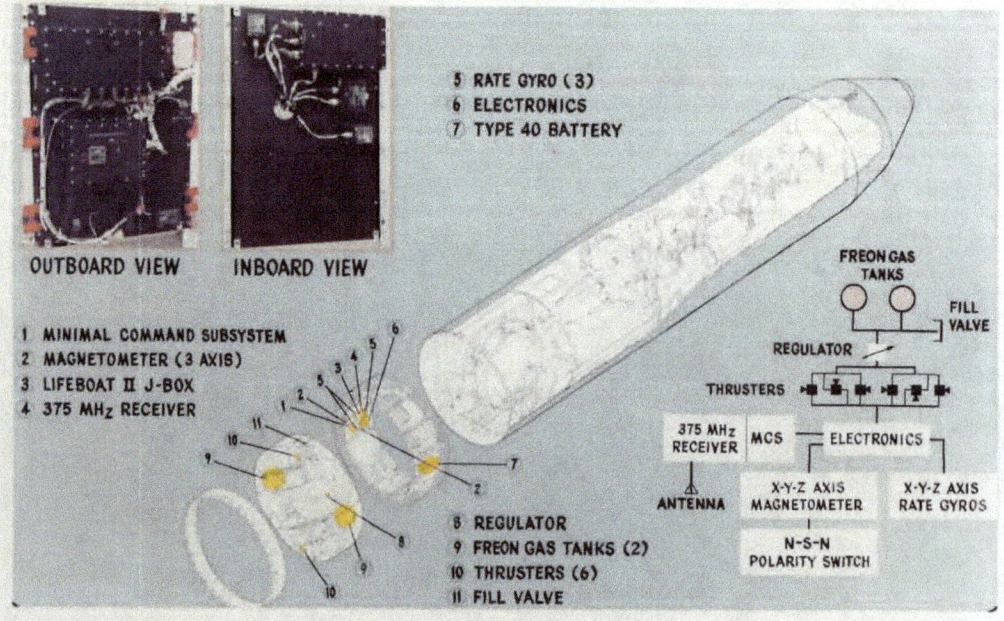

Fig. 1-12 — Lifeboat II

Search/Surveillance Cameras

The search/surveillance cameras (Figs. 1-13 and 1-14) provide high-resolution stereoscopic coverage of selected areas on the earth's surface by using two independently controllable panoramic cameras. The system provides a target resolution of 2.7 feet or better at nadir when operating at primary mission orbital altitudes with an apparent target contrast of 2:1, sun angles greater than 30 degrees, and using SO-208* film (SO-315 effective mission 1214 and subsequent).

The search/surveillance system has been designed with the following characteristics:

Optics	60-in. focal length, f/3 folded Wright (modified Schmidt) system (T 3.4 excluding filter factor)
Aperture diameter	20 in.
Field angle	±2.85°
Slit width range	0.91 in. (maximum); 0.08 in. (minimum)
Film	6.6-in.-wide (B&W) Type 1414 or SO-208 and others; currently SO-315. Also, SO-130 (infrared color) and SO-255 (natural color).
Resolution (2:1 contrast)	Center of format ≥200 l/mm; Elsewhere in format ≥160 l/mm
Film load	Currently (1982) 155,000 ft (per camera) mixed load of SO-315 and color. Total weight = 2,000 lb
Film stack diameter	68 in.
Scan modes	30°, 60°, 90°, and 120°
Center of scan	0°, ±15°, ±30°, and ±45°
Maximum scan angle	±60°
Stereo convergence angle	20°
Frame format (120° scan)	6-in. by 125-in.
Film velocity	200 in./sec (maximum) at focal plane
Image motion compensation range	0.018 rad/sec to 0.054 rad/sec for V_x/H,† ±0.0033 rad/sec for V_y/H‡
Weight (less film)	5,375 pounds

---

*SO-208 is a thinner base equivalent to Type 1414, used extensively for the first 13 missions.
† V_x/H: orbital angular rate (in-track)
‡ V_y/H: orbital angular rate (cross-track)

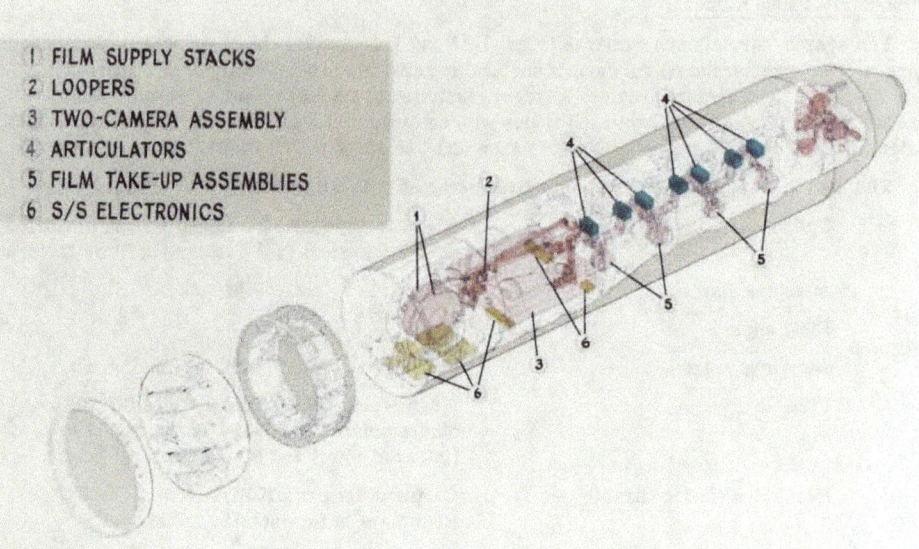

Fig. 1-13 — Search/surveillance cameras

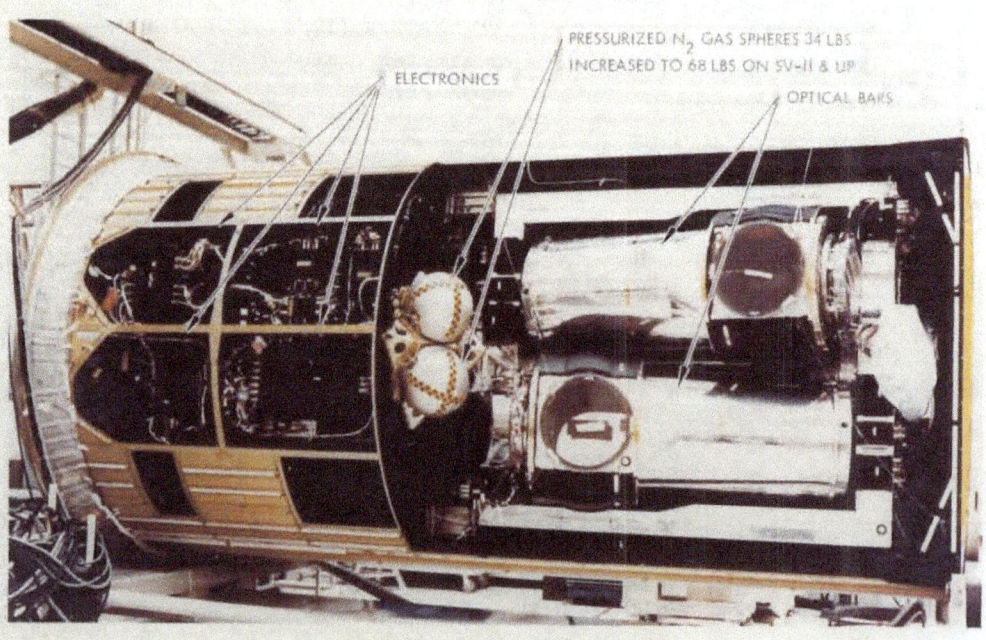

Fig. 1-14 — Two camera assembly

THE MARK 8 SATELLITE REENTRY VEHICLE (SRV)

The McDonnell Douglas Mark 8 SRV is used to recover exposed film from the search/surveillance cameras. Each of the four SRV's has a base diameter of $57\frac{1}{2}$ inches and is 85 inches from the heatshield nose to the retro-motor nozzle. Maximum total weight of the RV and film is 1,695 pounds. This consists of 956 pounds of RV and equipment, 239 pounds for film takeup assemblies, and 500 pounds of film.

When the takeups in an RV are filled, the next in-line RV is enabled and the full RV is ejected from the optimized pitched down SV at a 3-foot per second rate. The spinup to 10 radians per second is accomplished via hot gas generator to stabilize the RV during the retro-rocket motor burn. The retro-rocket provides a 1,623-pound thrust to slow the RV for reentry. The despin system then slows the spin rate to 1.4 radians per second, which provides the needed stability during the coast period and still permits the RV angle of attack with the flight path early in the reentry period. The drogue parachute is released and main parachute deployed upon the closure of a barometric pressure switch at about 50,000 feet. At 15,000 feet, the rate of descent is from 1,200 to 1,650 feet per minute, which is suitable for aerial recovery by USAF JC130 aircraft.

If aerial retrieval is not accomplished, water recovery becomes a backup phase. When sea water contacts a sensor, a relay closes the film canister vent valve and transfers vehicle power to the water recovery beacon. A salt water corrosion plug will sink the recovery capsule in 48 to 60 hours after water impact. This allows a reasonable time for location and pickup by Air Force and Navy forces.

If the RV significantly overshoots the specified impact point, it will be destroyed. This is accomplished by ejecting the heatshield and deploying the drogue chute if aero drag has not produced 0.003 g by a given time after RV separation. This results in the RV burning up when the atmosphere is encountered. It has not been necessary to utilize this provision to date.

In Fig. 1-15, the film is shown passing through the RV. Transfer of film to this RV consists of transferring takeup power, wrapping film on this takeup, cutting and sealing the film path on the exit side, followed by cutting and sealing the inlet film path on the forward RV. The bottom view shows the film on takeups A and B. The takeup drive motor and control electronics are contained mainly within the takeup hub.

Of the encapsulated volume inside the RV, 18 cubic feet is for the takeup assembly and 13 cubic feet is used by the RV equipment. The film stack diameter can be up to 35 inches.

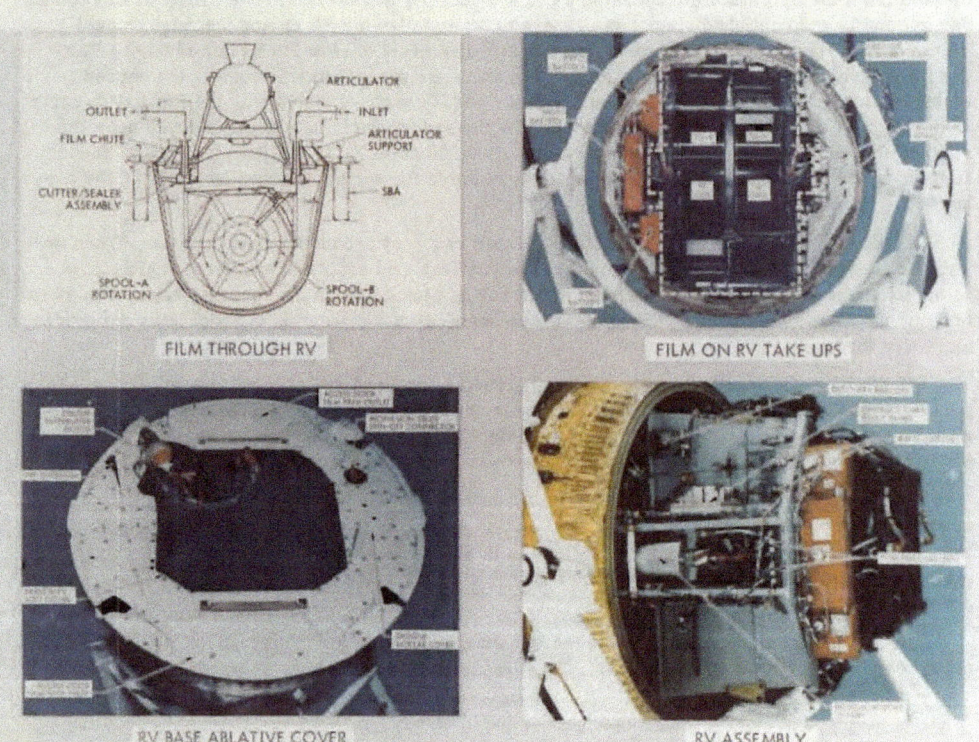

Fig. 1-15 — Mark 8 Recovery Vehicle equipment

MARK V SATELLITE REENTRY VEHICLE (SRV)

Introduction

In 1970 the Reentry Systems Division of General Electric Company (GE-RSD) was contracted to provide a modified version of the Mark V Satellite Reentry Vehicle (SRV) for recovering the exposed Mapping Camera Module film. The SRV mission was to provide a lighttight, environmentally controlled enclosure which had the capability to return the exposed stellar-terrain film from orbit for aircraft retrieval by parachute snatch with water recovery as the back-up mode.

This SRV, designated RV-5 on the HEXAGON Program, was an adaptation of an existing and proven design modified to be compatible with the Itek film cassette takeup, the required film paths, and the HEXAGON flight environments. A total of twelve RV-5's were flown successfully (Table 1-2), extending the maximum orbital duration from 43 days to 118 days in orbit; the longest duration in orbit of any Mark V SRV (Figs. 1-16a and 1-16b).

Table 1-2 — RV-5 Flight History

SRV	SV	Launched	Recovered	Days in Orbit
1801	5	9 Mar 73	20 Apr 73	43
1802	6	13 July 73	24 Aug 73	43
1803	7	10 Nov 73	7 Jan 74	58
1804	8	10 Apr 74	9 June 74	60
1805	9	29 Oct 74	27 Dec 74	59
1806	10	8 June 75	30 July 75	52
1807	11	4 Dec 75	2 Feb 76	60
1808	12	8 July 76	8 Sept 76	62
1809	13	27 June 77	17 Oct 77	112
1810	14	16 Mar 78	11 July 78	117
1811	15	16 Mar 79	12 July 79	118
1812	16	18 June 80	14 Oct 80	118

The Mapping Camera Module (MCM) involved three associate contractors:

- GE-RSD supplied the RV-5
- Itek supplied the camera and associated hardware
- LMSC supplied the APSA and was the integrating contractor.

As the integrating contractor, LMSC maintained the formal interfaces with Itek and GE-RSD. Documentation consisted of an Interface Control Document (ICD) which defined individual roles and responsibilities, supported by a mechanical ICD and an electrical ICD which defined the technical areas to be controlled.

General Electric built and tested each RV-5 at its Philadelphia facility and shipped it to LMSC for the following operations:

- SRV disassembly to subsystem level
- Film takeup installation and alignment
- RV assembly to the flight configuration, including the mass properties balancing and verification test

- Mating to APSA for tracking tests, acoustic tests, thermal vacuum tests, and final flight readiness tests
- Final flight functional verification, pneumatic fill and arming after the MCM module was mated to the parent SV.

The RV-5 was then flight ready except for installation of the retro-rocket igniter, accomplished at VAFB after the SV/Titan III spacecraft mate.

During the entire assembly and test flow at LMSC, the GE resident ensured that each RV-5 was processed in compliance with GE specifications.

System Description

The basic vehicle is approximately 33 inches in diameter, 42 inches long, and weighed 390 pounds with a full ($\approx$70 pounds) film load. Fig. 1-17a shows the interface between RV-5 and the APSA; Fig. 1-17b shows the interface between the payload-related hardware and the RV.

RV-5 was designed to comply with many requirements. The major ones were the mission flight envelope, on-orbit temperature control, controlled reentry dispersion, and the basic reentry environment.

The flight envelope consisted of a wide range of orbital parameters (see Fig. 1-18). RV-5 was compatible with this range of orbital operations, except for some de-orbit limitations in terms of dispersion and aft end heating. All missions flown had about 80 to 90-nm perigees, 150 to 160-nm apogees, 96-degree inclination (sun-synchronous), and solar angles (β) less than 20 degrees. These conditions did not result in any constraint. Only a payload malfunction, which never happened, would have caused an aft-end heating constraint.

Since dispersion and aft-end heating were the parameters most likely to effect RV-5 operations, detailed definition of their boundaries were made available to operations personnel during the mission. For contractual incentive purposes, an operational flight envelope was constructed defining the overall limits (see Fig. 1-19). This envelope consists of the locus of the outer limits of the dispersion and aft-end heating. The deep dip on the left is due to aft-end heating; the shallower, higher altitude is due to dispersion. A typical mission had an apogee of less than 160 nm and an impact anomaly (the angle between perigee and the impact point) of about 40 degrees. This condition lies in the acceptable zone, hence, it never constrained any operation.

In the launch mode, RV-5 was protected by the SV shroud, thus, no significant aerodynamic heating was encountered. The acceleration and acoustic noise levels realized were well within the RV-5 design capability.

On orbit, the RV kept the exposed film transported onto its takeup between 30 and 85°F while also ensuring that none of its major elements were exposed to excessive high or low temperatures.

Fig. 1-16 a — Mark V reentry vehicle

Fig. 1-16 b — Mapping Camera terrain and stellar takeup assemblies

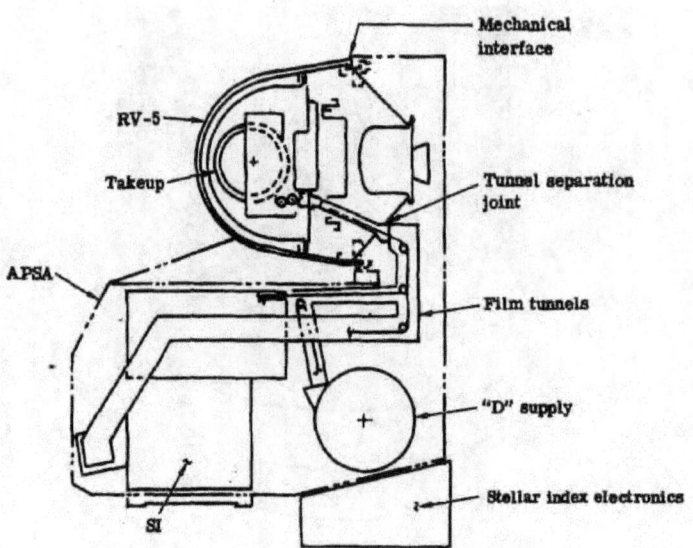

Fig. 1-17a — Mapping Camera Module

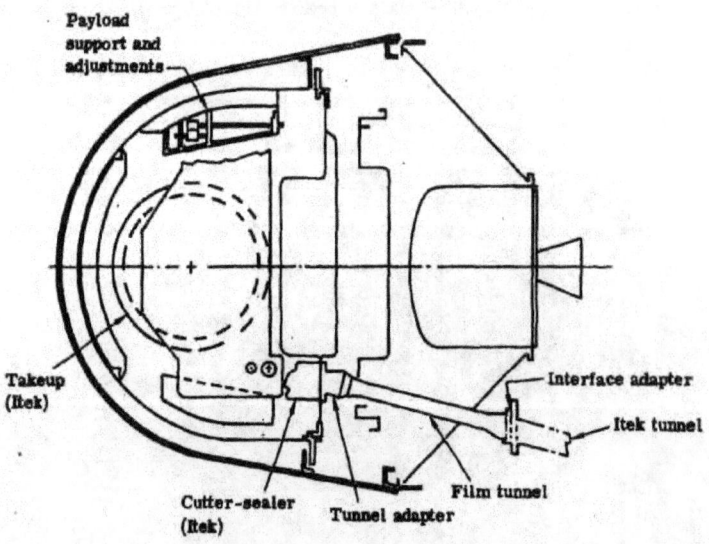

Fig. 1-17b — SRV inboard profile

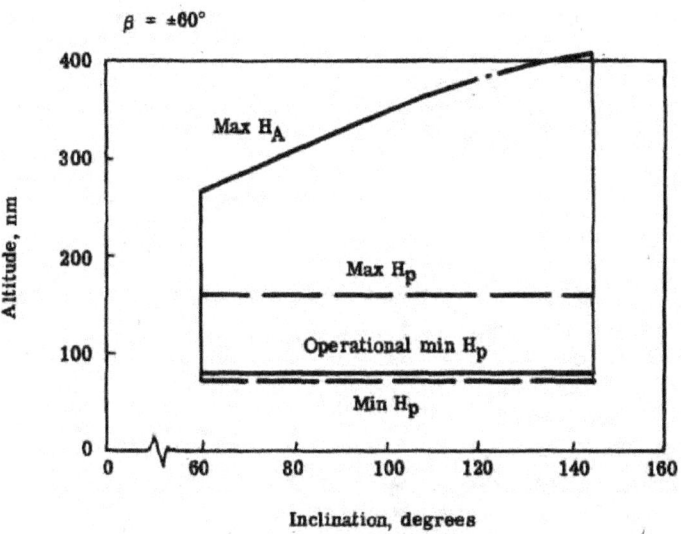

Fig. 1-18 — Mission flight envelope

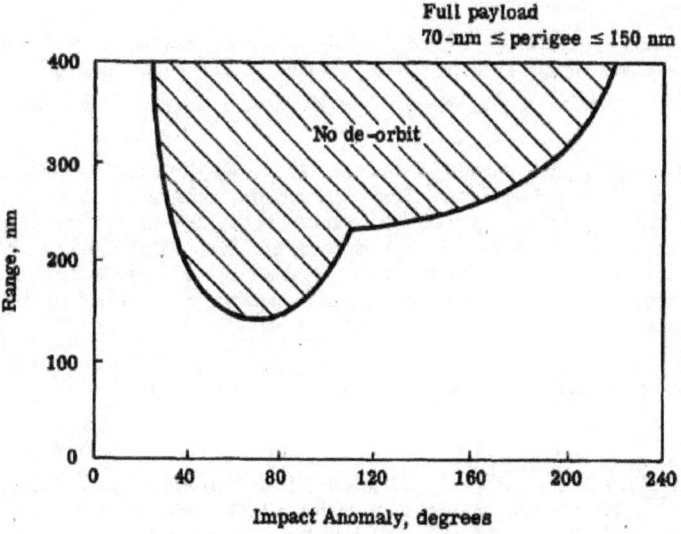

Fig. 1-19 — Operational flight envelope

This was accomplished by proper selection of thermal control coatings and thermostatically controlled heaters.

RV-5 dispersion requirements are summarized in Table 1-3. Deboost from the higher altitudes defined in the mission flight envelope could have violated these limits. In actuality, all RV-5 deboosts occurred from low altitude resulting in a mean dispersion of 3.8 nm uprange from the planned impact point.

Table 1-3 — Dispersion Requirements

	Full Payload	Payload Malfunction
In-track, nm	200	250
Cross-track, nm	±25	±30

Note: No dispersion requirement for Lifeboat

RV-5 consisted of five major subassemblies (see Fig. 1-20, a simplified unclassified exploded view of the recovery vehicle) which contained the de-orbit, recovery, electrical, and thermal control subsystems. These subassemblies, briefly, are:

1. <u>Thrust Cone (T/C)</u> — A self-contained subassembly which included the de-orbit subsystem. The T/C was ejected after the de-boost sequence was completed. Its forward ring provided the mechanical mounting interface with the MCM while an explosive in-flight disconnect provided the RV-5/MCM electrical interface.

2. <u>Thermal Cover</u> — This cover provided thermal protection to the parachute during re-entry. After reentry, it was ejected aft to initiate the parachute deployment sequence.

3. <u>Parachute</u> — A three-stage retardation system designed to be compatible with air retrieval.

4. <u>Recovery Capsule/Capsule Cover</u> — This assembly provided a lighttight, environmentally controlled enclosure that protected the exposed film. The recovery subsystem, which included a programmer, power supply, and retrieval aids, was contained in this assembly. If air retrieval attempts failed, the recovery capsule would land in the ocean and float. After a given time interval, if water retrieval had not been successful, it would sink. A destruct timer was included in the recovery system to preclude successful recovery if the parachute was not deployed within a given time period after de-orbit.

5. <u>Forebody</u> — This subassembly provided structural support to the capsule and contained the heat shield that protected the capsule and its contents from reentry heating.

<u>Hardware Description</u>

1. <u>Thrust Cone (T/C)</u> — The T/C and de-orbit subsystems provided all the functions necessary to deboost the RV. It contained thermal batteries, a dual ejection programmer, spin and despin pneumatics, retro-rocket harnessing (which includes carry-through wiring from the MCM to the capsule) and cable cutters for its separation. The entire subsystem was redundant. In addition to the subsystem components, a short section of film tunnel was mounted to the thrust cone and provided the film path from the MCM to the capsule.

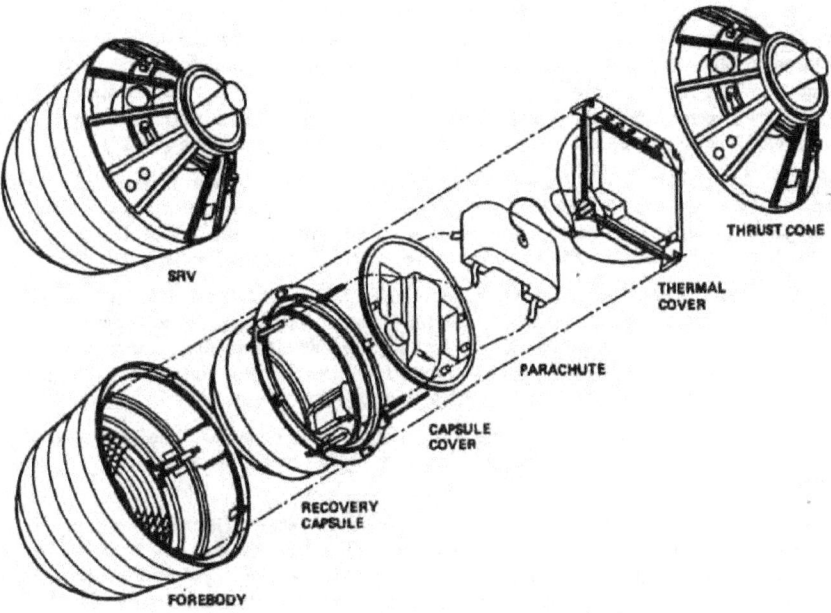

Fig. 1-20 — Mark V SRV (exploded view)

The basic thrust cone structure was a 45-degree truncated cone structure constructed of ring-stabilized thin walls with structural channels spot welded to the 0.020-inch-thick aluminum skin. Its forward ring provided the structural interface with the MCM and maintained electrical isolation by an epoxy/glass overlay on the interface ring. Mounting locations for trim ballast, required to maintain SRV mass property requirements, were provided on the T/C. The thermal environment for the thrust cone was maintained by APSA. After completion of the de-orbit sequence, the T/C was discarded.

2. Thermal Cover — The thermal cover was built-up from a pyro preg (phenolic glass) lay-up, aluminum corner fittings, and stainless steel frames as an unclassified assembly. It was then modified to the program's configuration and lined inside with fiberglass insulation. The thermal cover was one of the reentry thermal control elements and provided thermal protection to the parachute during reentry. After reentry it was ejected aft to initiate the parachute deployment system. Other reentry thermal control elements were the phenolic nylon heatshield, aluminized insulation, and ESM (a foamed elastomeric silicone heat protection material developed by GE) on the capsule cover.

3. Parachute — The retardation system used on RV-5 after reentry was the MK-5 B/C parachute system, which consisted of three stages of deceleration: (1) a 5.4-foot ribbon drogue, (2) a 29.6-foot ring slot main canopy reefed, and (3) the main canopy disreefed. Incorporated in the main canopy was a heavy line air retrieval structure designed to be compatible with JC-130 retrieval. This structure consisted of four laterals running around the circumference of the main canopy and six heavy suspension lines which went from the swivel through the canopy and back to the swivel.

Incorporated in the parachute assembly were two bagline cutters and four reefing line cutters. These mechanically-activated, time-delay cutters controlled the staging times. Drogue chute deployment activated the bag line cutters, which fired 10 seconds later. This released the top of the main bag and the drogue deployed the main canopy reefed. Four seconds later the reefing line cutters cut the lines to permit full canopy deployment at an altitude of about 50,000 feet.

4. Recovery Capsule/Capsule Cover — The recovery capsule was a ring-stiffened aluminum spinning. An external bellyband, spot-welded just forward of the cylindrical portion of the shell, rode on the forebody guides to facilitate a clean separation. The external surface of the shell was gold-plated to minimize heat loss by radiation to the forebody during orbital flight.

The aft end of the capsule was a ring from an aluminum forging which provided the interface with the forebody, the O-ring and the locking grooves for cover mating, standoffs to support the thermal cover, part of the interface with parachute fittings, the antenna mounting, and mounting points for the external nose ballast tension straps. Two recovery trays mounted in the capsule held the recovery programmer and the inertial switch/interval timer trays.

The film takeup was mounted at four points within the recovery capsule. This arrangement provided a five degree-of-freedom adjustment capability that could be used to align the takeup with the overall film path.

Two VHF beacons were located in the capsule, one aft of the other. The weight of these beacons balanced the weight of the recovery trays on the other side of the capsule. Each beacon was wired directly into one of the recovery battery sections. The beacon output was modulated to provide limited event information during flight, including retro-rocket firing, thrust cone separation, g-switch closure and opening, and thermal cover ejection.

The recovery programmer provided the switching and timing functions required for recovery subsystem events. It accepted power from two independent batteries (housed in a sealed case contoured to fit the nose dome of the capsule) and provided redundant signals for its primary outputs. The remotely activated silver-zinc batteries had ten cells each and a sump to retain excess electrolyte. Each battery provided between 14.8 and 17.0 Vdc.

Two 10-watt heaters, bonded to the inner capsule wall, provided the power necessary to maintain the capsule temperature on-orbit. They were thermostatically controlled at a setting of 35 ±4°F. A 67°F thermostat would provide system protection if the primary thermostat failed to close.

A sink valve that would operate between 69 and 107 hours after water impact was initially installed in the capsule. Starting with SV-11 (SRV 1807) the sink time was reduced to 40 ±5 hours.

The capsule cover was an aluminum weldment which sealed the aft face of the capsule. It was configured to provide a mounting plane for the cutter/sealer (GFE from LMSC), the film tunnel forward adapter, the parachute pack, and the following components:

- A pneumatic fitting that permitted capsule pressurization or evacuation during test.
- Three test connectors that were used to safe pyro circuits and test RV functions after final assembly.
- An in-flight disconnect which was a pyrotechnic separated connector.
- Two descent valves for pressure equalization during reentry and recovery. Ascent venting was through the open film tunnel.
- The pop-off valve which allowed entrapped air to escape when the sink valve opened, thus speeding up the sinking of the unrecovered capsule.

5. Forebody — The forebody/subassembly provided RV-5 with its reentry aerodynamic configuration, thermal protection, and recovery capsule structural support. The heatshield portion was a phenolic glass liner covered by a phenolic nylon heatshield. The heatshield was structurally stiffened by three rings. Capsule guides between the rings ensured a smooth capsule/forebody separation. A thermal blanket in the nose of the forebody isolated the capsule from the high temperature realized by the nose in flight. Three temperature sensors in the forebody monitored on-orbit temperature. Sensor data was retrieved by SV telemetry.

A magnesium ring provided the primary structural interface with the capsule. Four ejection pistons, mounted forward of this ring to the forebody, pierced the forebody and capsule rings and were fastened to the thermal cover corner fittings. Torquing down on the piston studs, therefore, clamped the thermal cover/capsule/forebody into a structural assembly. Electrically initiated charge adapters screwed into the base of the ejection pistons provided the impulse necessary to eject the thermal cover at the required velocity. This activation released the forebody from the capsule which allowed separation when the parachute exerted sufficient retardation force on the capsule.

The exterior surface of the heatshield was painted with D4D, an aluminum pigmented silicone-alkyd thermal control coating. Thermal control studies showed α_s/E_H of about 1 was required for capsule temperature control while a low emissivity coating was required to maintain the forebody temperature above an acceptable minimum. D4D limited the minimum temperature to $-110°F$, an acceptable level.

Mission Overview

Following launch, during the initial portion of ascent, the vehicle was protected from aerothermodynamic forces by the satellite vehicle shroud. The shroud was ejected prior to orbital injection only after the forces had reached an acceptable level.

On orbit the SV nominally flew nose forward, thus the RV longitudinal axis was parallel to the local horizon over the MCS. In this position, the RV did not have a direct line of sight to earth, which eliminated albedo as a thermal source to the RV's temperature control. Short time excursions from the nominal orbital flight attitude occurred for events like orbit adjust, MCS calibration, RV de-orbit, ▮▮▮▮▮▮▮▮▮▮▮▮▮▮▮▮▮▮▮▮ During these phases of flight, only the RV-5 capsule heaters, temperature sensors, and film takeups were active.

During orbit, the RV was essentially passive with only its film takeups, thermal control subsystem, and instrumentation functioning. After all the exposed film had been transferred to the RV's takeups and the cutter-sealer closure made, the RV was activated and separated via SV commands. The satellite vehicle yawed aft and then pitched down to achieve the correct orientation for RV de-orbit, thus minimizing dispersion. Following separation, the RV was autonomous. It spun-up, deboosted, de-spun, aerodynamically re-oriented itself to a nose-forward attitude, deployed the parachute, and was retrieved by the JC-130 recovery aircraft. The launch, through orbit, through recovery, and back to factory flow is shown in Fig. 1-21. More detail on the reentry and recovery events are covered in Section III "Mission Scenario".

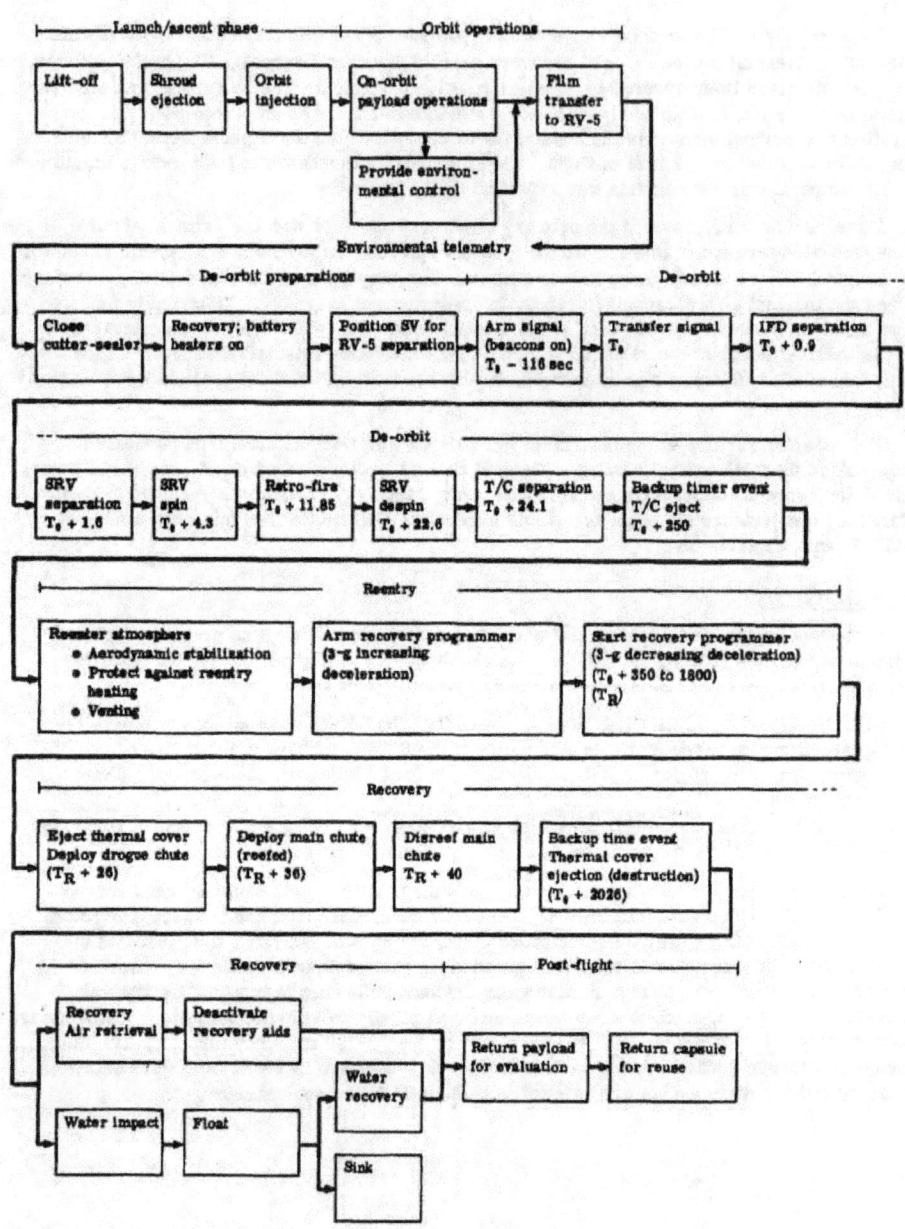

Fig. 1-21 — Mission flow

MCS HISTORY

SECTION 2

THE MAPPING CAMERA PROGRAM DEVELOPMENT

INTRODUCTION

Military Mapping, Charting, and Geodesy (MC&G) has employed satellite photography since 1960. With the aid of this photography, the Defense Mapping Agency (DMA) and its predecessor organizations have produced over 50,000 different maps and charts out of a requirement which exceeds 80,000 world wide, levied by the Unified and Specified Commands, the Military Services, and the Intelligence Community.[3]

Photographic coverage of metric accuracy [currently provided by the HEXAGON (KH-9) Mapping Camera System (MCS)] and medium to medium-high resolution (2 to 10 feet, such as that provided by the HEXAGON Panoramic Cameras) is indispensible at present, and will continue to be in the 1980's for the production and updating of these MC&G products to support operational needs.

The geodetic data derived from satellite imagery provides the military with tens of thousands of accurate point locations needed for operation of strategic and tactical weapon systems.

Operational or programmed weapon systems, including Minuteman II/III, Polaris, Poseidon, Cruise Missile, Pershing, B-52, and F-111, are dependent on this positional information and on maps and charts for navigation and target strike. Without the HEXAGON MCS these MC&G products would not be available today.

Twenty HEXAGON systems were programmed, the operational period starting on 15 June 1971 with the last system projected to be flown in 1985. The first four systems were flown without the MCS; the next twelve systems flown carried the MCS which produced calibrated photography for direct use and provided metric orientation for the panoramic photography.

A Doppler Beacon System (DBS) and a Navigational Package System (NAVPAC) provided ephemeral information which accurately establishes camera/vehicle position in space.

All of the twelve Mapping Camera Systems were successfully flown. The last four HEXAGON vehicles will be flown without a mapping camera system. To provide attitude determination for the panoramic photography on these missions, the government developed a Stellar Solid State (S^3) Camera Assembly under contract with the Perkin-Elmer Corporation. This will in effect allow panoramic imagery to be used for mapping purposes with the required metric accuracies.

The exploitation responsibilities were directed to the two major DMA organizations, the Hydrographic/Topographic Center (DMAHTC) in Washington, D.C. and the Aerospace Center (DMAAC) in St. Louis, Missouri. Also, following the downgrading of the product to SECRET in 1974, some of the material was made available to other organizations such as the Environmental Protection Agency (EPA), and the U.S. Geological Survey (USGS) for use in civil agency mapping and charting programs.

The organizations involved in the Mapping Camera subsystem of the HEXAGON program and their various roles are listed below. The degree of perfection achieved with this system, developed under priorities that were historically secondary to intelligence priorities, is evidence of outstanding contributions at all levels.

- The National Reconnaissance Office (NRO)
    - Budgetary
    - Contractual
    - Management
        - NRO Program A, the Secretary of the Air Force Special Projects (SAFSP) [including the Satellite Control Facility (SCF) and the Satellite Test Center (STC)]
            - Budgetary
            - Contractual
            - Management
            - Hardware and software development
            - Integration
            - Operations
            - Product evaluation

- The Defense Mapping Agency (DMA)
    - Requirement generation
    - Product evaluation and exploitation
    - Budgetary

- The Central Intelligence Agency (CIA)
    - Product evaluation and exploitation

- The Aerospace Corporation
    - Technical advisory
    - Product evaluation

- Itek Corporation
    - Hardware and software development
    - Integration
    - Operations
    - Product evaluation

- Lockheed Missile and Space Company (LMSC)
    - Hardware and software development
    - Integration
    - Operations

- General Electric
    - Hardware and software development
    - Integration
    - Operations
- Eastman Kodak
    - Photographic materials development and manufacturing
    - Product evaluation (sensitometry)
- TRW
    - Software development

MCS DEVELOPMENT TO FIRST FLIGHT

Events Leading to Contract Award

During the 1950's the mapping, charting, and geodesy (MC&G) community was called upon to prepare detailed line graphics on which the intelligence gleaned from the U-2 photography could be portrayed. These programs led to studies on how accurate and to what scales denied areas could be mapped from the U-2 photography, and, for future planning, the reconnaissance satellite systems being designed for the early 1960's. However, the precedent that the photographic collection systems would address MC&G requirements as secondary to the intelligence mission was deviated from only once—the ARGON (KH-5) system described in Part I.

The goal of the KH-5 geodetic system was to upgrade the world datums to a 750-foot accuracy, a very ambitious undertaking for the early 1960's. While this goal was not met in its entirety, the success of this program indicated that a combined program utilizing panoramic cameras for high-resolution wide-area coverage, and frame cameras for geometric control and indexing the panoramic coverage provided an efficient intelligence search system and an MC&G collection system.

During the evolution of the CORONA (KH-4) system in the 1960's, separate studies were being conducted by the MC&G community to design an optimum mapping and charting system. The Army requirements for positional accuracy led them to put primary emphasis on the frame camera approach while the Air Force targeting requirements caused them to investigate approaches to calibrate the panoramic cameras for geometry since targeting required higher resolution than the frame camera could provide. This two-pronged approach led to a dual effort to improve the MC&G collection program.

In 1964 the Army produced a study showing that a 12-inch-focal-length wide-angle frame camera lens could be designed and built that would provide sufficient resolution to meet the 1:250,000 mapping requirements and provide the control for 1:50,000 maps. It would also provide most of the planimetry for the 1:50,000 maps, but not all. Panoramic coverage would have to be used in conjunction with the frame to do the complete 1:50,000 job. This was considered an acceptable compromise since the intelligence community required the panoramic coverage for search and a larger frame camera would be probibitively expensive to build and operate as a separate collection system.

The Air Force concepts for calibrating the panoramic cameras continued during 1963-1968, with several approaches tried on the CORONA (KH-4).

MCS HISTORY

Also during this period, the advanced programs office (SP-6) of SAFSP developed a prototype 12-inch-focal-length wide-angle lens under the direction of Colonel Charles Ruzek, USA. This was the direct result of the extensive design study and lens development program for the proposed Geodetic Orbital Photographic Satellite System (GOPSS). In 1965 the KH-4 system had evolved into a very efficient system but was no longer completely adequate for either the intelligence community or MC&G. Looking toward the future, Dr. Flax, the DNRO, started studies looking at meeting the 1979-and-beyond requirements. This led, amongst other actions, to the United States Intelligence Board (USIB) revalidating both the intelligence and the MC&G requirements. The MC&G requirements compiled by COMIREX were approved by the USIB on 4 December 1967. These requirements were the basis for the final design of the 12-inch HEXAGON Mapping Camera System.

The improvements in the KH-4 system and in exploitation techniques developed by NPIC indicated that a stellar index (SI) camera was no longer needed for the intelligence mission. The question remained, however, how to best meet the MC&G requirements: a short focal length SI camera coupled with the panoramic camera or a longer 12-inch-focal-length camera which could stand alone for MC&G? Studies conducted by the NRO indicated that both approaches could be accommodated by the HEXAGON vehicle and that both were technically feasible.

In August 1966, SAFSP had issued a request for proposal (RFP) for the Stellar and Terrain Camera (SI) for the Photographic General Search and Surveillance Satellite System (HEXAGON). RFP's were sent to several camera contractors, but proposals from only the Fairchild Camera and Instrument Company (FCIC), and the Itek Corporation wound up in the final evaluation. FCIC had proposed a 7.5-inch lens for the terrain lens and solid-state devices for the stellar sensors, while Itek had proposed a camera system—based on the GOPSS development program—with a 12-inch-focal-length terrain lens, and two 10-inch-focal-length stellar lenses, using photographic film throughout. The Source Selection Board scoring summary showed the Itek Corporation proposal[4] to be the more responsive to the requirements stipulated in the RFP.

The proposal covered the design, development, fabrication, test, and qualification of the Stellar and Terrain Camera and associated aerospace ground equipment, system analysis, integration and evaluation, test planning and operational support, and program management and administration functions during the acquisition phase of the program. This phase was to include the launch, operation, recovery, and post-flight evaluation of six flight systems.

The Stellar and Terrain Camera performance, when coupled with that of the Sensor Subsystem panoramic cameras, was to be capable of meeting the mapping and charting requirement presented in Table 2-1. The SI photography was to be capable of providing adequate control to the panoramic photography to enhance its geometric fidelity sufficiently to enable utilization of the Sensor Subsystem panoramic cameras for large scale map compilation. It was the goal of the SI photography that it alone should be capable of meeting as many of the mapping and charting requirements as feasible, consistent with reasonable SI Subsystem size, weight, angular coverage, etc.

In order to ensure this level of performance, the terrain camera was to provide a ground resolved distance of not less than 80 feet at any point in the format with MIL-STD-150A three-bar targets, at a 40-degree sun angle from an altitude of 90 nm; however, the goal for ground resolution should not be less than 40 feet at any point. The cross-track coverage from this altitude should be at least 65 nm. Forward coverage should provide at least triple overlap exposure ($33\frac{1}{3}$ percent new coverage per exposure), with a goal of quadruple coverage (25 percent new coverage per exposure). The stellar camera(s) would be required to measure the absolute attitude of the terrain camera with respect to the stellar coordinate system at the mid-point of the period the terrain shutter was open.

Table 2-1 — Mapping and Charting Requirements

Type of Map	Scale	1σ Accuracy Requirements (including datum degradation)			
		Local Horizontal		Vertical	
		Accuracy, feet	Distance, statute miles	Accuracy, feet	Distance, statute miles
Large scale topographic	1:50,000	39.6	25	10-20	10-20
Medium scale	1:250,000	193.5	125	49.8	20
Medium scale aeronautical charts	1:200,000	155	100	30-45	20-30
Significant features	—	—	—	12-30	10

The Stellar and Terrain Cameras would carry sufficient film to permit operation in conjunction with Sensor Subsystem (panoramic) cameras which at that time were stipulated to photograph a total forward coverage of 3,000 miles per day for 30 days. The SI reliability was to be 0.97 for a 30-day mission, based upon a 50 percent confidence limit.

Although the SAFSP Mapping Camera competition had been concluded, a contract award was not immediately forthcoming. The Deputy Secretary of Defense, Mr. Paul Nitze, on 22 December 1967, issued a memorandum requesting a study to evaluate the potential of the 12-inch camera, and a 3-inch SI, which was also under consideration for the KH-9.[5] The memorandum stated, "It is essential that a study be made to provide a cost-effectiveness comparison of these two cameras and to identify any other factors which may bear on the selection of one or the other for inclusion in the new system."

"The study should include consideration of the comparative cost of the two options, including satellite vehicle and SI camera development, procurement and operations, as well as the costs of map production and other exploitation of the product of the SI cameras. The incremental value to possible military operations of the increased map accuracy which would result from the 12-inch SI camera should also be assessed. In addition, the need for either SI camera to support the exploitation of the primary mission camera products (i.e., mensuration and accurate location of the targets) should be analyzed."

"The following alternatives should be assessed among those studied:

1. Proceed with development and integration of the 3-inch SI camera and discontinue all effort on the 12-inch camera.

2. Proceed with the development and integration of the 3-inch camera and also fund the minimum camera and vehicle engineering and development efforts needed to provide the option for later inclusion of the 12-inch SI camera.

3. Discontinue all effort on the 3-inch SI camera and fund the level of development engineering effort on the 12-inch SI camera and vehicle integration to permit incorporation of the 12-inch SI camera into the new search system in a block change subsequent to the early flights of the system which would include only the primary mission camera (this alternative choice assumed that the SI cameras were not necessary for NPIC exploitation of the product of the primary mission cameras)."

"In light of the above, I request the Assistant Secretary of Defense (Administration) to establish a group chaired by a representative of his office and composed of representatives of the offices of the Director of Defense Research and Engineering and the Assistant Secretary of Defense (Systems Analysis) to conduct such a study. The NRO, NPIC, DIA, the JCS and the Department of the Army may be asked to provide such information and support for the study as may be required."

"I would appreciate receiving a preliminary report of the study results by February 1, 1968. This preliminary report should, as a minimum, clearly indicate whether the probable need for the capabilities of the 12-inch SI camera is great enough to warrant immediate action to protect the option for later inclusion of this camera in the new search system vehicle without major redesign and requalification. In addition, recommendations should be included as to any additional study needed to make a final determination on the SI camera selection, taking into account the acquisition schedule for the new search system."

Mr. Sol Horowitz, the Assistant Secretary of Defense, transmitted the completed study,[6] conducted by Colonel ███████ to the Deputy Secretary on 15 February 1968. Based on review of pertinent facts derived from reports, briefings, and research, the group had arrived at the following conlusions and recommendations:

"a. An SI camera is not required to support the primary intelligence mission of the HEXAGON.

b. All three alternatives are technically feasible.

c. An SI camera on HEXAGON would not interfere operationally with the primary mission.

d. The size and cost of the HEXAGON booster is not affected by the inclusion or exclusion of an SI camera.

e. There is a requirement for photographic coverage of 13 million square miles for medium scale mapping.

f. There are unfulfilled mapping production requirements for 20,000,000 square miles of medium scale maps and 10.4 million square miles of large scale maps.

g. Accurate mapping from satellites requires frame photography.

h. Use of conventional means to satisfy the outstanding mapping collection requirements is not feasible from either the cost or operational viewpoints.

i. 3-inch SI photography (or better) correlated with panoramic photography generally is required to produce medium scale maps.

j. The 12-inch SI photography from the HEXAGON will permit medium and large scale mapping without correlated panoramic photography.

k. Map production costs for the medium and large scale maps can be reduced using the 12-inch rather than the 3-inch materials at a rate of $.45 million per million square miles for medium scale and $3.0 million per million square miles for large scale.

l. Approximately $20 million associated with conventional acquisition techniques can be removed from the CIP during the FY 71-73 time period.

m. Option II (3-inch SI initially with 12-inch incorporated later) is least desirable of the three from both cost and engineering viewpoints.

n. Maps of foreign areas produced from satellite photography are classified SECRET.

o. Security restrictions on satellite photography preclude full utilization of this material to satisfy cooperative mapping agreements with foreign countries. Relaxation of these restrictions could result in further cost reductions in the DoD MC&G program.

p. Other federal agency programs can benefit from 12-inch SI photographic materials."

The study then recommended that all effort on the 3-inch SI camera be discontinued and that funds be provided for the level of development engineering effort on the 12-inch SI camera and vehicle integration needed to permit incorporation of the 12-inch SI camera into the New Search system in a block change subsequent to the early flights of the system which would include only the primary mission cameras (Option III).

The recommendation that the third option be exercised was approved by the Deputy Secretary of Defense on 11 March 1968. On 12 March 1968 Dr. Alexander Flax, the Director of the NRO, issued instructions to the HEXAGON program office to proceed with the development of the 12-inch SI camera and to terminate all work on the 3-inch camera. To support this development, the Deputy Secretary of Defense directed the Secretary of the Army to transfer ▓▓▓▓▓ of RDT&E funds from the Army to the Air Force in each of fiscal years 1968 and 1969. Lt. General William Cassidy, Chief of Engineers, delivered a check for ▓▓▓▓▓ to Lt. Colonel William Williamson of the NRO staff on 15 March 1968, which completed the birth pains of the 12-inch mapping camera.*

---

*As an aside, the process of obtaining program funding produced an interesting vignette. From meetings involving representatives from the Bureau of the Budget, OSD/Administration, OSD/Systems Analysis, and DIA/MC&G, came the concensus that there was opportunity to trade off existing conventional mapping collection to help pay for the 12-Inch Mapping System. The USQ-28 High Altitude Program employing two RC-135 aircraft based at Forbes AFB, Kansas, became the tradeoff candidate. As events materialized, the USQ-28 Program terminated before the first Mapper flew on the HEXAGON.

Another tradeoff came later. By 1969, the costs associated with the 12-inch camera, along with other components of the HEXAGON System, had increased for known and explainable reasons. In a 19 May 1969 memorandum from Lt. Colonel Williamson to Dr. J. L. McLucas, DNRO,[1] Williamson stated that "even with the projected total costs through FY 74 of about ▓▓▓▓▓ this approach is still more efficient and cheaper than the alternatives of considering a greatly reduced mapping and charting capability at considerably more cost. The decision of the DDR&E and DIA to stop the Rapid Combat Mapping Acquisition System (RACOMS) development at an estimated cost of ▓▓▓▓▓ has, in itself, offset the cost of continuing the 12-Inch Mapper Program."

System Configuration Changes

During the approximate 18-month time span between the proposal competition and the contract award, both government and contractors continued to study program requirements and approaches. The major deviations between the configuration initially proposed and the first flight hardware were:

The film recovery concept. Itek had initially proposed a dual recovery vehicle (RV) configuration. This was looked at briefly but not pursued. The SAFSP program office next considered two approaches: (1) feeding the MCS film into each of the four panoramic camera recovery vehicles; (2) having a separate recovery vehicle as an integral part of the mapping camera system. As the study progressed, the advantages seemed to be heavily in favor of the separate RV, these advantages being reflected in many areas—design, development, integration, test operations, film processing, and product evaluation. Upon deciding in favor of the separate RV approach, and after due consideration, SAFSP proposed using the General Electric Mark V Satellite Reentry Vehicle (SRV) already proven on the GAMBIT program. In a message to Major General John L. Martin on 12 September 1968, Dr. Flax approved the use of a modified Mark V SRV to return the SI film.

Film selection. The film proposed for the terrain camera was the relatively high-speed type 3401 which, with the lens proposed, would enable achievement of the minimum GRD (80 feet) across the entire format without the requirements to incorporate image motion compensation, thus permitting the 1.0-cm grid reseau plate to be mounted as an integral part of the lens and ensuring the stability of the metric calibration. Upon further evaluation it was decided that the incorporation of forward motion compensation (FMC) would provide growth potential for the use of higher resolution films as they became available, so the FMC design was incorporated on the first flight enabling the use of type 3400 film.

Filter selection. The lens was to be protected from axial thermal gradients by an integral thermal window which also would incorporate the Wratten 25 color and anti-vignetting filter. This was changed to a Wratten 21 filter effective with the first flight system.

Stellar camera lens selection. The lens proposed for the stellar cameras was the 250-mm-focal-length, Wild f/1.8 Falconar. After additional study, it was determined that Itek had the in-house capability to produce a lens that would be more suitable for the job. This resulted in Itek producing in-house 25 stellar lenses to accommodate the entire program (12 systems, two each system, 1 spare).

Establishing First Launch Schedule

As these contractual and technical details were being formalized, the subject of first launch for the MCS was receiving much attention. Up to this point in time it had been generally postulated that the earliest a mapping camera could be launched in the HEXAGON program would be on Satellite Vehicle Number 7 (SV-7). The Defense Intelligence Agency (DIA) had expressed a desire to advance the availability date of the first HEXAGON MCS coverage by interchanging the launch schedule for vehicles number 6 (SV-6) and 7 (SV-7). DIA fully understood that this action, while advancing the availability of initial MCS coverage by 3 months, would mean a 6-month period between this first coverage and subsequent coverage which would start with vehicle number 8 (SV-8). It was Dr. Flax's understanding "that SV-7 was scheduled as backup to SV-6 and therefore it might be feasible to launch SV-7 in place of SV-6, and then SV-6, without serious impact on the program, or incurring substantial cost increases." In a message to General John Martin on 20 February 1969, Dr. Flax asked for a detailed analysis on the impact and cost of this proposed schedule change for the HEXAGON program.

The best approach for scheduling the first MCS flight was not immediately obvious. In the ensuing months other combinations were looked at with respect to their impact on cost and schedule, and the question was still unsettled in mid-1971. In July 1971, SAFSP asked Lockheed for cost estimates for "advancing the HEXAGON mapping camera launch by:

1. Installing the Mapping Camera on vehicle 6
2. Interchanging vehicles 6 and 7."

Taking Lockheed inputs, SAFSP advised the DNRO that their review indicated it would be cheaper to modify SV-6 to accept the MCS than the other approach of interchanging vehicles 6 and 7. On 14 October, SAFSP advised Itek that "your first units are now scheduled to go on the 6th HEXAGON launch which is currently scheduled for December 1972." But by October, due to increased interest to obtain mapping camera photography at the earliest possible date, it was determined that the MCS could in fact be launched on vehicle number 5 (SV-5) in January 1973 and without any omission in the subsequent schedule of HEXAGON missions. The DNRO agreed to this approach.

NPIC Statement of MCS Product Requirements

As early as June 1969, the National Photographic Interpretation Center (NPIC) had formally notified the DNRO that they planned to use the terrain product from the MCS to supplement the search phase of intelligence production in areas outside the pan photography area, and it was expected that some stellar reduction would be required. Their plans indicated that two copies of the terrain material would be needed and one copy of the stellar would be required to satisfy intelligence requirements at NPIC.

MCS Calibration Responsibilities and Procedures

As to the responsibility for calibration of the MCS, during a meeting held at Itek on 18, 19 June 1969, it was agreed by all participants that the U.S. Army Topographic Command (TOPOCOM)* be tasked with this responsibility. DIA concurred with this decision and formally tasked TOPOCOM to perform the function, requesting that DIA, NRO, and the HEXAGON SPO be provided a description of the proposed calibration data reduction procedure by 15 August 1969.

During the ensuing months there were many informal and formal discussions on the approach to calibration. Since the actual requirement was considerably downstream, the question was explored thoroughly before the calibration plan was announced.

By September 1971 agreements had been reached on action items (from previous meetings relative to calibration film processing and procedures). In a message (23 September 71) to TOPOCOM (Serenus Dossi/Ralph Smith) and DNRO (Lt. Colonel Robert A. Schow), SAFSP made reference to agreements reached at a 20, 21 September meeting at TOPOCOM:

A. DMA's recommendation to process the film exposed at the calibration test facility, on site, rather than at DMA, was concurred with, providing the following conditions could be met:

1. The FE 102 processor and chemistry must be qualified prior to use.
2. Delivery of the FE 102 processor and ancillary materials would be compatible with the arrival of the qualification camera at the calibration site.
3. The adequacy of the stellar images for camera calibration would be determined by on site inspection of the processed film by DMA personnel.
4. Calibration film processing would be accomplished by DMA personnel.

---

* TOPOCOM merged into DMA in 1972.

B. In order to expedite the qualification of the FE 102 processor and chemistry, SAFSP would insure that sample film from system testing would be delivered as soon as possible.

C. Vacuum window distortion data and error ellipse data would be transmitted to DMA as soon as possible.

D. Film specifications, to include the math model and mensuration techniques, which would be used by DMA for all HEXAGON frame calibrations, would be furnished to SAFSP, Itek, and DMA on/about 1 October.

The referenced message then scheduled another meeting at SAFSP on 14-15 October to affect final coordination on the following items:

A. Specifications for the mensuration of stellar images obtained at the calibration site.

B. Math model to be used for the 12-inch camera calibration.

As to the possible methods of collecting data for MCS calibration, there were from the beginning two basic options under consideration. These were:

1. Preflight: collection of star images with the flight systems from a suitable observatory site.
2. Inflight:
   a. Photograph a controlled test range
   b. Pitch the vehicle and photograph the stars with all three cameras simultaneously.

There was never any doubt as to the wisdom or the necessity of the preflight data collection; it was just a matter of site selection and preparation, and working up the procedures.* Nor were there any arguments against photographing a controlled test range, although it was recognized that some additional preparation would be required. In May 1972, TOPOCOM was in the process of service testing the aerial SMAC program. This program was designed to solve for rigorous calibration of all interior and exterior relative orientation parameters. Since the interior relative orientation parameters (XP, YP, and focal length) are highly correlated to the vehicle position, it would be necessary to have an extremely precise (2 to 5 meters) ephemeris over the controlled range. The necessity to deploy six geoceivers strategically located around the Bar XC range was considered highly desirable in order to accurately track the HEXAGON missions. This number of trackers had been verified with simulations conducted by Naval Weapons Laboratory (NWL). In full concurrence with the necessity of photographing the Bar XC range (equipped with geoceivers), SAFSP in a July 1972 message to the NRO requested approval of funds to purchase the geoceivers for the Bar XC test range and additional funds for visual edge match comparators and associated equipment for TOPOCOM.† Other details (not insignificant) to be worked out were the software preparations to assure proper execution of the Bar XC photography.

The third option, i.e., pitching the vehicle in flight to obtain star photography with all three cameras simultaneously, did, as one might readily suspect, receive the most deliberation.

---

*Details of site selection, preparation, and procedures are discussed in "Preflight Calibration," further in this section.

†On 3 August 1972, the NRO approved the request to purchase the geoceivers, but disapproved the purchase of VEM equipment on the understanding "that the visual edge match comparators are not mission essential and arrangements can be made by TOPOCENTER to use existing equipment at other organizations."

The pitch option offered an excellent potential, since the data for calibration would represent the system configuration after having experienced the launch environment. But there was an element of risk—there were significant technical unknowns (at the time) and potential operational problems associated with the maneuver. It was finally decided, however, that the potential attractiveness of this option warranted at least the incorporation of the capability in the camera system to handle the program. Since this involved both hardware and software additions requiring early incorporation into the program, the decision was made to develop the capability, and delay the decision to actually do the maneuver until a later date after more experience with the HEXAGON vehicle had been gained.

MCS Block I Procurement

The first buy of mapping cameras, commonly referred to as Block I, was for six flight systems, a prototype (or engineering model), and a qualification system. But on 19 May 1979, SAFSP formally directed Itek "not to undertake the procurement necessary to fabricate the sixth flight mapping camera. In lieu of fabrication of the sixth flight unit, Itek shall refurbish the qual unit as required for the sixth mapping camera launch."

MCS Software Development

By November 1970, Itek had defined the MCS hardware/software limitations and advised SAFSP via a 19 November message as noted below:

A. Operating time constraints (violation may cause degradation of thermal control).

  A1. Four hours maximum in ascent mode (A-mode). Allow one-half hour OFF after four-hour operating period.

  A2. Fifteen ON commands maximum per rev and forty ON commands maximum per day.

  A3. Twenty-five minutes maximum operating time per rev and fifty minutes maximum operating time per day. The maximum operating time per rev shall not be repeated on successive revs.

  A4. Main bus power shall not be removed during pre-launch or orbital time phases.

B. Camera response constraints. (Violation may cause loss of frame following command change or loss of synchronization.)

  B1. R factor (FV/H)* update increment shall not exceed one unit. Update rate shall not exceed three units per frame period.

  B2. Overlap changes shall not be commanded while camera is operating.

  B3. R factor (FV/H) shall not exceed 82 when 78 percent overlap is commanded.

C. Reliability constraints. (Violation may cause loss of redundant operational capability.)

  C1. Do not command B-mode except in event of failure.

  C2. Do not command secondary controls (ones in 4 bits VSPC) except in event of failure.

  C3. Do not command EDAP to secondary except in event of failure.

  C4. Do not command emergency enable, thermal shutter cap except in event of failure.

---

*FV/H: forward velocity divided by height.

D. Mission planning constraints. (Violation may cause unnecessary loss of film or degradation of photographic performance).

D1. Command operate OFF when leaving day light segment of each rev.

D2. Avoid operating when sun is within 15 degrees of either stellar lens field.

D3. Do not command material change detector (MCD) search ON except for final one inch of terrain film supply radius, and unless calibration maneuver is planned.

D4. Do not command calibration mode (C-mode) except in calibration attitude. Calibration maneuver shall be performed only on dark side of earth with moon outside all fields of view.

D5. Do not command 3-millisecond terrain exposure option.

D6. Run film to depletion at end of mission before cut/seal.

D7. Do not operate camera outside the V/H range 0.0165 to 0.0566 radians per second.

D8. Optimum terrain resolution is obtained in north to south pass, near polar orbit.

Incorporation of Wobble Roller

On 12 February 1971 SAFSP informed the DNRO that recently they had determined that the GE RV structure did not provide adequate rigidity to maintain the film path alignment through the launch environment. It had become necessary, therefore, for Itek to incorporate a wobble-roller mechanism to maintain the required alignment and dimensional tolerances in the film path. Two solutions had been possible to maintain the film path alignment: (1) make the RV structure more rigid, or (2) have Itek design a wobble-roller to accommodate the increased tolerances. The latter decision was made.

Early Plans for Block II Procurement

Long before the first mapping camera had been launched, plans were already in process for a follow-on buy of six additional units.

In a message dated 17 May 1971, the SAFSP contracting officer, Colonel Charles C. McBride, Jr. advised Lt. Colonel Schow and Mr. ▇▇▇▇ in the NRO of the following planning factors for the next buy of mapping cameras from Itek:

Procurement: Four complete systems plus two sets of components

Configuration: Same as Block I

Schedule:
1 July 71	Request for proposal
1 Sept 71	Proposal due
1 Oct 71	Contract negotiations
1 July 73	First item delivery date (7th mapping camera)
15 June 74	Flight date 7th Mapping Camera (13th HEXAGON vehicle)
1 Oct 74	Flight date 8th Mapping Camera
1 Feb 75	Flight date 9th Mapping Camera
1 June 75	Flight date 10th Mapping Camera
1 Oct 75	Flight date 11th Mapping Camera
1 Feb 76	Flight date 12th Mapping Camera

As to funding, the fiscal 1971 financial plan made provisions for initial long-lead procurement actions for raw glass related to follow-on HEXAGON flights 13 through 18 (MCS flights 7 to 12). These funds were to be used primarily for glass procurement from Schott, glass polishing, and developing a second source for encoders to replace Sequential Corporation which had gone bankrupt.

On 10 June 1971, Itek was authorized to procure long-lead time items in parallel with the preparation of the Block II proposal. Following the usual fact-finding and negotiations, SAFSP issued a contract to Itek on 7 December 1971 for the six additional systems.

Proposal for MCS Improved Design

In the early months of 1971, dialogue between DMA, SAFSP, and Itek had revealed several areas of possible improvements to future HEXAGON systems. Itek had performed preliminary studies that revealed the possibility of increasing the dynamic resolution by a factor of 2 (94 lines/mm AWAR). This required a new lens design of the same focal length (12 inches) with an aperture of f/4. The new lens would permit the use of slower but higher resolution film than was presently planned and thereby significantly improve the resolution of the system. Itek believed that a lens could be designed with such increased resolution and yet maintain or reduce the total distortion in the lens system. The option to include color correction was also available.

DMA encouraged a feasibility study for a primary lens design, providing at least a two-time increase in resolution, coupled with studies for reducing the blur rate. The physical dimensions for the new lens should remain near current values to avoid major system revisions as a result of weight and space limitations. The preliminary analysis indicated that with the increased resolution the photography would support the 20-meter relative contour interval with a significant safety margin. The doubled resolution would permit:

1. The photographic compilation to be performed at a greater enlargement than previously planned. This would allow, theoretically, twice the compilation scale and thereby half the ground error in the final map product. Therefore, DMA should approach the capability of compiling a 10-meter interval as specified in the approved USIB requirements.

2. The photographic compilation of the 20 meters case would be expedited since the input photography would not have to be enlarged to meet the accuracy requirements. This reduced enlargement factor would allow more ground area to be compiled per unit time.

3. The dependence on the panoramic photography for additional detail would diminish and in many areas be eliminated entirely with the increased information content that the new lens would provide. This would lower the MC&G requirements for panoramic camera coverage which should be of distinct benefit to the intelligence community.

By August 1972 the feasibility study for incorporating an improved lens system into the HEXAGON Mapping Camera had been completed. The study indicated that an f/4 lens could be incorporated which would increase the dynamic resolution of the camera by approximately 70 percent for the AWAR and 40 percent for the minimum resolution, to AWAR/min values of about 85/50 for black and white imagery on EK 3414 film. This would significantly reduce the amount of map detail which must be obtained from panoramic photography and would improve the vertical mapping accuracy. In addition, the f/4 lens would have a spectral band-pass capability which would be compatible with SO-242 color film. If the users had a requirement for color, the f/4 lens would provide color photography with about the same resolution (50/36) which the current lens provided using black and white film.

In a message on 16 August 1972, SAFSP suggested to the NRO that it was time to decide whether this development should be incorporated into the system, or dropped. The optimum schedule for the incorporation would be:

"Now through early 1973—contractor study specific configurations, performance, manufacturing techniques, and the effect on existing Aerospace Ground equipment (AGE).

Early 1973—Government evaluation

Mid-1973—Initiate contract for next lot of Mapping Cameras to include the improved lens

Mid-1975—Delivery of first f/4 camera

Mid-1976—Flight of first f/4 camera."

SAFSP stated that a firm decision could be deferred until FY 74, but that in order to protect the option, they should "now be conducting the specific configuration studies."

It is not clear from the records, or from recollection, exactly what transpired from this time forward in a step-by-step sequence relative to this proposal. The important thing from a historical point of view is that the improvement program, as proposed, was not activated. However, as described in Section 8, system capability was increased significantly (approximately doubled in film footage and resolution) as the program progressed, with relatively minor and inexpensive modifications being made to the camera system to accommodate the use of higher resolution and thinner base films.

Pressure Makeup System (PMS)

The compartmenting of space within contractor facilities to physically separate classified programs, together with communication restrictions imposed (for security reasons) on personnel within contractor facilities and government program offices, sometimes causes the wheel to be invented more than once. This might have been the case in the early design and development days of the mapping camera, had it not been for the continuity of experienced personnel in both government and Itek. One example was the method of addressing the problem of electrostatic discharge, commonly called "corona marking," which can result when film is transported over rollers in a near-zero pressure environment. The MCS contractor team had initially proposed that the marking problem could be controlled satisfactorily via passive means only, i.e., by the selection of suitable roller material and adequate testing. To those who had lived through the nightmare of electrostatic marking problems on previous space programs, specifically CORONA and DISIC, the thought of using the passive approach alone was not only extremely risky, but unnecessary. Although, through the brute force approach of roller material selection and arduous testing, Itek had learned to eliminate or minimize to an acceptable level the marking in the CORONA system,* in the latter years of the CORONA program a backup system was provided by the integrating contractor, Lockheed, which provided high assurance that the processed film would be free of this potentially catastrophic problem. Basically, the pressure makeup unit,† known generally as PMU, included a storage supply of dry nitrogen gas, orifices of varying sizes and the controls necessary to dispense the gas upon command. Testing for corona marking was done in a vacuum environment in three

---

* The technique developed on the CORONA program was later made available to the DISIC program.

† In the CORONA program, this installation was known as "pressure makeup unit (PMU);" on the HEXAGON program, "pressure makeup system (PMS)."

steps: first, at the subassembly level to obtain the selection of rollers which provided minimal or no marking, then at subsystem level in the Itek vacuum chamber under simulated PMU conditions, and finally, at the full system level in the integrating contractor facility. Pressure sweeps were made using various size orifices which provided a range of chute pressures varying from a few microns to 100 microns. Since each system had its own personality, the orifice which provided the pressure range best suited to the individual system was selected and installed for flight.

After this information had been brought to the foreground by the "old timers" (such as Grant D. Ross, Field Operation Manager for Itek, and Lt. Colonel Albert W. Johnson, SAFSP), the decision was made to incorporate a pressure makeup system for the mapping camera.

It was not immediately clear as to the best approach for developing and installing the PMS. The initial thought was to make this the responsibility of the camera contractor, and pursuant to this approach, SAFSP sent a message to Itek on 22 July 1971 directing the incorporation of a pressure makeup system into the prototype and flight systems. But after further discussions between SAFSP and the contractors (Itek and Lockheed), this responsibility appeared to fall more naturally to Lockheed, since they had the experience of the pressure makeup units on the CORONA program and also, the obvious location for the PMS gas storage spheres and dispensing system was in the Lockheed Auxiliary Payload Structure Assembly (APSA).

On 10 September 1971 SAFSP requested via message that Lockheed submit an ECP "to provide and install pressure makeup hardware in the mapping camera module." In response, Lockheed submitted ECP's for the design and qualification of a PMS, and modifications of the APSA to accommodate installation of the hardware. The PMS would "provide for the storage of high pressure nitrogen and the ON/OFF control of the release of the nitrogen in response to commands from a using system." To develop and qualify the PMS, a prototype unit and a qualification unit would be assembled for the test program and acceptance tests and qualification tests would be performed. Lockheed emphasized that the qualification tests must be expedited to support delivery of the flight units, stating "the authorization to proceed is needed no later than 1 November 1971 to support already tight schedules for APSA delivery."

To summarize this subject, with both contractors vigorously pursuing their areas of responsibilities in fabrication, qualification, test, and analysis of pressure sweep data to select the optimum orifice for flight, the pressure makeup system was ready for the first MCS flight, and performed successfully throughout the program.

Two Additional Mapping Cameras

In early 1972, the NRO was giving some thought to the procurement of two additional mapping camera systems to be flown on vehicles 17 and 18. In March 1972, responding to an NRO request, contractors submitted planning estimates for this proposed follow-on program.

The idea was kept alive and by 1975 definitive actions were being taken to formalize this follow-on effort. On 14 March 1975, via message to SAFSP and DMA Headquarters, the NRO advised "it has been decided to procure two more Mapping Camera Systems for HEXAGON SV-17 and 18, in order to encourage the contractor to maintain a bidding capability through October 1975." The message further stated that this did not as yet have congressional approval and that a firm commitment and contracting for these units must be delayed until approval was given (expected in October 1975). That if congressional approval were obtained, the NRO would be committed to buy the additional units (one in FY 76 and one in FY 77). The message emphasized that SAFSP was not authorized to commit NRO funds to support the plan pending congressional approval. Then on

1 April 1975 a message was sent from the NRO (Hill/Wheeler) to SAFSP (General Bradburn) stating "in view of the limited availability of FY 75 funds and the overriding requirement for the capability to acquire the follow-on mapping cameras for HEXAGON vehicles 17 and 18, you are authorized to apply minimum FY 75 NRP funds not to exceed ▮▮▮ to meet this requirement." The intent was to limit the application of these funds primarily to long-lead items since a definitive contractual acquisition of the mapping cameras would have been delayed until congressional action took place on the FY 76 budget.

On 11 April 1975, the contractor was authorized to proceed as directed by the message. By October 1975, however, authorization to proceed was not forthcoming and all efforts were stopped. In short, there were no more mapping cameras built for the HEXAGON program beyond Block II.

MCS Testing to First Flight

Following the formal program turn-on in 1969, all associate contractors had proceeded along individual time lines of engineering, hardware procurement, subassembly, and test to bring all of the major subassemblies together in a timely schedule for system testing and launch. As stated previously, Lockheed, the HEXAGON integrating contractor, would build the Auxiliary Payload Structure Assembly (APSA) to which would be mounted the Itek Mapping Camera Assembly, the Lockheed pressure makeup system, the Doppler beacon system built by Applied Physics Laboratory (APL) and supplied GFE to the white contract, and the General Electric recovery vehicle. The modified Mark V RV provided by General Electric would house the Itek film takeup assemblies.

By January 1972, three units, having reached significant stages in the schedule, were being tracked closely on the Itek project office weekly status reports. The prototype in the APSA was essentially complete. A minor change to the terrain transport had been required to allow access to the upper areas of the camera without removal from the APSA. Initial film tracking runs were scheduled for the last week in January. The qualification unit was through system updating. Distortion boresight test unit (DBTU) tests had been conducted, assembly and retracking on the system dolly was underway, and Main Instrument System Electrical Assembly (MISEA) temperature tests were scheduled to start on 23 January. The first production unit, known in Itek circles as "P-1," had completed a limited functional test, with installation of an updated rotary shutter. Engineering debug and checkout of the main instrument was expected to be completed by 21 January. A customer buy-off meeting of the first production unit was scheduled to start on 1 February 1972.

On 16 February, a failure in the Itek dynamic resolution test (DRT) chamber put serious perturbations in the schedule, resulting in the necessity to interchange flight units. The DRT failure resulted in back-streaming of oil during the testing of P-1, coating the unit with oil to the extent that it was necessary to initiate a complete rebuild plan. (Needless to say, the chamber vacuum system was also rebuilt and modified so that this failure could not recur.) In addition to the flight unit perturbations, the failure caused a vacuum chamber bottleneck in the program qualification testing. This problem was resolved by moving the 60-day thermal vacuum qualification test to another chamber, i.e., the "E" chamber in the Perkin-Elmer Danbury, Conn. facility.

By 12 May 1972 the prototype unit had been shipped to the Lockheed facility building 156, being set up for thermal vacuum testing. The P-1 unit (now S/N-003), having completed calibration tests at Cloudcroft, was delivered to Lockheed on 15 May 1972 to start functional testing. The qualification unit was in orbital thermal vacuum (OTV) testing scheduled to be completed during the third week in May. At this time another system, P-2 (S/N-004) began to show up on the weekly status reports. DBTU testing and main instrument vibration was completed on 17 May, and the system was being set up for functional test prior to MISEA temperature test.

During the last week in June the prototype was installed on the HEXAGON SV-5 unit in building 156, with electromagnetic interference (EMI) testing scheduled to begin the first week in August. (The SV-5 vehicle was to go on a 6-day-work-week, 10-hours-a-day, double-shift schedule starting on 19 July through 20 October.) P-1 (S/N-003) was experiencing stellar transport problems, thus delaying the completion of tracking tests. The qualification unit (S/N-001) had completed the main instrument hot shake and second photobaseline test, and was scheduled for shipment to the Perkin-Elmer facility the last week in June. P-2 (S/N-004), having completed the second baseline on 16 June, was in the MISEA hot shake, scheduled to be completed by 23 June. The fifth unit, P-3 (S/N-005), now coming into prominence, was in MISEA temperature test scheduled for completion on 24 June.

As of 21 August the prototype was on hold awaiting the start of EMI testing. P-1 (S/N-003) had completed acoustic and post-acoustic functional testing and was scheduled for thermal vacuum testing, which was to start on 25 August. P-2 (S/N-004), with platen centering and FMC problems resolved, was in altitude thermal vacuum (ATV) testing, scheduled to be shipped to Cloudcroft on 25 August. P-3 (S/N-005), undergoing engineering evaluation of technical problems with the terrain transport, forward motion compensation (FMC) assembly, and the rotary shutter was scheduled to start ATV tests on 23 August. The qualification unit started the 60-day simulation test in the Perkin-Elmer chamber on 8 August, but the test was halted due to chamber problems (leaking cold panel feed lines). The problems were resolved and the test resumed on 3 August with a projected completion date of 12 October.

By late December 1972, P-1 (S/N-003) had completed both the A-1 and A-2 vacuum chamber tests in the system configuration at Lockheed, and was scheduled for a launch date of 15 February 1973. P-2 (S/N-004) was scheduled to undergo acoustic testing at the module level starting on 2 January to be available for a system mate date of 9 January. P-3 (S/N-005) was undergoing various module level testing at Lockheed and scheduled for a system mate date of 29 January. P-4 (S/N-006) at the Itek factory had completed main instrument vibration and the second photobaseline tests. The qualification unit completed the 60-day simulation in the Perkin-Elmer chamber on 12 October. It was then shipped to Cloudcroft on 29 November for final calibration, and upon completion of these tests was returned to the factory for refurbishment. The prototype, having completed scheduled testing, had been on inactive status at Lockheed since September.

By 26 January 1973, P-1 (S/N-003) had completed final loading and pre-ship confidence testing. All work at Lockheed had been completed and the system buttoned up, still scheduled for launch on 15 February. P-2 (S/N-004) was experiencing problems with the pressure makeup system. It was anticipated that these problems would be resolved to enable start of the A-1 thermal vacuum testing on 7 February. P-3 (S/N-005), having experienced problems with the terrain camera transport was rescheduled for system mate on 20 February. P-4 (S/N-006), following engineering investigation and testing of rotary shutter and stellar camera transport problems, was scheduled to resume thermal cycle testing on 26 January. P-5 (S/N-002) had completed DBTU testing and was scheduled to start photobaseline testing in the Itek DRT on 26 January. The qualification unit refurbishment of major subsystems was in progress at Itek. The prototype was shipped from Lockheed back to the Itek factory on 8 January.

During February, the APSA with P-1 (S/N-003) had to be demated from SV-5 to make a repair on the recovery vehicle. This necessitated installing a new flight load and repeating the final performance test. It was shipped to Vandenburg on 21 February for a newly scheduled launch date of 6 March. The P-2 (S/N-004) test progress was still being gated by associate contractor problems with the A-1 testing scheduled two or three times; however, as of 23 February the start of A-1 test preparations was still to be determined. P-3 (S/N-004) completed vacuum testing on 8 February and it was anticipated that this unit should be ready well in advance of the then scheduled 2 April mate date. As subsequent testing progressed smoothly, a new mate date of 15 March was anticipated. Thermal cycle testing of P-4 (S/N-006) was delayed due to additional stellar transport problems. Following engineering investigation, rework, and retest of the transport, the thermal cycle test was started (or resumed) on 12 February and completed 14 February. The system was next scheduled into the DRT for altitude thermal vacuum tests. P-5 (S/N-002) completed first and second photobaseline testing in February and was scheduled to complete thermal cycles tests on 28 February. Refurbishment of the qualification unit was still in progress.

P-1 (S/N-003) was successfully launched on 9 March 1973, two HEXAGON systems earlier than was originally thought possible. The orbital performance of this system and subsequent systems, along with major hardware or program changes, are covered in Section 7, "Operational Considerations and Statistics."

MAPPING CAMERA SYSTEM (Fig. 2-1) HARDWARE AND TEST CYCLE

Configuration

The wide variety of MC&G products required for Department of Defense (DoD), military, and intelligence community users led to establishing requirements for the HEXAGON Mapping Camera System (Table 2-2). These requirements in turn dictated the MCS basic design.[8]

Table 2-2 — HEXAGON Mapping Camera Requirements

Coverage
1. 16 million square nautical miles of denied areas
2. World-wide mapping coverage of free world at a rate of 10 million square nautical miles per year

Accuracy
Sufficient accuracy and resolution to permit compilation of large and medium scale topographic maps and aeronautical charts

Reference Orbit
92.5-nm perigee

Mission Duration
45 days for the first two flight units
60 days for remaining flight units
(As the program progressed, mission duration increased beyond requirements, achieving 118 days on each of the last two missions.)

MC Ground Coverage
70-nm width at 92.5-nm altitude
Triple overlap photography with quadruple overlap at altitudes over 100 nm
(70 and 78 percent forward overlap per exposure, respectively)

Terrain Camera Resolution (92.5-nm altitude)
GRD: 45 ft maximum, 35-ft area-weighted average (3400, W-21)
26 ft maximum, 20-ft area-weighted average (3414, W-12)

Camera Geometry
Terrain camera calibration precision: 2 micrometers
Angular relationship between terrain and stellar cameras (knee angle) calibrated to precision of 3 arc-seconds
Base-to-height ratio of 0.9 to 1.0 (triple or quadruple overlap)

Terrain Mensuration
Object point locations accuracy: 4 micrometers

Terrain Exposure
Selectable: 3, 6, or 12 milliseconds (3400, W-21)
6, 12, or 24 milliseconds (3414, W-12)

Reliability
0.97 for 30-day mission
0.997 for initial operation on-orbit

Table 2-2 — HEXAGON Mapping Camera Requirements (Cont.)

Environmental Control
    Active temperature control of camera unit at 73 ±1°F within passively controlled
        APSA spacecraft of 38 to 70°F

Telemetry
    PCM/FM telemetry data providing complete coverage from launch through
        completion

Timing
    Processing and distribution of time codes and sequencing functions
    1.0-millisecond absolute accuracy from interface clock
    0.1-millisecond relative accuracy between frames from internal MC clock

The vertically mounted 12-inch-focal-length, f/6 terrain camera, providing a 74-degree in-track and 40-degree cross-track field angle, furnished the base-to-height ratio, scale, and accuracy necessary to compile the desired 1:50,000 scale map manuscripts. The 10-inch-focal-length, f/2 stellar cameras provided the short exposure and point-to-point separation necessary for angle recovery from star imagery.

The orientation of the stellar lenses, selected after a careful tradeoff study, represented a best compromise of such considerations as all-attitude orientation sensitivity, minimization of horizon flare, and minimization of direct view of the sun (for north to south passes). The use of two stellar cameras ensured that the required attitude determination would be obtainable under all expected starfield population conditions. In the event of a failure of one of the stellar cameras, sufficient data could be acquired from the remaining camera to provide degraded metric information.

Together, the Main Instrument (MI), Doppler beacon subsystem (DBS), satellite reentry vehicle (SRV), and the Auxiliary Payload Structure Assembly (APSA) made up the mapping camera module (MCM). The major assemblies of the MC system are shown in Fig. 2-2.

Physical Characteristics

Dimensions and weight for the MC System are presented in Table 2-3. Since few of the assemblies are rectangular in shape, the box of maximum rectangular dimensions is given for the various assemblies.

Lens Systems

Terrain Lens

The prototype terrain lens produced by Itek during the Geodetic Orbital Photographic Satellite System (GOPSS) contract met the GOPSS mission requirements including ground resolution and wide field angle. This lens system utilized a total of eleven elements and three aspheric surfaces. Maximum radial distortion of the image was over 1 mm.

Design improvements were made to optimize the lens performance for the MC mission:

1. Readjustments were made which resulted in a decrease in on-axis resolution, but an increase in AWAR and minimum resolution. This adjustment provided more uniform image quality throughout the format since the accuracy of stereoscopic height measurements is greater

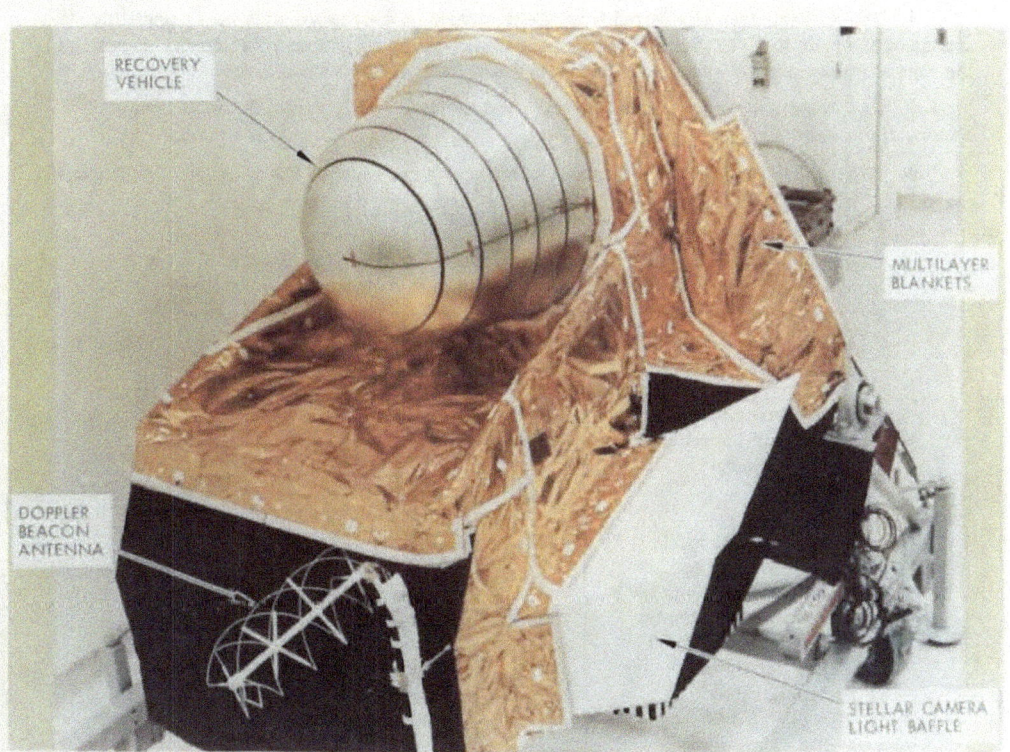

Fig. 2-1 — Mapping Camera System

when the conjugate images are of the same resolution level. Additionally, the error in measurement is nearly proportional to the difference in resolved distances between the two conjugate frames.

2. Geometric distortion was reduced drastically to simplify map compilation procedures and increase the accuracy of MC&G products.

3. The spectral band of the lens was broadened to permit photography with shorter exposure times for reduced dynamic smear.

4. The aspheric was eliminated from the underside of the reseau plate thus permitting motion of the plate for FMC without introducing second-order optical misalignments. In addition, the reseau plate material was changed to fused silica for thermal insensitivity.

5. The length and weight of the lens were reduced to simplify packaging and to permit a more rigid assembly.

6. The number of elements and aspheric surfaces were reduced for economy of production.

Table 2-3 — Physical Characteristics of the MC System
(dimensions in inches, weight in pounds)

Assembly	Height (Z Axis)	Width (Y Axis)	Length (X Axis)	Weight
Main Instrument	34	34	31	339.0
Stellar and Terrain Takeup (empty)	21	16	14	20.8
MISEA	16	17.5	27	96.0
EDAP	5	17.5	22	31.5
+Y stellar light baffle*	27	26	52	20.0
−Y stellar light baffle*	27	26	52	20.0
Terrain thermal shutter	3	12	21	5.7
Terrain supply (empty)	17	18.5	17	19.0
Stellar supply (empty)	14	6	17	19.5
Stellar supply servo amp	6	2.5	2.5	1.0
System harness	–	–	–	30.0
Terrain 9.5-inch film on spool	13.3	9.8	13.3	57.0
Stellar 70-mm film on spool	11.7	3	11.7	13.2
Chutes (9.5-inch and 70-mm)	–	–	–	19.2
MC System	83	87	86	691.9

*Light baffle dimensions are given in maximum rectangular box dimensions not keyed to the system axes.

MC System dimensions are given as a totally integrated configuration into the APSA. Height given is from bottom of MISEA to top of SRV. Width given is from extremities of light baffles in final configuration.

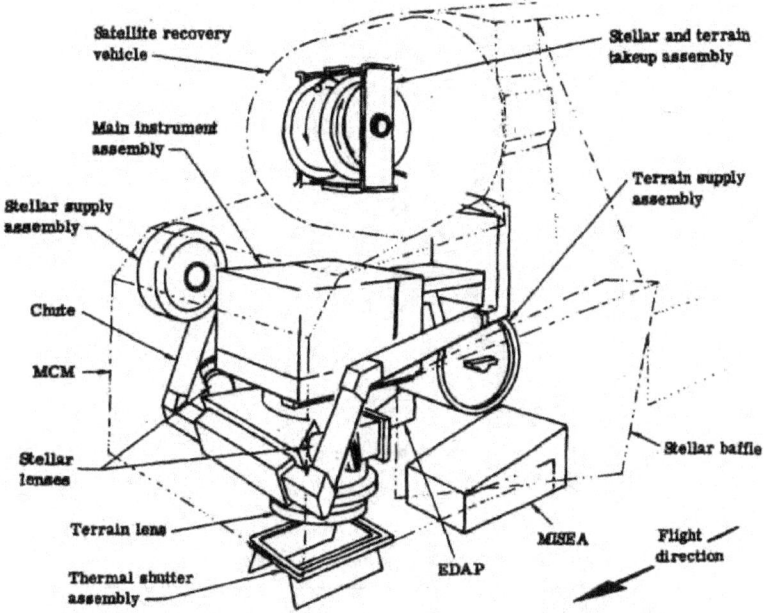

Fig. 2-2 — Mapping Camera major assemblies

The evolution of lens performance from the GOPSS design to the MC capability is shown in Table 2-4.

Table 2-4 — Evolution of Terrain Lens Design

	GOPSS	MC	
1. Resolution (1/mm, 2:1 contrast)			
Film	3400	3400	3414, SO-315
Filter	21	21	12
AWAR	50.2	53	100
Minimum	25	40	70
2. Distortion (maximum), micrometers	1,000	26	—
3. Relative weights	1.0	0.92	—
4. Number of power elements	11	8	—
5. Number of aspherics	3	1	—
6. Reseau	Asph.	Flat	—
7. Total length, inches	28.2	25.2	—
8. Number of air-glass surfaces (no window)	14	12	—
9. f-number	6.0	6.0	—
10. Focal length, inches	12.0	12.0	—
11. Field of view, degrees full	80.0	80.0	—

The final lens design is shown in Fig. 2-3 and the lens specifications are listed in Table 2-5.

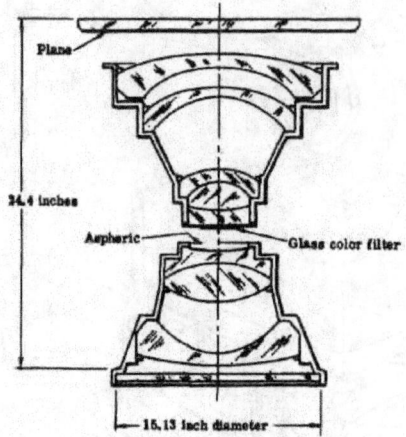

Fig. 2-3 — 12-inch metric lens

Table 2-5 — Terrain Lens Specification

Focal length, inches	12
Aperture	f/6, T/14*
Format, inches	9 × 18
Field of view, degrees	80
Distortion, micrometers	
Radial, max	100
Tangential, max	20
Static resolution, ℓ/mm	
AWAR (EK-3400 film)	50
Minimum (EK-3400 film)	38
Film type/filter	EK-3414/W-12
Weight, pounds	186
Overall length, inches	24.4
Number of elements	8 powered plus window, filter, and reseau plate

*T/number does not include filter factor

Stellar Lenses

Since the stellar cameras determined the angular position of the MC at the instant of exposure, theoretically they must acquire at least two known stars. In practice, 10 to 20 stars of known position were necessary to allow determination of pitch, roll, and yaw attitudes with requisite accuracy. In order to acquire this number of stars, the stellar cameras must have sufficient sensitivity to image stars of 6th magnitude and brighter. Mensuration errors and star catalog errors were the principle limitations in accuracy determination. Since these errors were independent and random, they could be minimized by imaging a large number of stars in the stellar format. A wide-angle lens was desirable to increase the number of stars imaged, and to furnish accurate determination of all three orientation angles in cases where one of the two lenses was inoperative due to automatic capping against solar impingement. A large aperture lens was required to record the necessary stellar magnitude in as short an exposure time as possible. The lens focal length should be similar to the terrain lens focal length so that equivalent mensuration errors would result in equivalent contributions to attitude determination error from all three lenses.

The initial lens considered for the stellar cameras was the Wild f/1.8 Falconar with a 248-mm focal length and effective field angle of 25 degrees. Initial investigations predicted that the Falconar lens would be capable of furnishing the desired accuracy and sensitivity for the required stellar acquisition. However, subsequent tests revealed that the lens would not provide the imagery required and, in effect, it did not perform to expectations.

Itek then initiated a design effort which resulted in the stellar lens that was implemented in the MC. Since the camera design was advanced at the point this decision was made, the mechanical interface of the new lens was constrained to be similar to the Falconar. The final lens design is shown in Fig. 2-4.

The mounted stellar lens envelope (see Fig. 2-5) was 5.5 inches outside diameter, and 13.8 inches in length. The mounted lens weighed 19.3 pounds.

The stellar lens assembly consisted of a two-part cell; the main section was basically cylindrical and provided the flange interface with the terrain camera structure. A rear section accommodated the rectangular reseau element, and was machined at final assembly to position the reseau surface at the optimum vacuum focal plane. Spacer rings were machined so the air gap between elements could be adjusted as required to optimize photographic performance for variations in indices of refraction and fabricated element dimensions. The optical elements were retained in the cell by means of circumferential mylar shims and threaded retaining rings. The cell material was fabricated from beryllium, which provided lightness and high structural rigidity. The thermal expansion coefficient of beryllium closely matches that of the optical elements, thus it prevented undue strain on the glass elements through the anticipated thermal excursions.

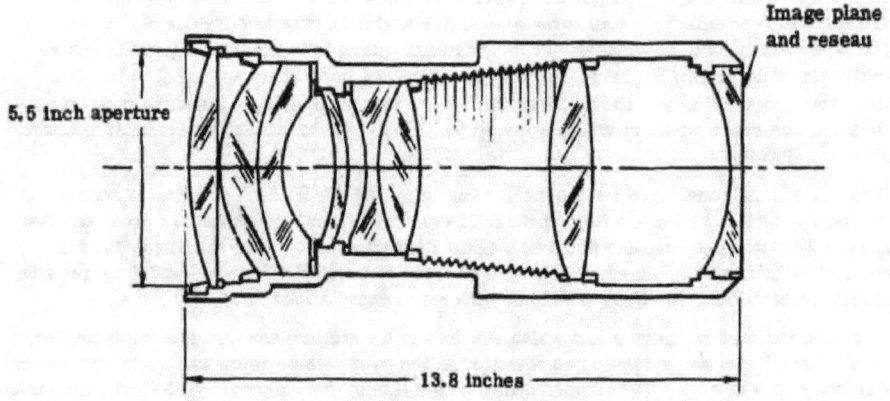

Fig. 2-4 — 10-inch stellar lens

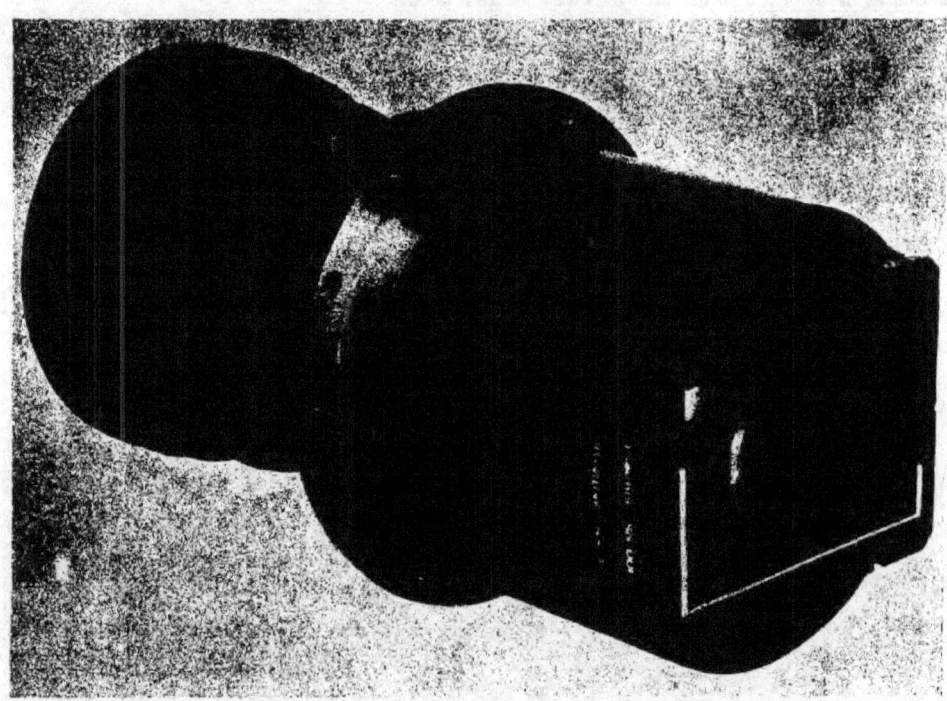

Fig. 2-5 — Stellar lens structure

MCS HISTORY

MECHANICAL DESCRIPTION

Inherent to the mapping camera design was the ability to maintain the overall metric error to 0.8 micrometer between two frames over a 60-day operating period. To maintain the knee angle error to 2.6 arc-seconds, it was essential that the mechanical mating of subassemblies was accurate and reproducible, particularly the terrain and stellar lens systems. To ensure this integrity, the mapping camera structural design concentrated on the center section of the camera assembly (see Fig. 2-6) to which the terrain and stellar cameras were attached. The center section thus became the integrating structural element of the mapping camera. A rectangular magnesium baseplate attached to the upper end of the center section (as shown in the figure) provided the actual interface between the mapping camera and the APSA through three kinematic mounts which isolated the camera from APSA deformations.

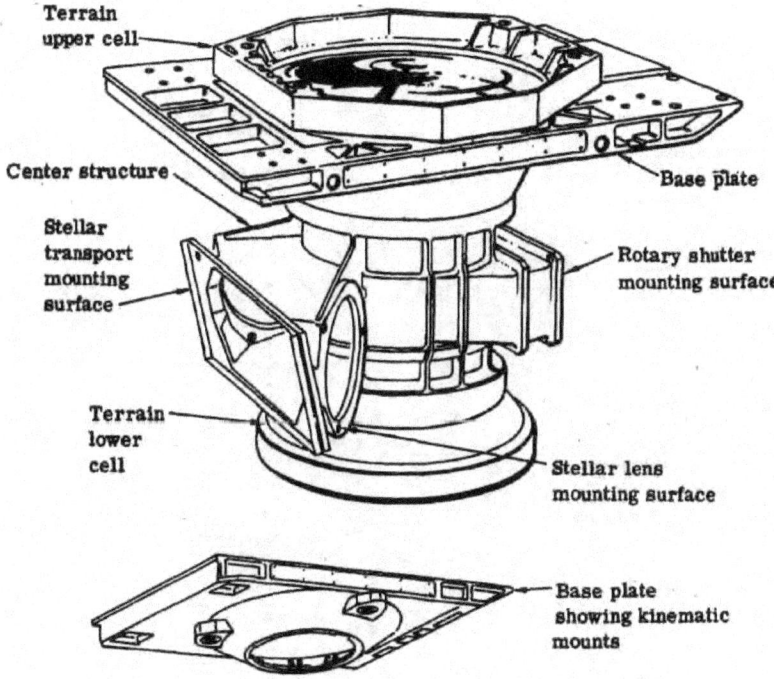

Fig. 2-6 — MC center structure and mechanical interface

Terrain Camera

The terrain lens pointed downward on orbit and imaged the ground scene onto the 9- by 18-inch format (18-inches in-track). Fig. 2-7 shows the mechanical arrangement of the terrain camera. The transport and platen press, located above the image plane, advanced the 9.5-inch film into the exposure station and clamped the film to the flat rear surface of the lens (reseau plate) during exposure. The reseau plate contained a matrix of tick marks (+) which were used to compensate for film distortion. The reseau plate was attached to the lens body through flexures which maintained precise axial and lateral alignment but allowed the plate and film to move along the flight path for FMC. Four projectors in the lens housing imaged fiducial marks at the corners of the format at the instant of exposure. These marks precisely located the moving terrain image with respect to the calibrated reference frame of the optics. The time at which the fiducials flashed

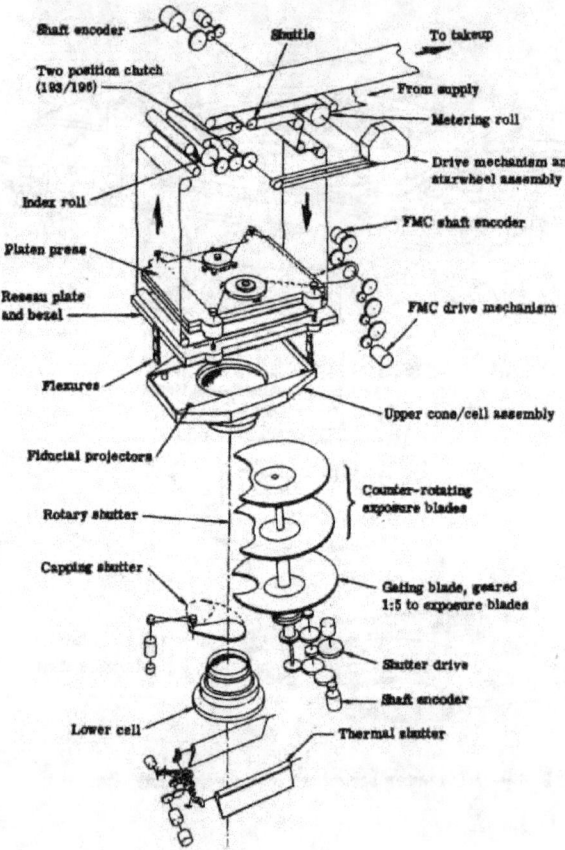

Fig. 2-7 — Terrain camera mechanical arrangement

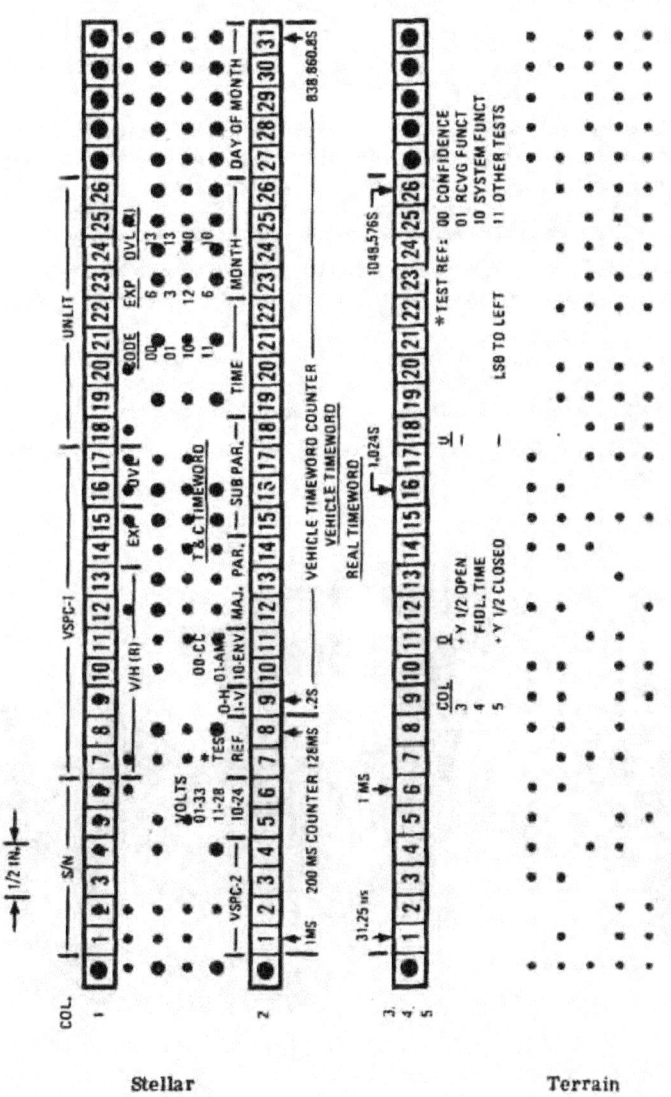

Fig. 2-8 — Data blocks with readout overlay superimposed

was recorded in binary code on the data block at one edge of the format (Fig. 2-8). This recorded time was used in determining the spacecraft position through the orbital ephemeris. The high speed rotating disc shutter and capping blade were located within the lens (between the lens) to control the exposure.

The actual terrain camera photographic system was comprised of the center section, upper cone, lower cone, and the reseau plate.

Major structural elements (see Fig. 2-9) were beryllium, which provided light weight, high rigidity, and a coefficient of thermal expansion which closely resembles that of the glass elements. The center section billet was fabricated from a large sinter of I-400 alloy with properties that included a high rating of precision elastic limit (PEL).* The upper and lower cones mated to the center section on steep conical surfaces which assured radial stability.

The reseau plate and bezel assembly was attached to the upper cone by three flexures which allowed for FMC motion in the direction of flight (x direction) but introduced restraint in the y and z directions.

The terrain rotary shutter and capping shutter were installed between the lens cones through a large cutout in the center section. The terrain platen press area included the platen press, reseau bezel, and FMC drive mounted to the upper cone and lens assembly.

The terrain film transport mounted to the baseplate, and the stellar film transport mounted to the center structure.

Fig. 2-9 — Terrain camera structural elements

---

*PEL is the stress at which 1×10^{-6} inch/inch of nonrecoverable deformation occurs

Stellar Camera

The stellar camera (see Fig. 2-10 for mechanical arrangement) included two 10-inch-focal-length, f/2 lens systems served by a common 70-mm film supply. Two adjacent 70- by 100-mm starfield images made up the stellar format. The image of the stellar data block was exposed between the two starfield images. The stellar transport and platen press assembly advanced the film for each frame and pressed the film to the flat rear lens surfaces (reseau plates). Double-bladed shutters were located in front of each lens to control the exposure and limit the thermal load on the lenses between exposures. Since the starfield did not provide sufficient background illumination to image the reseau ticks, lamps, installed on the inside of the shutter blades, pre-exposed the formats during clamping, thus raising the background density except where shadowed by the reseau.

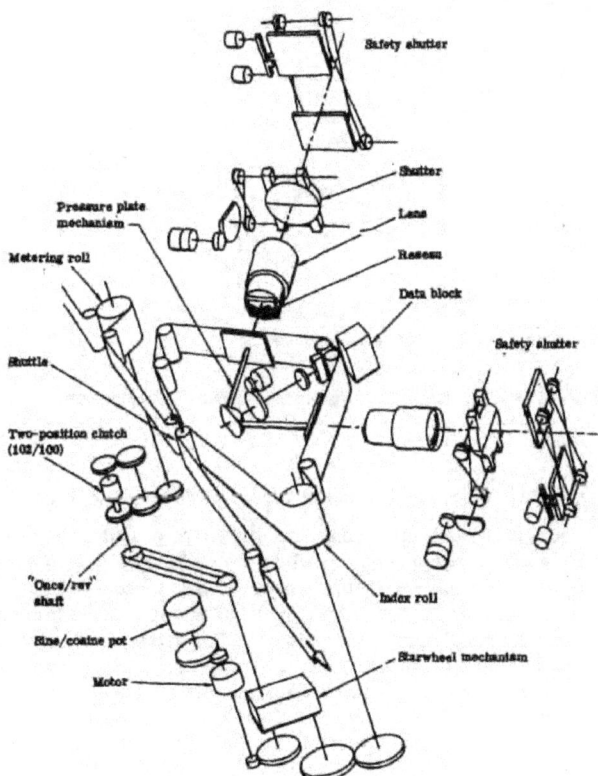

Fig. 2-10 — Stellar camera mechanical arrangement

Terrain Primary Shutter

The primary shutter subsystem (see Fig. 2-11) was introduced between the upper and lower lens cones and mounted to the center section, and provided exposure control to the terrain optical system. A capping shutter was mounted to the center section just behind the rotary shutter as shown in Fig. 2-11.

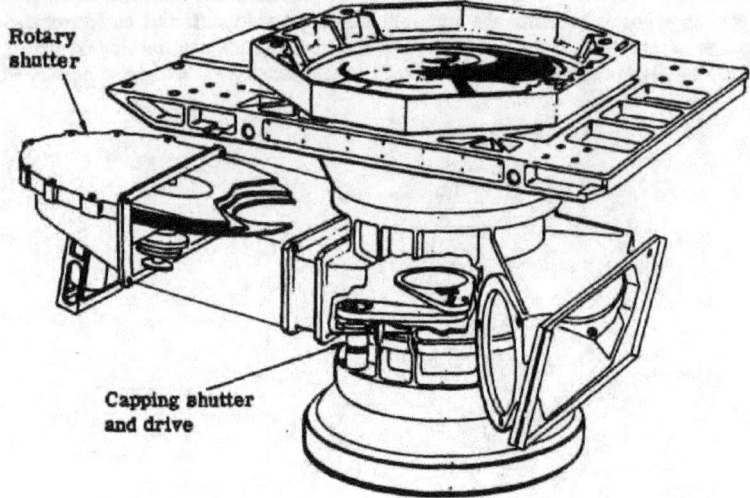

Fig. 2-11 — Terrain camera primary shutter arrangement

Rotary Shutter

Three motor-driven rotary discs controlled the exposure of the terrain camera.

Two discs rotating at equal speeds and in opposite directions provided the actual terrain exposure. A slower rotating disc acted as a gate to allow the capping shutter to open and close without causing double exposure. One of the motors contained a dc tachometer used for servo stabilization. An encoder was geared to rotate at the same velocity as the slow speed disc to provide position and velocity information for servo control. The drive motors were also geared to a sine/cosine potentiometer which processed fixed rate input data from the terrain transport for B-mode operation. During C mode the rotary shutter was commanded to remain open, and the capping shutter was used to control exposure.

Capping Shutter

The capping shutter shown in Fig. 2-11 was an electro-mechanically positioned blade which was synchronized to correspond to selected rotary shutter exposure openings. A torque motor and chain and sprocket drive positioned the capping shutter upon appropriate command from the control and synchronization (C&S) electronics subsystem. A cam operated switch was activated upon shutter closure and instrumentation indicated the closed condition.

During C-mode operation, an electroluminescent panel on the surface of the capping shutter pre-exposed the terrain reseau plate onto the terrain format. The reseau was artifically exposed in this manner since there is not sufficient background illumination available in the calibration attitude, i.e., all cameras facing the star field.

Stellar Shutter and Light Baffle Subsystems

The operation and physical appearance of the stellar shutters (+y) and (-y) and light baffles (+y) and (-y) were essentially identical; the primary difference being symmetrical left (-y) and right (+y) configurations (see Fig. 2-12). Each shutter and its associated light baffle was mounted so that the circular aperture saw a rectangular field of view of 16 × 25 degrees, providing the basic format on the 70-mm film.

Mechanical Operations (Shutters)

On each stellar shutter assembly there were two shutter doors; a drive door and a slave door. Each door opened outward by pivoting about a hinge shaft. The drive door was opened or closed by a sector gear mounted on one end of the drive shaft. The servo motor drove the sector gear. The slave door was operated by a mechanical link of two pairs of steel belts wrapped around pulleys on the end of the drive shaft and slave shaft. Of the four steel belts, two were used for opening and two were used for closing the slave door. One of each pair was redundant in case of failure of the other belt.

Terrain Pressure Plate and Platen Press

The pressure plate and platen press (see Fig. 2-13) was attached to the reseau plate bezel; the reseau being the last element (or focal plane) of the terrain lens assembly. Four ball nuts, supported by the bezel, mated with the ball screws on the pressure plate. Each ball screw was driven by separate motors and power amplifiers which clamped the pressure plate in response to the C&S command. A microswitch indicated when the platen was in its clamped position. A sprocket chain coupling synchronized the ball screw motors, and also drove a spiral spring which loaded during clamping and unwinded when the clamping signal was removed, and returned the press to its unclamped position. A potentiometer ganged to the sprocket supplied positional information for telemetry.

Forward Motion Compensation (FMC)

The FMC drive mechanism mounted to the base plate as shown in Fig. 2-14 drove the bezel in the flight direction at the velocity selected to compensate for forward motion. The terrain bezel was attached to the upper cone through a three-point flexure arrangement which allowed the bezel to move for FMC, while the platen press was in its clamped position in the exposure sequence. The FMC drive moved the bezel approximately 0.13 inch through a gearless direct drive. The FMC drive motor turned a crankshaft which translated the reseau through a connecting rod between the crankshaft and the bezel bracket.

During photography, the four fiducials were exposed onto the terrain format to precisely record the platen position with respect to the bezel (hence, the camera optical axis at the time of exposure). The FMC mechanism was aligned at assembly to ensure that top dead center of the crankshaft throw positioned the center reseau tick nominally on the optical centerline.

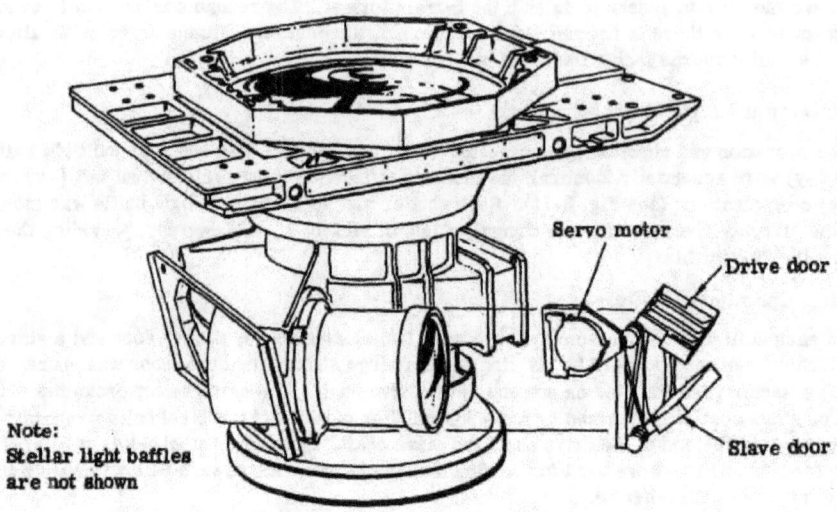

Note:
Stellar light baffles are not shown

Stellar shutter arrangement

Lens focal length = 10.0 inches
f/2.0
Sensitivity = 6th magnitude stars or brighter
Boresight stability = 2 arc-sec in operation
Field of view: 16 by 25 degrees

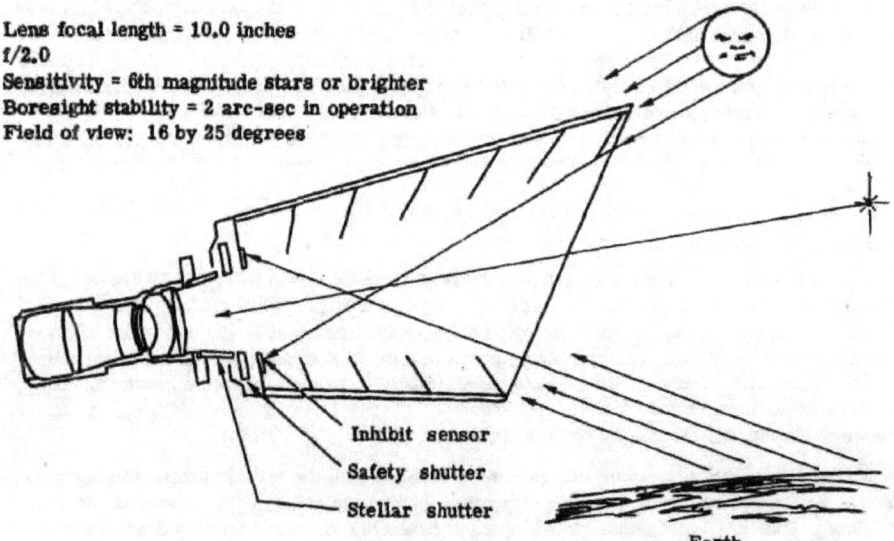

Stellar baffle configuration

Fig. 2-12 — Stellar shutter and light baffle arrangement

MCS HISTORY

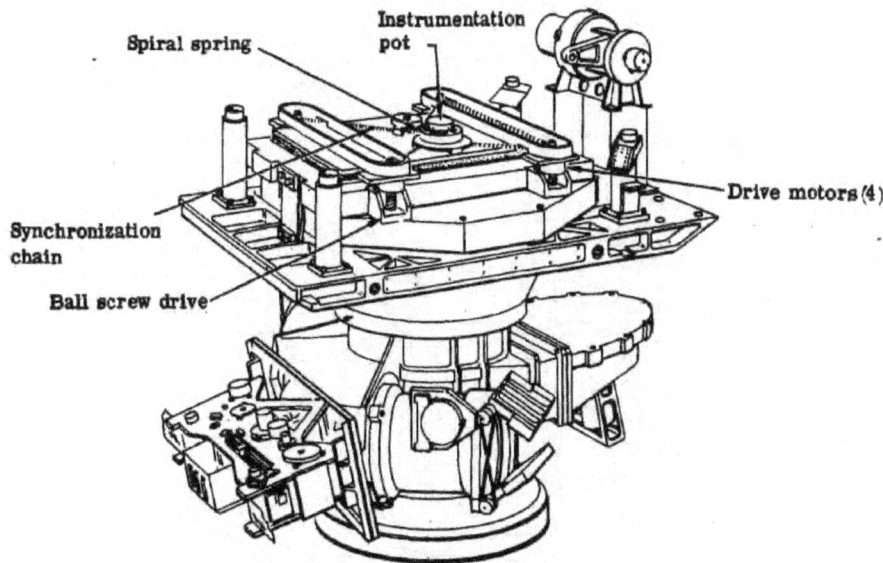

Fig. 2-13 — Pressure plate and platen press arrangement

Terrain Film Transport

The terrain film transport was mounted to the base structure by means of four mounting posts, as shown in Fig. 2-15. This arrangement allowed the platen press and pressure plate, which was mounted to the upper cone and cell assembly, to operate through the excursions necessary for film clamping. The function of the terrain transport was to continually meter film from the terrain film supply, intermittently frame the required film to the terrain pressure plate and platen press area, and to guide the exposed film to the exit chutes for retrieval by the takeup. The terrain transport film flow is shown in Fig. 2-16.

During operation, film was pulled off the supply and fed into the takeup at a continuous rate determined by the particular framing rate selected. The transport was the mechanism that provided this function. The metering roller turned continuously at the selected rate determined by the required framing cycle. The indexing roller turned at 4/3 the angular rate (of the metering roller) for 3/4 of the selected framing cycle period. Thus, the average velocity of the indexing roller over one complete cycle was nominally equal to the velocity of the metering roller. The shuttle mechanism provided the means for storing the film, fed by the metering roller, during the 1/4-cycle dwell period of the indexing roller. During the remaining 3/4 of a framing cycle the shuttle reversed direction and paid out the stored film while the indexing roller was rotating at 4/3 the metering roller angular velocity. Thus it can be seen that the shuttle was simply a mechanical differential device whose output is the algebraic sum of two inputs. The two inputs were the film passed by the metering and indexing rollers. The dwell portion of the cycle was provided so that the film was completely stopped during the time the platen press clamped the film to the reseau during the photographic exposure.

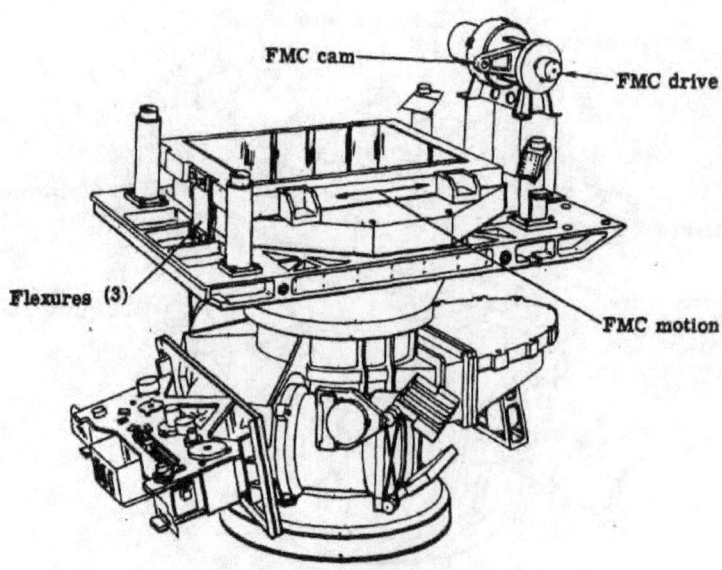

Fig. 2-14 — FMC drive arrangement

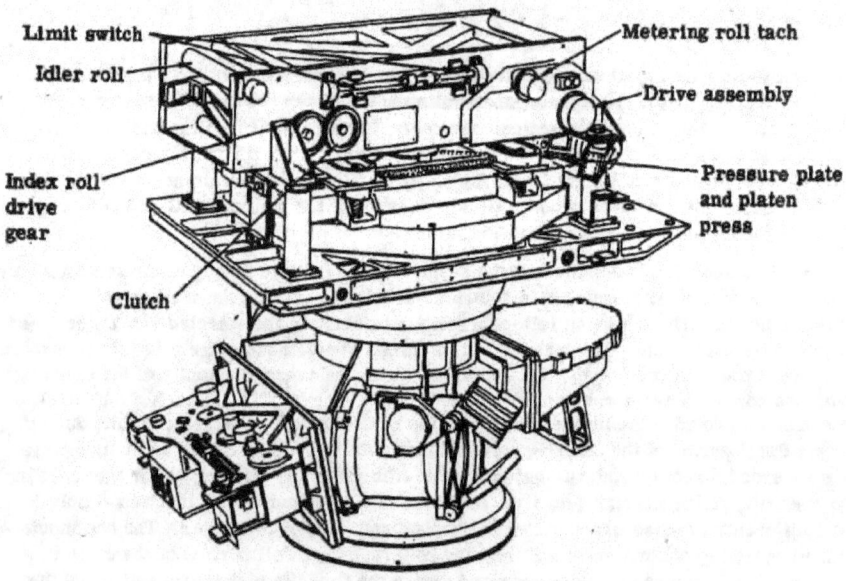

Fig. 2-15 — Terrain film transport

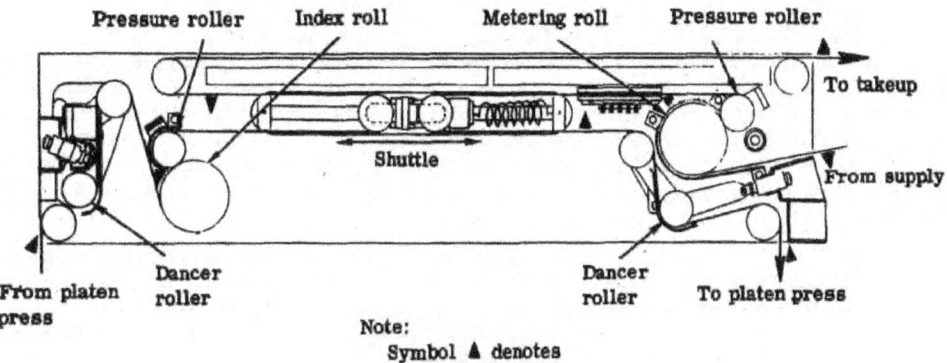

Fig. 2-16 — Terrain transport film flow

Both metering and indexing rollers were made such that their diameters were nominally equal. However it was not possible to manufacture two rollers with exactly the same diameters. Therefore it can be seen that over many cycles of operation the roller having the larger diameter would have transported more film into the shuttle than the roller with the smaller diameter. If nothing was done to alter this condition then the shuttle would slowly drift in a direction toward the roller having the smaller diameter. Since space limitations prevented the necessary total shuttle travel to accommodate this possible condition, other means had to be devised. Such means were provided by driving the indexing roller at slightly higher or lower velocities than nominal. The selection of the sign of the value of the velocity differential was controlled by limit switches actuated by the shuttle. The limit switches controlled the operation of a clutch mechanism that determined which of the two gear ratios was to provide the velocity difference value.

Frame Advance and Mechanical Operation

The terrain film transport advanced approximately 19 inches (a frame) of 9.5-inch film through the pressure plate and platen press subsystem at a variable interval of 7.789 to 87.38 seconds per frame. The film was transported during the platen unlock segment of the timing cycle.

Other Rollers

Film was guided from input to format to output in the terrain transport by idler rollers. The alignment of some of the rollers could be adjusted by end-bearing eccentrics.

Spring-loaded dancer rollers were necessary to provide compliance where the film left the transport to the platen press and where the film entered the transport from the platen press.

Stellar Film Transport and Platen Press

Although the stellar film transport and platen press were separate subassemblies, they were mated during the subassembly operations, and were functionally integrated. The combined transport and platen press assembly mounted to the mapping camera center structure as shown in Fig. 2-17. The film flow through the stellar transport is shown in Fig. 2-18. Film from the stellar supply was fed through the metering-pressure roller arrangement and through the shuttle. The stellar transport and platen press operation was synchronized to corresponding terrain camera operations.

The stellar film transport subsystem accepted film from the stellar supply assembly, allowed the proper length of film to be fed into the stellar platen press subsystem, and stored exposed film for retrieval by the takeup subsystem. The stellar film transport framing cycles were synchronized with the framing cycles of the terrain film transport. Mechanical operation of the stellar film transport was essentially the same as the terrain film transport and is not repeated here except where differences apply.

Frame Advance and Mechanical Operation

Though the configuration or packaging is different, the stellar film transport employed the same film advance approach as used in the terrain film transport. The metering and index rollers were 2 inches in diameter. With 1.72 revolutions per frame, these rollers advanced the film 10.80 ±0.1 inches. The frame interval was approximately 7.0 to 80.0 seconds, coinciding with the terrain film transport. The stellar transport, with identification of the shuttle and rollers, is shown in Fig. 2-18. The duplex clutch selecting one of two gear ratios drove the metering roller in the stellar film transport in the same manner that the clutch in the terrain film transport drove the index roller.

The stellar platen press was housed within the stellar film transport to clamp the 70-mm film for exposure by the stellar camera. Because the stellar camera employed two shutter and lens subsystems at right angles to each other, it follows that the stellar platen press must clamp the film when it is at right angles in the stellar film transport. In addition, the data block for the exposure was centered between the images on the 70-mm film format and thus required its own platen. The data block platen was in between the two stellar platens, 45 degrees from each of them.

The platen press mechanism consisted of three pressure plates, all of which were covered with a foam latex pad where they came into contact with the film. Two of the pressure plates were the main or stellar lens plates driven by one of two cams. The other pressure plate, for the data block, was a narrow strip plate which was driven by the second cam. Both cams were mounted on a common shaft and operated by one of two motors through a pinion gear and a sector gear. Because the index roller was stopped at the platen clamp command, the film could have been scratched if all three platens clamped simultaneously. The pressure plates were clamped in the following sequence: the pressure plate closest in the film path to the index roller clamped first, followed by the data block pressure plate, which in turn was followed by the platen nearest the metering roller. The sequence was determined by the configuration and phasing of the cams on the shaft.

MCS HISTORY

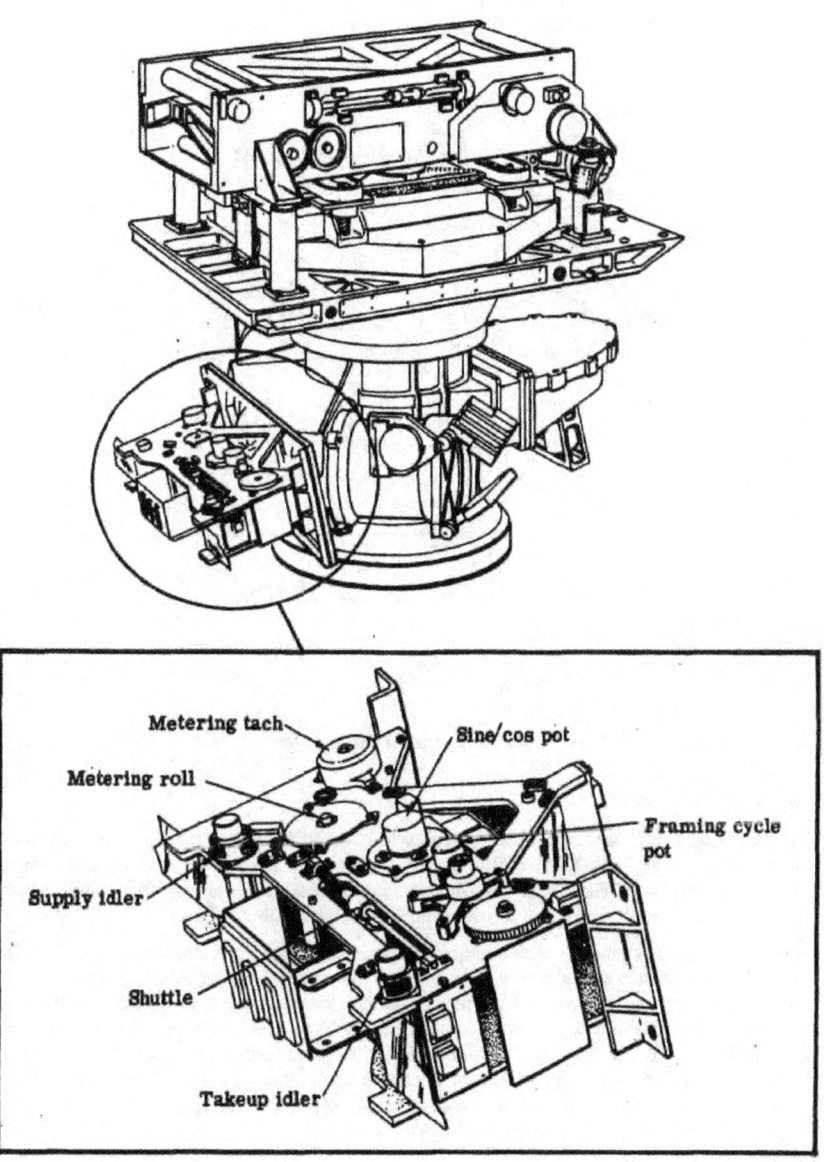

Fig. 2-17 — Stellar camera film transport

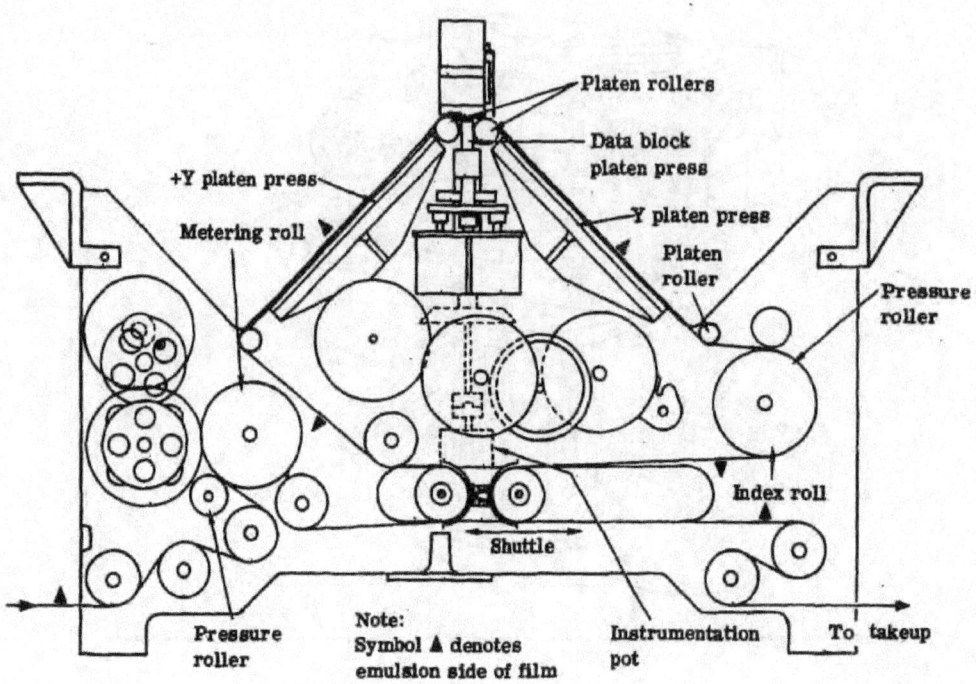

Fig. 2-18 — Stellar transport film flow

Terrain Film Supply Subsystem

The terrain film supply assembly was mounted to the APSA below, behind, and in alignment (as viewed from the +z axis) with the terrain film transport assembly; it provided approximately 6,300 feet of 9.5-inch-wide UUTB film to the terrain instrument. The terrain film supply assembly consisted of a film spool, a sensor arm, a brake, and associated electronics.

The supply electronics was a voltage-controlled current source. A sensor arm measured the radius of the spool and a potentiometer connected to the sensor converted its position to an analog signal. This signal was converted to current which flowed through the motor and provided the torque necessary to maintain 5 pounds of tension on the film at any radius. The current was measurable at an analog data point at the MC/SBA (Satellite Basic Assembly) interface. The brake prevented off-spooling when the system was shut down. Upon application of power (Operate ON and A mode), the brake was released and tension was applied to the film.

Stellar Film Supply Subsystem

The stellar film supply assembly was located along the +y side of the APSA at approximately the same level as the terrain film transport. The stellar film supply assembly provided 70-mm film to the stellar instrument, and accommodated approximately 4,000 feet of UTB film.

Operation of this subsystem was similar to that of the terrain supply. The tension applied during normal and A mode operation was nominally 2 pounds. The stellar supply current was measurable at an analog telemetry point at the MC/SBA interface.

Stellar and Terrain Takeup Subsystem

The stellar and terrain takeup assembly was mounted inside the satellite reentry vehicle (SRV) for the purpose of retrieving the exposed film from both cameras for eventual recovery. The takeup assembly was a terrain spool and a stellar spool, each identical to the corresponding supply spool. Each spool had a sensor arm, antibackup device, and electronics. Operation of each film takeup was essentially identical. The prime difference between the subsystems other than film and spool size was that the spools rotated about the same axis, but in opposite directions. Since the width of the film and the radii of the two systems varied, the takeup torque required by each was different.

For A-mode operation, the antibackup mechanisms were released and tension was applied to the terrain and stellar films.

Terrain Thermal Shutter Subsystem

The terrain thermal shutter assembly was mounted to the APSA directly below the terrain lens at the interface with the main thermal shroud. Its purpose was to maintain the integrity of the thermal enclosure surrounding the camera at all times except during a photographic exposure. This reduced the power requirements for maintaining the proper temperature of the critical lens elements. An "emergency open" command opened the shutter in the event of failure in both the primary and redundant electronics. With the terrain thermal shutter open, the terrain lens saw a field of view of 74 by 41 degrees and provided the basic format on the 9.5-inch film.

Mechanical Operation

The terrain thermal shutter had two overlapping blades, each of which contained 30 layers of aluminized mylar on the outside surface for protection against thermal radiation to the outside. The inside surface of the blades was covered with magnesium sheets which serve to evenly distribute heat from heaters sandwiched between the inside blade surface and the magnesium sheets. Each blade had a separately controlled heat zone.

Each blade was pinned to a hinge which was driven by a worm gear train. In the event of a servo malfunction, a mechanical limit stop was provided to prevent the door from hitting a portion of the frame or thermal shield when opening or closing.

Chutes

The chute assemblies provided lighttight passage of stellar and terrain films from the two supply assemblies to their respective film transport assemblies and to the respective takeup spools. A gas flow, provided from the APSA pressure makeup system (PMS) maintained a pre-determined pressure to preclude corona discharge marking.

Light Baffle

A light baffle was mounted externally to each lens of the stellar camera. The baffles absorbed stray light incident from outside the specified angular field of view during all modes of operation. In addition, the baffles contained over-illumination photo cells to temporarily inhibit the associated stellar shutter should the sun come into its field of view. Each baffle had a lighttight safety shutter which permanently capped the lens aperture upon command should a stellar shutter fail to open.

Main Instrument System Electronics Assembly

The Main Instrument System Electronics Assembly (MISEA) contained the electronics which initiated and synchronized camera operations, and provided conditioned instrumentation outputs. All command and telemetry interface connections to the SV or the test and checkout (T&C) console were made at the outboard side of the MISEA. Internal connections to all camera assemblies through the system harness were made on the top surface of the MISEA.

Electrical Distribution and Power

The Electrical Distribution and Power Assembly (EDAP) received dc power from the SV and distributed the power to the various MC subsystems. Distributed dc power consisted of + and −5 volts regulated, + and −15 volts regulated, +38 volts regulated and unregulated, and all associated ground returns. Eventual chassis ground was made external to the camera system.

System Harness

The entire system was electrically interconnected by a system harness. Signal wiring was separated from power wiring.

Physical Orientation

The MC and structure, comprising the MCS, were all referenced to one common coordinate base. This arrangement provided uniformity and consistency among program personnel and supporting documentation. Fig. 2-19 shows the MC system components in two views for the final configuration in normal orientation. In the normal orientation, all three axes (x, y, and z) were mutually perpendicular. The x and y axes were horizontal and the +z axis was vertical down. The front of the system was at the +x end, the rear at −x. Although some dollies and handling equipment used with the system permitted the system to be rotated about the y axis, all references in this book are made to the normal orientation unless otherwise noted.

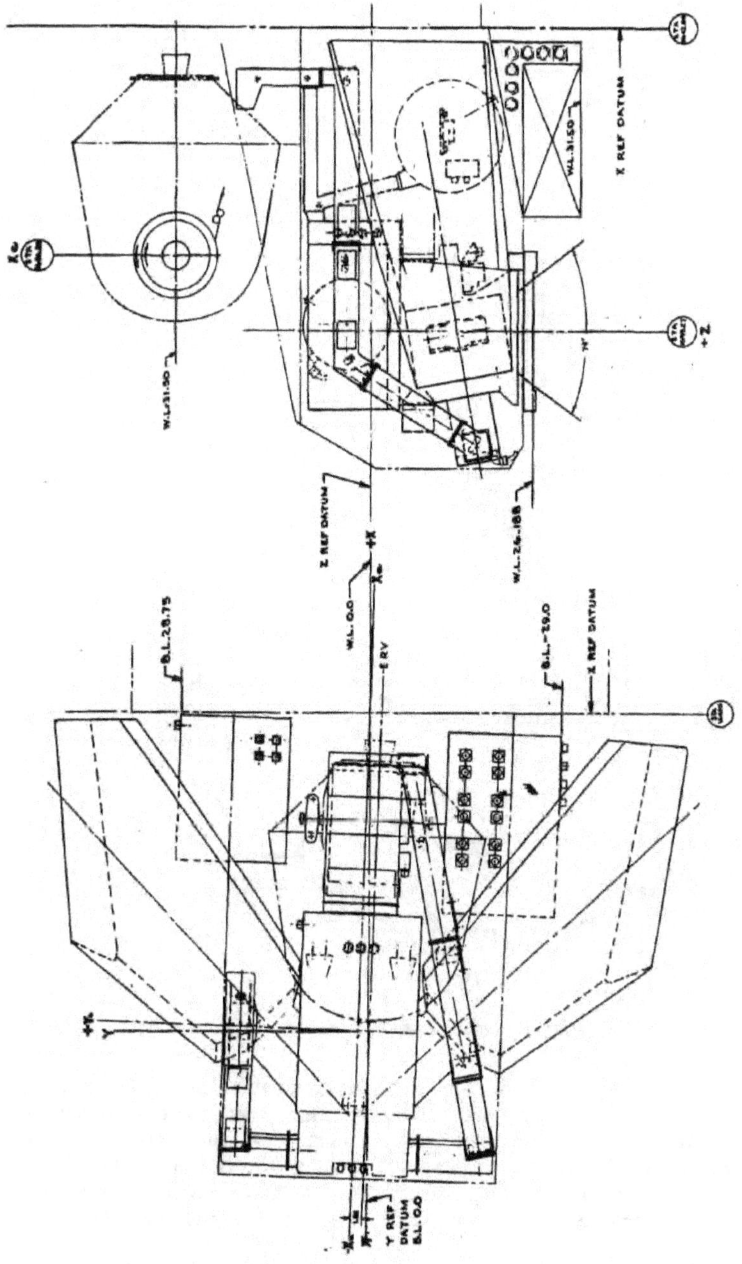

Fig. 2-19 — APSA and MC coordinates and assembly identification

TEST AND INTEGRATION

The sequence of events from MC subassembly buildup and test at the Itek facility through integration of the Mapping Camera Module and orbital mission is shown in Fig. 2-20.

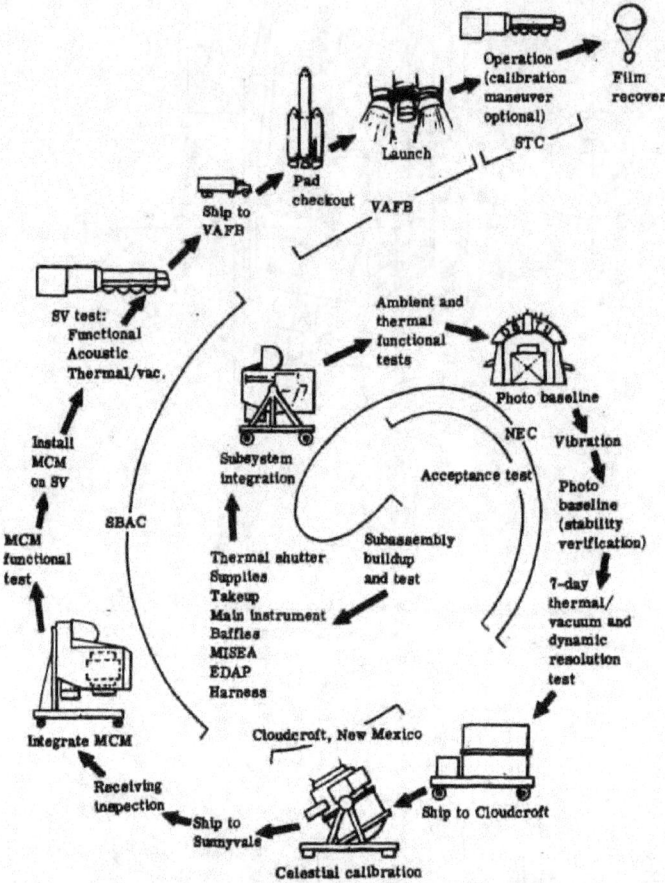

Fig. 2-20 — Mapping camera integration and operational flow

All subassembly integration operations and acceptance testing of the MC were conducted at Itek. Acceptance testing included functional baseline testing under ambient conditions, and a photographic baseline test in the Itek distortion and boresight test unit (DBTU) under controlled temperature (Fig. 2-21). The DBTU, comprised of a rotary camera mount, photo theodolite, special cassettes, and a multicollimator array, provided the necessary image sources for both stellar and terrain cameras to determine lens distortion, stability, and resolution.

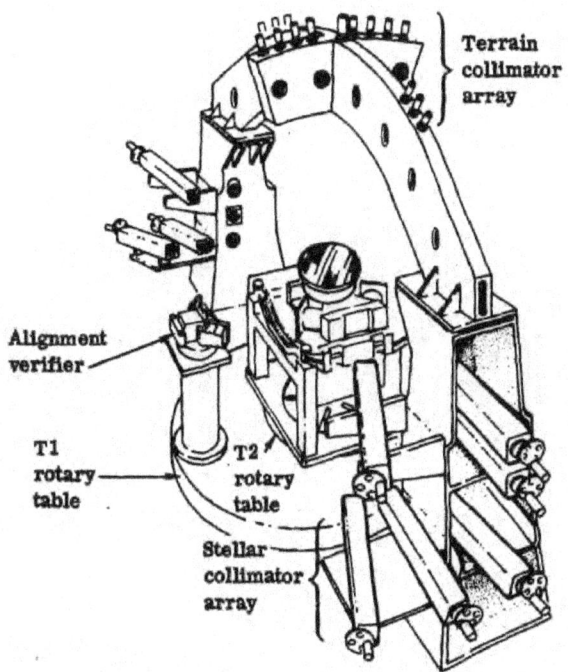

Fig. 2-21 — Distortion boresight test unit (DBTU)

Delivery of the twelve systems from Itek to the integrating contractor was established by contract as required to meet launch dates as scheduled at the time of contractual agreement. As the program progressed and the HEXAGON mission duration capability increased, launch schedules were relaxed from three, to two systems each year, with the last three Mapping Camera Systems being launched on approximate 1-year centers. However, in the interest of efficiency and economy, systems were delivered from Itek as contractually scheduled, the last system to fly, mission 1216, being delivered on 15 January 1977, 3 years prior to launch.*

Preflight Calibration

Following formal customer acceptance of a system at Itek, the "main instrument" was installed in an exotic shipping container and transported via Air Force C-141 aircraft from Hanscom Field, Massachusetts, to Holloman AFB, New Mexico, and then transported via truck to an Air Force celestial calibration site in Cloudcroft (CC), New Mexico. The CC calibration site is located at an altitude of 9,200 feet above sea level, at a latitude of 32° 58' 46" N, and a longitude of 105° 43' 58" W. As seen in the aerial photograph (Fig. 2-22) the site is completely surrounded by pine forest. This minimizes surface light which would be objectionable, particularly during periods of full moon. The 9,200-foot altitude places it above most of the haze, and there are no urban lights on the horizon to preclude the capture of stars at low elevations.

---

*Upon delivery of the last system, program management was transferred to the Itek West Coast Facility for program completion, approximately 4 years.

Fig. 2-22 — Air Force Celestial Calibration Site Cloudcroft, New Mexico

MCS HISTORY

In addition to the natural assets of the site, two major factors influenced the choice of CC as the celestial calibration site for this program. First, the Air Force Electro Optical Facility at the CC site had been in operation since 1965, thus providing an "established base" in this otherwise remote area. Secondly, another Air Force observatory, known as the AN/FSR 2 site and located about 1/4 mile from the Electro Optical Facility, was not being used at the time. Following deactivation from its original purpose, this site had been used during the 1966 to 1969 time frame in the calibration of two satellite cameras, the Dual Improved Stellar Index Camera (DISIC), CORONA program, and the Astro Positioning Terrain Camera (APTC), GAMBIT program. Only minor modifications to the building had been required to accommodate calibration of these two cameras, since they were relatively small and could be handled easily by two men. But calibrating the MCS presented a more complex requirement by several orders of magnitude.

To provide calibration with the required 2-micrometer accuracy, a critical choice had to be made between operating the camera in air or in vacuum. Operation in air would eliminate errors associated with bending and thermal gradients in the viewports of a vacuum chamber. However, the cameras were focused for optimum imaging in vacuum, so the star images in air would be much larger than in vacuum, and consequently mensuration errors would be larger. In addition, the camera temperature would be more difficult to control and film conformity to the reseau surfaces would be more difficult to achieve. On the other hand, sensitivity analyses indicated that the chamber viewports could be accurately calibrated, so the choice was made in favor of camera calibration in vacuum.

This decision helped to define the magnitude of the support equipment design and manufacturing tasks, and the building modifications. Design and development included the vacuum chamber with acceptable three-optical-window viewports, the vacuum system, and handling/support equipment. Extensive modifications to the building itself were required to accommodate the installation of the system into the vacuum chamber, hoisting it to the dome, etc. The facility already contained a precise synchronous drive equatorial mount, but it had not been used for some time and would require refurbishment and checkout.

The SPO issued a contract to Itek for the design and manufacture of the vacuum chamber plus handling and test equipment. Concurrently arrangements were made through Air Force channels for nearby Holloman Air Force Base to handle the building and dome superstructure modifications. To play an important role in this effort was the Electro Optical Site resident manager, Mr. Edmund Tyson, who had participated personally in building the EO observatory in the early 1960's. Mr. Tyson was also familiar with the calibration requirements in general, as he had provided support to the calibration activities previously mentioned.

Along with the fabrication of hardware, site modifications and handling procedures, the security aspect of this operation had to be considered. In a message addressed to the NRO, DMA, and Itek on 6 January 1972, Subject "HEXAGON security procedures and ground rules for Cloudcroft," SAFSP provided the following direction:

"1. It is considered essential to protect the following information at Cloudcroft.

    A. The fact that a camera is being calibrated at the facility.

    B. The identity of Itek.

    C. The fact that TOPOCOM or other mapping or intelligence agencies are involved.

    D. The fact that SAFSP is involved.

2. Therefore the following procedures should be adhered to:

   A. Under no circumstances should room reservations be identified with the person's organization. In practice, it is desirable that reservations be made by SP.

   B. No credit card identifying the corporation or organization will be used in the CC area.

   C. All phone calls to Itek or TOPOCOM will be made on commercial lines. Autovon may be used to call the Pentagon or Los Angeles AFS.

   D. Air Force uniforms may be worn at the site. If pressed, Air Force personnel may identify their home base as SAMSO. Under no circumstances should army uniforms be worn or army personnel identify themselves with an army organization."*

Summarizing several months of intensive effort in the building modification and production and installation of the vacuum chamber and other new equipment, suffice it to say that the site was ready to receive the first system on schedule. A schematic of the completed facility is shown in Fig. 2-23. A more detailed view of the vacuum chamber and supporting structure is shown in a sketch (Fig. 2-24).

As one can well imagine, there were some rough spots associated with getting the first system through. The second system went considerably smoother. By the third system, equipment had been wrung out, organizational responsibilities had been clearly resolved, and procedures had been refined.

The routine for the remainder of the program went about as follows. The organizations actually involved in the CC task were the SPO, DMAHTC, and Itek. Upon being advised of the shipment date from Itek in Lexington, the DMA team (comprised of four members) and the Itek team (comprised of four members from the Itek West Coast facility and two from Lexington) would get their gear together and make travel plans. The Itek team would arrive at CC two days prior to system arrival to make ready all test equipment. An SPO officer who had participated in the system buy-off at Lexington would serve as courier on the flight from Hanscom AFB to Holloman AFB and would then remain at CC to supervise the operation. Members of the Itek team would meet the aircraft at Holloman and assist in transporting the camera up the hill and into the site.

Baring complications, personnel would have the system ready for stellar photography in two days—it was considered a "no-no" to miss a good clear night. The DMA team, having arrived on the same day as the system, would have their processing chemicals and equipment ready to process the first stellar photography collected. The stellar photography would commence as soon as it was dark—this would vary from about 1800 to 2200 depending upon the time of the year. The summer months offered only a few hours of operation as it would again be too light to operate by 0400. The collection routine was varied somewhat throughout the program to cope with special requirements or adversities, but basically the operational flow followed the sequence shown in Fig. 2-25.

---

*These were good initial groundrules, but as usual, groundrules and procedures laid down in advance are never all inclusive. As the program progressed, on the spot decisions had to be made as occasions presented themselves, e.g., how to protect security in case of sudden illness, injury, preparation and submission of insurance claims, etc.

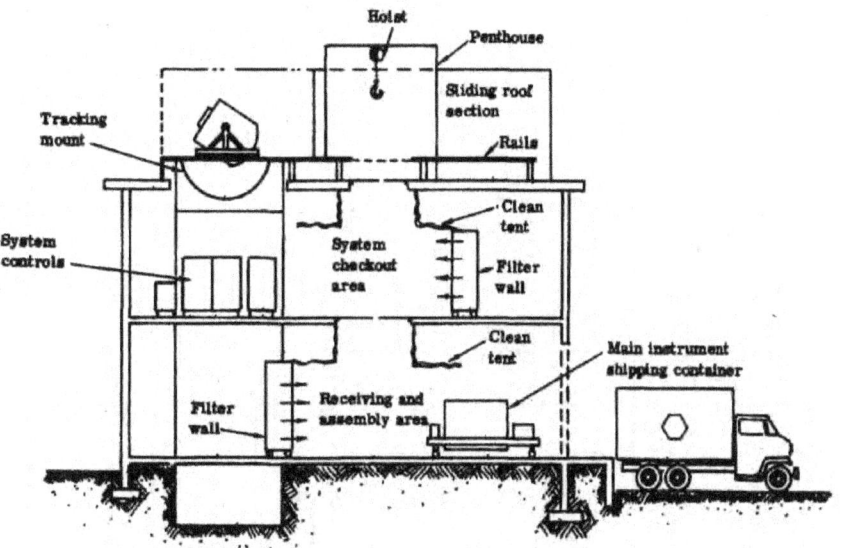

Fig. 2-23 — Cloudcroft Facility

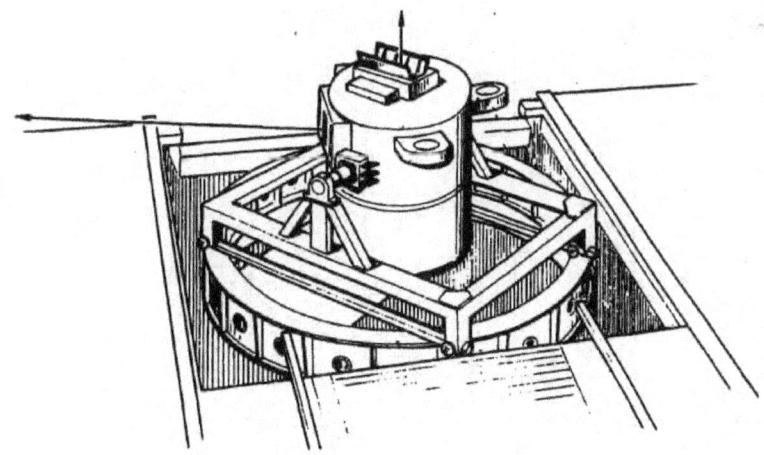

Fig. 2-24 — Cloudcroft calibration enclosure

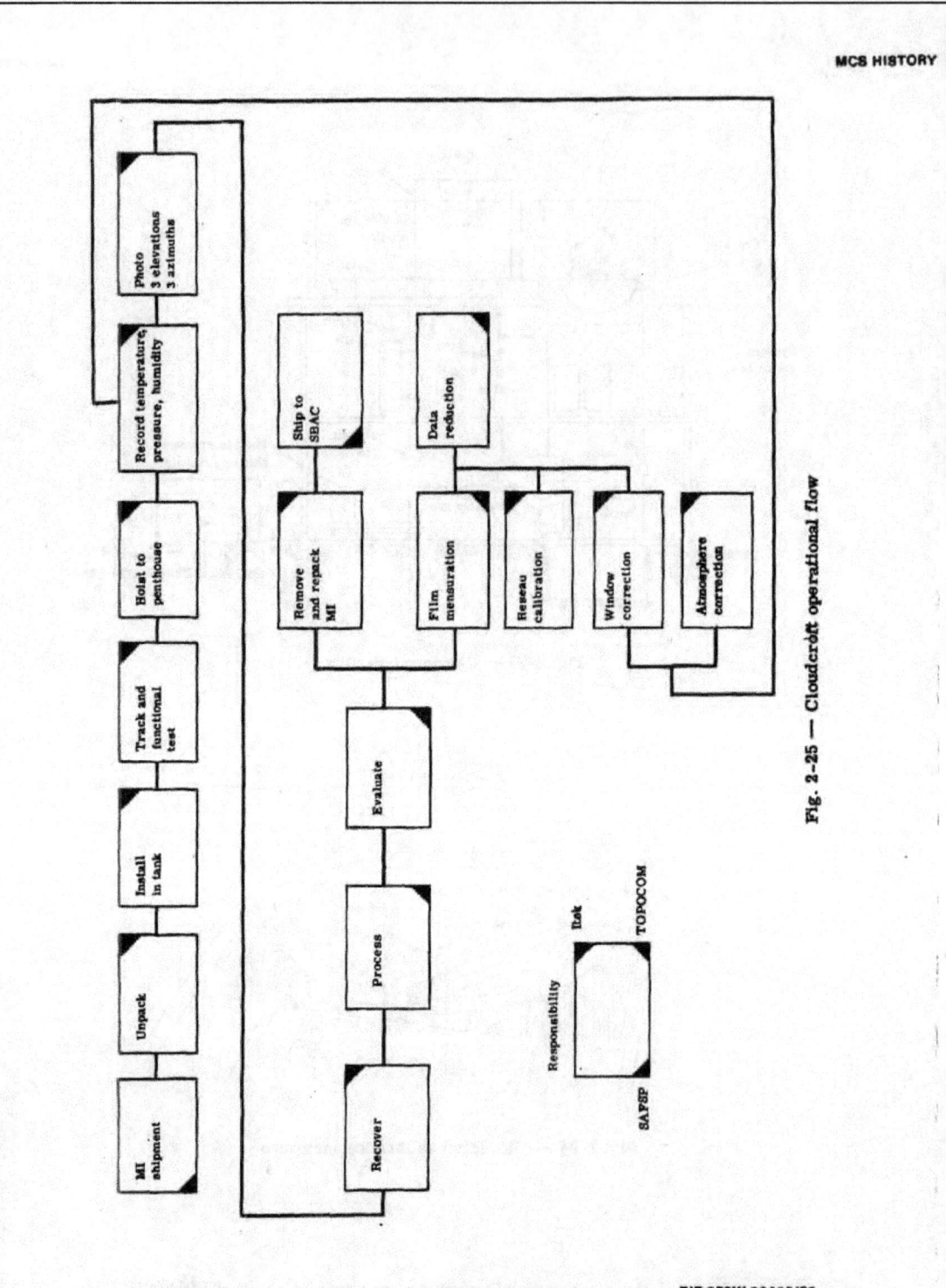

Fig. 2-25 — Cloudcroft operational flow

MCS HISTORY

A considerable amount of ancillary data also had to be obtained. Meterological conditions were monitored continuously. Vacuum tank temperature and pressure were monitored—exposure times were recorded. For the last five units at Cloudcroft, DMA had a team nearby for the purpose of observing known stars with a Wild T-4 theodolite in an effort to obtain a better atmospheric refraction model.

The film was processed immediately upon the conclusion of a night's photography. It was then inspected for critical defects, i.e., critical in terms of usability for calibration. Star imagery had to be sufficient in quality, quantity, and distribution throughout the format. Fiducials, crucial not only to calibration but also to production use of flight material, were checked for proper exposure. Reseau intersections, used for film distortion corrections, were also checked for proper exposure. Format background density, a key element in determining proper exposure, was checked. Other characteristics, such as cosmetic and physical damage which could render a calibration effort useless, were looked for.

When the inspector was satisfied that he had sufficient quantity and quality of exposures, it was time for people to pack up and go their separate ways. The DMA team returned to Washington to begin the calibration task, (addressed in Section 5), the Itek crew removed the MC from the vacuum chamber, reinstalled it in the shipping container, and assisted in transporting it down the hill to Holloman to meet an awaiting C-141 for transport to the Naval Air Station, Moffett Field, Sunnyvale, California. The SPO officer again would ride "shotgun" on the flight to Moffett. The reusable shipping container (Fig. 2-26) included insulation, refrigeration units and heaters to maintain the contents at a relatively constant temperature (72°F ±5°), and a suspended inner frame to protect and isolate the contents from the harmful effects of rough or abnormal handling or from the effects of vibration encountered in normal transit. From the calibration phase forward throughout ground integration, test, and flight, a system was never allowed to be outside specified temperature and humidity limits.

Two aspects of the CC operation are worthy of special mention. In the early program planning, three weeks had been allocated in the schedule flow for this phase; however, as the result of good management and a super team effort under the direction of Stephan Herman of Itek, the average time required was only 7 days. Secondly, and of unestimatable value, there was no equipment damage or injury to personnel during this 6-year period of potentially hazardous operations.

<u>System Integration and Test</u>

When the MCS subsystems arrived at the Lockheed facility they were put through various subassembly testing and buildup operations that were designed to produce flight-ready systems having a high degree of reliability. This activity was divided into three major phases as follows:

- Phase I   Receiving to APSA mate
- Phase II   APSA mate to SV mate
- Phase III   SV mate to final use.

All testing and operations conducted during each phase was certified by Quality Assurance personnel. Itek Field Engineering was responsible for evaluating and dispositioning any discrepancies or anomalies that resulted from the performance of the operations. A brief description of each phase follows.

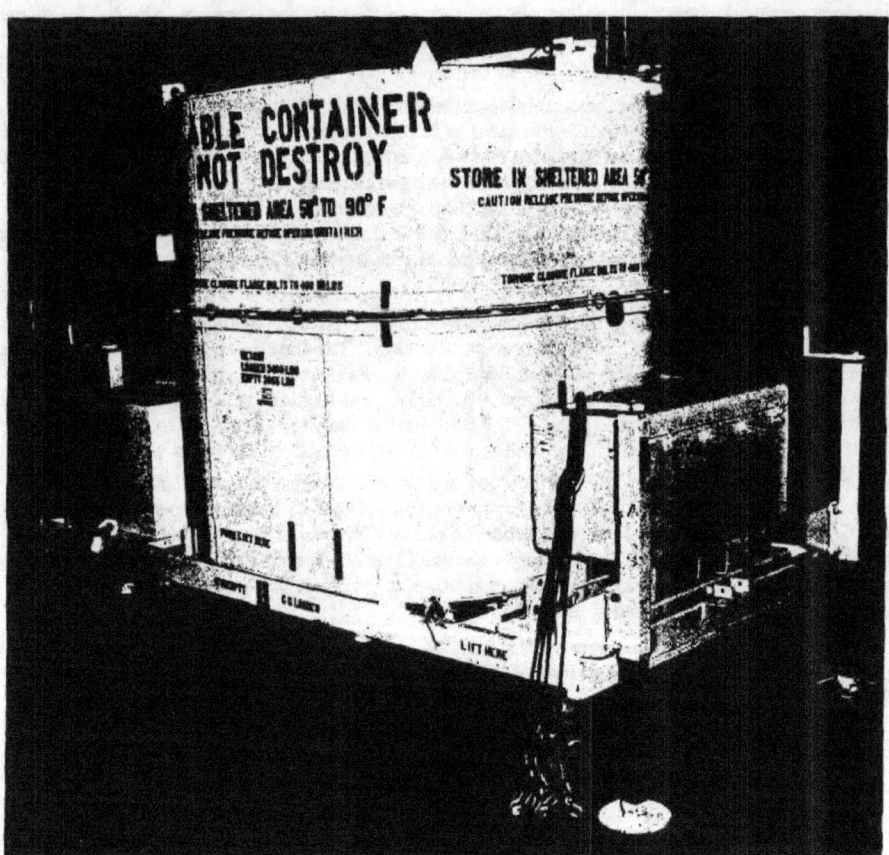

Fig. 2-26 — Main instrument shipping container

Phase I — Receiving to APSA Mate

The mapping cameras arrived at Lockheed disassembled into subsystems convenient for shipping. Prior to assembly, incoming acceptance testing was performed on the stellar supply assembly, terrain supply assembly, takeup assembly, viewport baffle assemblies, and the terrain thermal shutter assembly. The film path chute assemblies were tested for being light tight.

The systems were then assembled on the A&M (assembly and maintenance) dolly and run through a receiving functional test. Data from this test served as a baseline to determine if variations in performance resulted from shipment. This test checked interface resistance, heater zones, thermal sensors, normal and redundant electronics, ascent mode electronics, calibrate mode electronics, redundant electronics, backup mode electronics, material change detector, shuttle timing, sun sensor inhibits, light tightness, and film formats.

Phase II — APSA Mate to SV Mate

After the system successfully passed all tests in Phase I, it was removed from the A&M dolly and reassembled into the APSA structure. This was the final installation of the system and great care was used to assure proper mounting and alignment of component parts, such as film path alignment, thermal shields, light shields, etc. When system performance was verified, the film path interfaces were pinned in position. After the APSA structure assembly was complete, the recovery vehicle was removed for final buildup and reinstalled. A pressure makeup system and Doppler beacon were installed; the system then received a light leak test and a complete functional test. The functional test was essentially a repeat of the receiving functional test described in Phase I. After successfully completing all tests, the stellar and terrain supplies were loaded in preparation for mate to the SV. Fig. 2-27 shows the MCS in mated configuration with the SV in the high-bay system test area.

Phase III — SV Mate to Final Use

After mating with the SV, the mapping camera module (MCM) was tested in conjunction with the other systems on the SV. All commands and the monitoring of system responses was under the control of the Automatic Data Processing and Control System (ADPACS) computer. After the completion of a complete functional test, the system was subjected to an acoustic test, a 7-day thermal vacuum test, and an A-2 vacuum test (Fig. 2-28). Four systems also received electromagnetic interference testing. Between each of these tests a functional test was performed to verify that the MCM successfully survived the environments to which it had been subjected.

The SV was then moved to the vertical position (see Figs. 2-29 and 2-30) where flight film loads were installed and final use preparations were performed, including installation of safety wire, shorting of limit switches, and pressurizing the pressure makeup system. Following the final shipping preparation run, the SV was then installed in the transporter and shipped to the base, completing Phase III.

(It should be noted that many modifications, retrofits, and repairs had to be performed by the field personnel during all three phases. These were usually the result of one or more of the following: (1) a generic failure found in some component requiring the replacement of all like components; (2) a defect revealed due to environmental testing, time duration, or usage time; and (3) requested changes in system requirements, such as film types, change in overlap, and extension on end use dates affecting limited life items. Also, there appears to be an inherent increase in sensitivity to system performance deviations by personnel who are the nearest to the end-use operations.)

The mapping camera modules, mounted on and electrically connected to the Satellite Vehicle (SV), were transported to the base approximately 13 days prior to launch.

On arrival at the pad, temperature, humidity, and G-loading recorders were removed from the transport trailer and evaluated to assure that all conditions were in tolerance during the trip from Sunnyvale.

The cover was then removed and the transporter positioned at the base on the gantry. Immediately, an extension cable was connected to the APSA from the umbilical to supply heater power to the Mapping Camera and to monitor internal temperature from auxiliary sensors.

During SV hoisting, the MC temperatures were recorded on strip recorders. After mating to the booster vehicle, the extension cable was removed and ground heater power was supplied directly from the umbilical.

During the entire period of time at the base, the temperature was continually monitored via the auxiliary sensors, or by telemetry when vehicle power was on.

On L-11 day, the SV functional test was conducted during which the mapping cameras were exercised a total of 7 frames for a confidence test to assure no damage during transport.

On L-7 day, the Aerospace Vehicle systems tests were performed, which included a simulated launch preps, countdown, and launch. Mapping cameras were not operated, but A mode (ascent mode) torquer power was applied during terminal count, simulating conditions at and after launch.

On L-5 day, mapping cameras were advanced two frames to verify integrity.

On L-1 day, two more frames were run, this being the final operation prior to launch.

On L-0 day, Itek personnel continuously participated in the countdown, particularly monitoring temperatures after the environmental enclosure was removed. At T-30 minutes, A mode power was applied to the supply and takeup torquers to assure no film slack loops during launch and orbital insertion. A mode power also functioned to hold the terrain and stellar shutters in closed position.

After launch, monitor and control activities were transferred to the Satellite Test Center (STC) for orbital operations.

Thus, a total of 11 frames were advanced during base activity; the data collected was minutely analyzed and compared with previous runs to verify launch readiness.

Flight Readiness Assurance Program (FRAP)

Late in the program, in 1976 to be more precise, the last two systems and spare subassemblies were approaching the LOL (limited operating life), causing concern for the possible degeneration of acceptability of the hardware.

In order to maintain confidence in the integrity of the hardware, a system of automated checkout was developed which tracked trends in order to spot insidious deterioration. The system was called TAC (Trend Analysis Console), and consisted of a Hewlett Packard (HP) 1000 computer system comprised of two HP 2100 MX computers, tape deck, disc memory, control console with CRT readout, a printer, and plotter.

Great quantities of data were acquired and stored on tape, forming the basis for trend analysis by comparing present with previous data. All telemetry and hardline data points were recorded during short runs of the flight hardware. Any desired instrumentation points could be displayed on the CRT, permitting routine and diagnostic data analysis.

Analysis consisted of timing and amplitude measurements, (including rms, average, and standard deviation) plus PSD (power spectral density) analysis of tach and error signals in cases of dynamic problems.

TAC testing was repeated periodically on the MCM systems and all spare subassemblies.

As an example of TAC utilization, one significant possible problem concerned the suspicion that a particular error signal was exhibiting a trend upward in average value. Further analysis revealed that the changes were a result of hardware test configuration differences which were causing a shift in zero level as a result of ground loops in the test setup. Thus, what at first appeared to be a hardware problem finally was resolved as an Aerospace Ground Equipment (AGE) setup characteristic. Corrective action was taken, directly resulting from comprehensive TAC analysis.

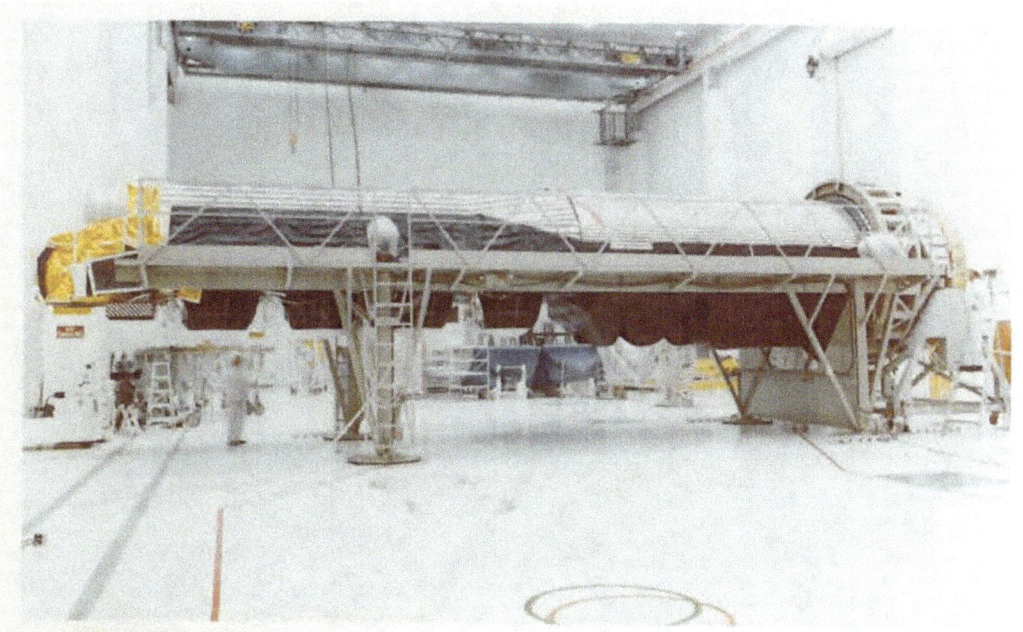

Fig. 2-27 — SV in high-bay test area

Fig. 2-28 — SV entering vacuum chamber

TOP SECRET/RUFF/GAMBIT/HEXAGON

BIF-059W-23422/82
Handle Via
BYEMAN/TALENT KEYHOLE
CONTROL SYSTEMS JOINTLY

Fig. 2-29 — SV going vertical

Fig. 2-30 — SV in position for MCS film loading and vertical tests

SECTION 3
MISSION SCENARIO

This section describes how MC&G requirements were generated and how they were fulfilled using the HEXAGON hardware/software and procedures.

OPERATIONAL SUMMARY

The major operational events of a HEXAGON mission (Fig. 3-1) are launch, orbit maintenance/payload operations, and RV recovery/SV deboost. It cannot be said that one event is more important than another, since total success depends upon proper operation of all events, each in the progression of events having its own spot in the limelight.

The sequence of launch events are:

0.0 sec	SRM ignition
0.2	Lift off
40.0	Transonic passage
54.0	Maximum dynamic pressure
113.9	Core I start burn
125.3	SRM separation
262.0	Core I shutdown and Core II start burn
262.7	Core I separation
276.0	Shroud separation
460.6	Core II shutdown
472.6	Core II separation (injection)

The solar arrays are deployed after SV stabilization on rev 1 with payload operations starting on rev 5. Orbit adjusts to correct period, altitude, and perigee location occur every 2 to 4 days. All control of the SV telemetry data is processed through the Air Force Satellite Control Facilities and associated remote tracking stations.

In addition to the normal operational events, the incorporation of the Mapping Camera on Missions 1205-1216 placed another maneuver requirement on the satellite vehicle to accomplish the MCS in-flight calibration photography. At a convenient time following completion of the mapping mission, the vehicle was programmed to position itself, usually by pitching down approximately 120 degrees, on the dark side of the orbit, so that the terrain and stellar cameras would all be pointing at a strong starfield for a series of simultaneous photographs.

The SV is pitched down to a specified angle for each RV ejection. After the last RV is ejected and all engineering requirements are satisfied, the SV is deboosted for ocean impact. (More details of the recovery sequence are given at the end of this section.)

DEVELOPMENT OF REQUIREMENTS

Requirements for imagery are based primarily upon programmed production with secondary consideration given to developing a worldwide data base. The mapping community, composed of the Defense Mapping Agency (DMA) and its associated production centers, the U.S. Geological Survey, and the Civil Application Committee of COMIREX (Committee on Imagery Requirements and Exploitation) define the collection requirements of the mapping photography prior to each mission. These requirements vary from long-term requirements in support of worldwide map production to short-term requirements of high priority in support of MC&G production programs connected with the Cruise Missile and Single Integrated Operational Plan (SIOP) target areas.

Specific overlap modes of camera operation are assigned depending upon the stringency of the accuracy requirement. Single-overlap, dual-overlap, or triple-overlap modes determine how many times a specific target area is included in successive frames, overlap being necessary to produce stereo photography from a single camera (see Fig. 3-2). If, for example, a map is required for the orientation of an island in a group of islands in the Pacific, a single-overlap mode of camera operation (mono) would satisfy the requirement. However, if the requirement is an aim point for ICBM's or Cruise Missiles, very precise positioning with vertical information would be acquired by triple overlap mode operation.

Area requirements for approved products are submitted to DMA/PR by the Services and U&S Commands annually in July. Following a validation process, the requirements are submitted to Director, DMA, and to Deputy Under Secretary of Defense (Policy Review) for approval. The results are published in the annual document "Department of Defense Mapping, Charting and Geodesy Area Requirements," the "Gray Book," issued in January. The "Gray Book" contains the total requirement for each product—regardless of whether or not the requirement has been satisfied.

Upon receipt of the requirements, DMA Production Centers determine the adequacy of existing products to satisfy them and develop a program of new production and maintenance. Resources are obtained through the POM* process. The approved production program is the basis for establishing requirements for source material. A summary of the results of this analysis and the planned production program is provided in the publication, "Defense Mapping Agency Requirements Status and Programs," otherwise known as the "assessment graphics," published annually in July following program development.

MISSION REQUIREMENTS VALIDATION

Defense Mapping Agency Validation

The announcement of a future KH-9 launch date starts the imagery requirements preparation machinery. The DMA collection plan is due to ICRS (Imagery Collection Requirements Subcommittee of COMIREX) 65 days prior to launch.

Collection requirements are requested by DMA from USGS (consolidates all Civil Applications Subcommittee members' requirements), Canada, United Kingdom, and Inter-American Geodetic Survey (IAGS). Production plans are requested from the DMA Production Centers. The Centers also submit imagery requirements for system calibrations.

---

*Program Objective Memorandum

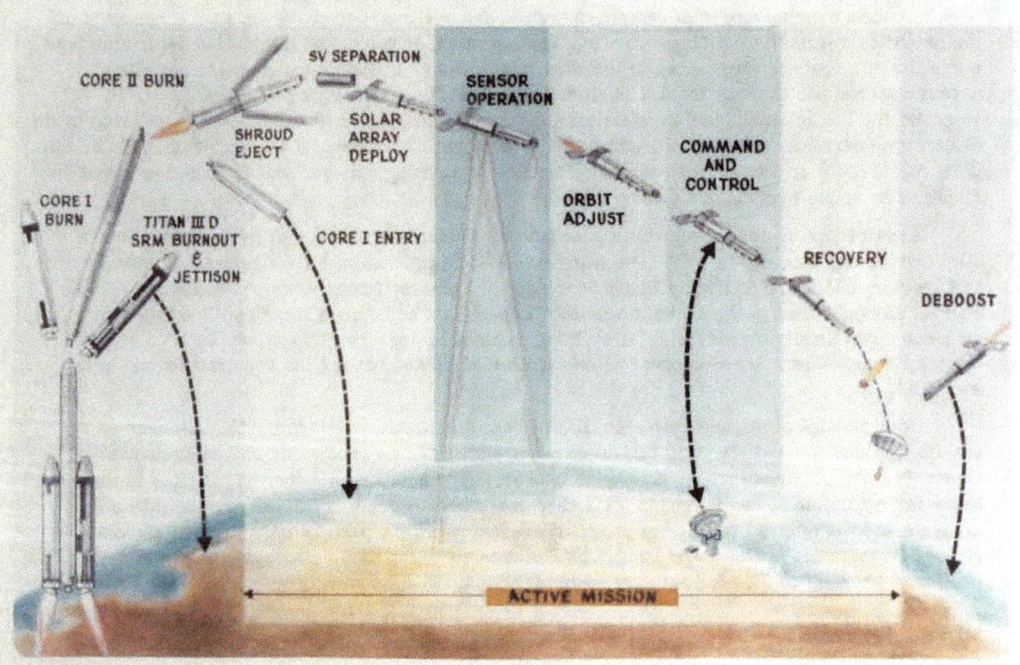

Fig. 3-1 — Operational events

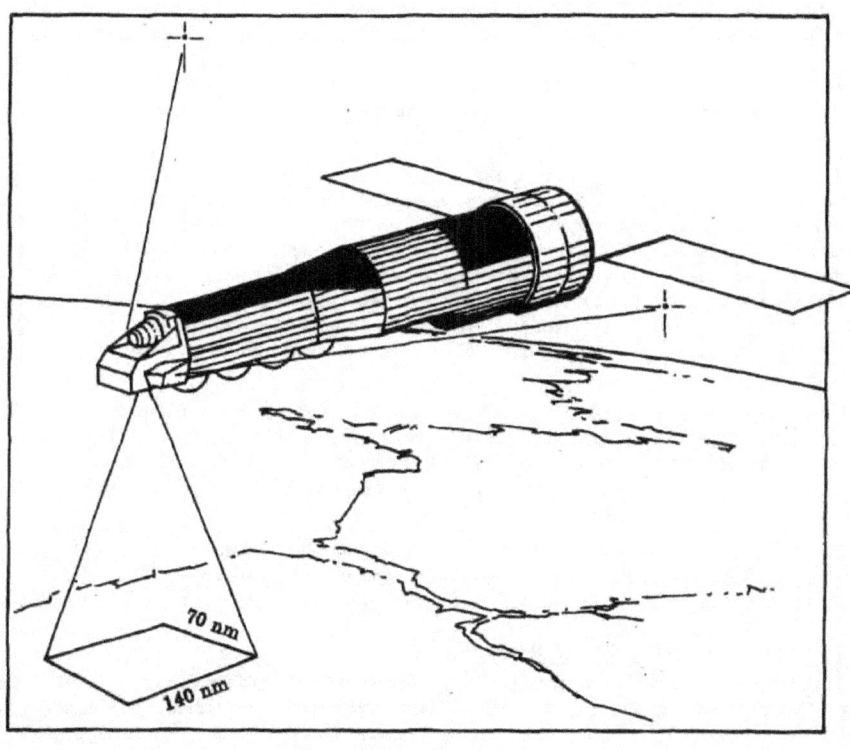

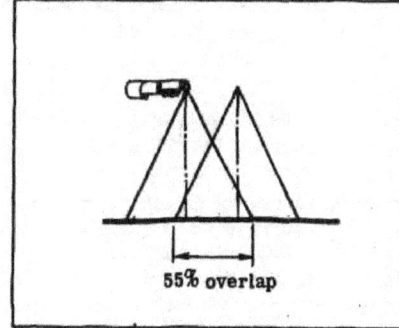

(a) Bilap photography

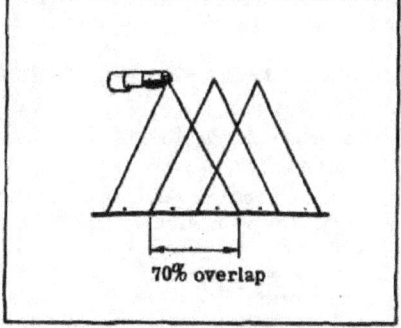

(b) Trilap photography

Fig. 3-2 — Mapping camera operations—92-nm altitude

One-hundred and fifty days prior to launch, 1:18,000,000 scale plots are prepared which contain the following information: DMA planned production programs by product and class (new production or revision), and collection requirements of civil agencies, UK, Canada, and IAGS. A second set of graphics is also provided, similar to the first, but from which have been removed all areas for which existing imagery is satisfactory in terms of mode (stereo/mono), currency, and cloud-freeness for each product. (Note: Civil Agency requirements are accepted as submitted.)

An initial review is made within HQ DMA. At this stage, any changes in area requirements or production plans are reflected by additions/deletions to the imagery requirements. Other sources of data are reviewed for application, including: HAP (Interagency High Altitude Photography program), U-2, other conventional aerial photography, map sources, KH-8, KH-11, and LANDSAT. Where data which meets the production requirements for particular products are available, the KH-9 plan is correspondingly reduced. Generally, the collection requirement is further modified by considering the time of year and duration of the mission, date of following launch, number of missions remaining in the KH-9 program, and intelligence requirements in the same areas.

Necessary changes are defined and 1:35,000,000 plots are then made which portray the requirements in the following ways: by type (topo/aero/hydro), mode (stereo/mono), priority (high/low), map scale (small, medium, large), and by category of product. The consolidated requirement is then reviewed. Following DMA staff-level review, the requirements are briefed to the Director for approval.

During the requirements definition process, a prioritization scheme is developed by broad area/product. The priorities are also presented for approval by the Director. The priorities are used to guide the partitioning of the overall requirement into sub-areas and the allocation of film.

Director Central Intelligence (DCI) Validation

Following the Director's approval, the MC&G Working Group and then the full COMIREX are briefed. The requirement is then sent to ACRES (Area Collection Requirements Evaluation System) for entry to the mission plan. The basic working document is a charter granted to the MC&G Community by the USIB (U.S. Intelligence Board), now NFIB (National Foreign Intelligence Board), in 1972, which endorsed MC&G participation in the satellite imaging collection programs.

MISSION OBJECTIVES SENT TO SATELLITE TEST CENTER

Once the preflight mapping objectives are finalized, the cells and their mapping categories are transmitted to the STC ▓▓▓▓▓ for incorporation into the data base Missions Objectives (MOB) File. The MOB file normally arrives one week prior to launch.

Mapping camera and panoramic camera orbit payload operations consisting of one or more exposures are dynamically interactive, with all mapping camera operations selections being manually implemented. All payload operations are optimized by an operations officer prior to implementation, whether automatically generated via the selection algorithm or manually generated by mandatory requirements from DMA. Thus, although the operation selection algorithm may have selected an MCS operation, the operations officer can manually override the selection by deleting or modifying the selected operation based on a prior factor such as weather, economy of film, or changes in area coverage priorities. For 120-day missions, patience can be exercised to access cloud-free coverage of required areas.

HEXAGON (KH-9) Data Base

The HEXAGON mission activity is supported by the TUNITY software system. The purpose of the TUNITY software system is to determine the specific mission profile which best meets stated intelligence collection requirements, to optimize photographic payload and vehicle operations, to command and control the vehicle, and to report mission results.

TUNITY fulfills its purpose by performing the following tasks in a sequence that permits the timely satisfaction of the mission requirements.

- Mission Operations
  - The purpose of TUNITY in mission operations is to determine the maximum possible satisfaction of mission area targeting requirements, taking into account predicted cloud cover weather information as well as the vehicle/camera/operational constraints and capabilities. TUNITY determines the specific optimum sequence of vehicle and camera operations intended to maximize the selected photography and maintains information pertaining to the degree of satisfaction of mission requirements.

- Command and Control
  - The purpose of TUNITY in command and control is to translate the sequence of required vehicle and camera events into conflict-free command messages to be transmitted to the orbiting vehicle. TUNITY also maintains a current status of the vehicle including the command memory image and predicted state of the vehicle subsystems.

- Mission Reporting
  - The purpose of TUNITY in mission reporting is to provide information to the intelligence community pertaining to the relative satisfaction of mission requirements. TUNITY also reports on the actual photography accomplished during the missions.

- Data Base Validation
  - Prior to each mission, the data base is validated to assure all data base items are set correctly to ensure the mission objectives will be satisfied. Validation is accomplished by having each associate contractor verify the data base items they are responsible for and by actually testing the software in a development rehersal during which actual mission activities are simulated.

SATELLITE TEST CENTER OPERATION

Purpose of the Satellite Test Center (STC)

Overview of STC Operations

Fig. 3-3 is a block diagram depicting the various interfaces required to support the HEXAGON mission. Premission activities include mission requirement finalization, operational software verification, telemetry processing verification, and Satellite Control Facility coordination verification through various development and dress-rehearsal support.

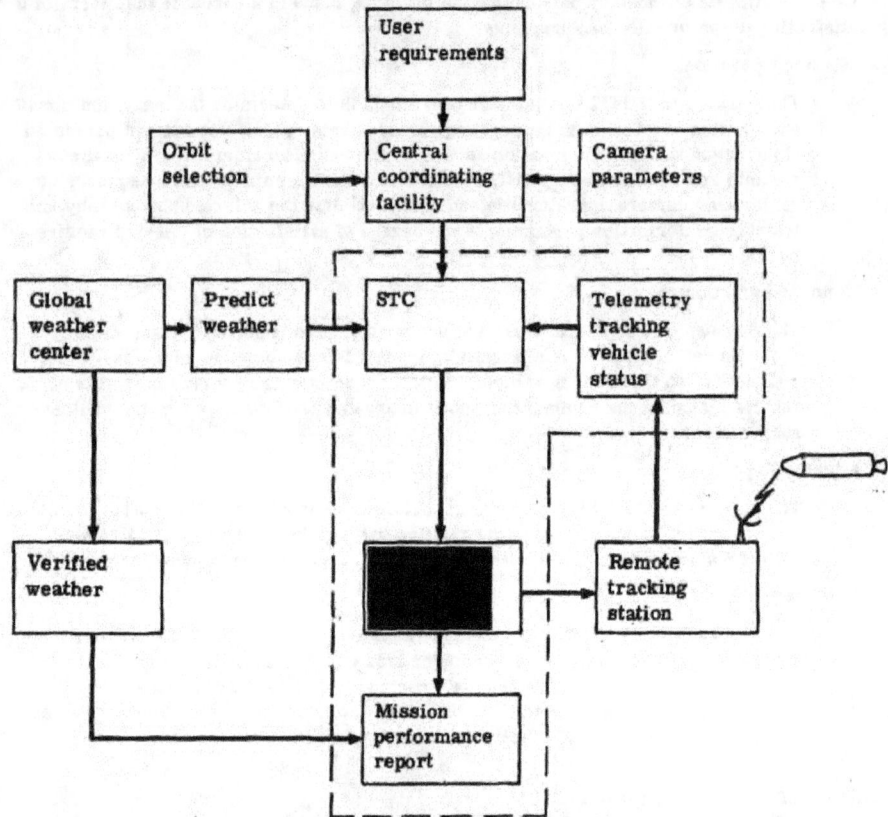

Fig. 3-3 — Operational support overview

STC Staffing

Staffing requirements to support the HEXAGON mission include representation of several NRO and Air Force program offices and various vehicle and payload associate contractors. All payload operations are optimized by an operations officer prior to implementation whether automatically generated via the selection algorithm or manually generated by mandatory requirements from DMA.

MCS HISTORY

USAF TRACKING NETWORK

The Sunnyvale Satellite Test Center (STC), part of the Space Division (SD) Satellite Control Facilities,

The remote tracking stations, acronyms and locations are as follows: (see also Fig. 3-4)

- Vandenberg Tracking Station (VTS) or COOK at Vandenberg Air Force Base, California
- Guam Tracking Station (GTS) or GUAM on Guam Island
- Hawaii Tracking Station (HTS) or HULA at Kaena Point on the island of Oahu
- Indian Ocean Station (IOS) or INDI in Seychelles Island group on Mahe' Island
- New Hampshire Station (NHS) or BOSS near New Boston, New Hampshire
- Thule Tracking Station (TTS) or POGO at Thule Air Force Base, Greenland.

The Global Weather Center (GWC) provides the latest weather information over the areas of interest for each orbit. By the use of weather satellites, the most recent weather forecast is made available to the STC to help determine the payload activities on an orbit-by-orbit basis. After specific payload operations are completed, the GWC provides a weather verification to assist in determining if the area of interest was accessed under acceptable weather conditions.

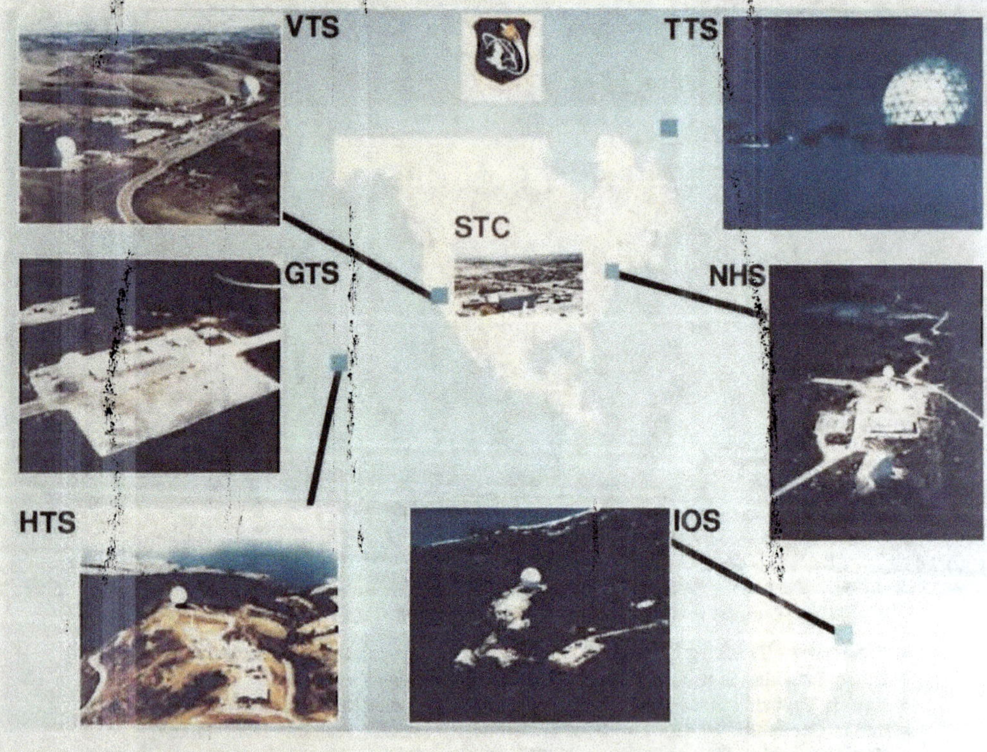

Fig. 3-4 — USAF tracking network

Associate Contractor Support

Associate contractor support is maintained throughout each mission activity to help ensure maximum probability of mission success. Representatives from each associate contractor assist the SCF in determining vehicle and payload health, in maximizing operational efficiency, and recommending corrective action in the event of a system malfunction. Contractor support is provided on a 24-hour basis from vehicle launch to vehicle deboost.

The associate contractor support, representatives of the Aerospace Corporation, and members of the SCF comprise a support group identified as the Technical Advisor Staff (TAS). The TAS is located within a central area to ensure proper interfacing and coordination.

Listed below is a summary of the various associate contractors that provide HEXAGON mission support. Contractors are listed in alphabetical order.

- CSC
    - Command System Contractor. General Electric, Utica, New York is the associate contractor for the command subsystem. The command system receives and stores commands sent by an RTS. Each command is time labeled and executes when the command search time equals the command label time.

- HTC
    - Home Town Crew. TRW is the associate contractor that provides the HEXAGON software programs. The HEXAGON software program is named TUNITY and is comprised of a multitude of programs and routines that control the vehicle and payload operations, requirements management, and several reporting functions.

- MWC
    - Mid-West Contractor. McDonnell Douglas, St. Louis, Missouri is the associate contractor responsible for the reentry vehicle (RV) for the panoramic camera film. The film, after it is exposed, is moved into the RV. The RV returns the recoverable payload to the earth for an air recovery by the Air Force.

- NEC
    - North East Contractor. Itek Corporation, Lexington, Massachusetts was the associate contractor responsible for the mapping camera.

- OPC
    - Our Philadelphia Contractor. General Electric, Philadelphia, Pennsylvania, was the associate contractor responsible for the reentry vehicle (RV) for the mapping camera.

- SBAC
    - Satellite Basic Assembly Contractor. Lockheed Missiles and Space Company, Sunnyvale, California is the associate contractor responsible for the space vehicle. The space vehicle provides the stabilized space platform to support both camera systems and all ancillary equipment.

- SDC
    - System Development Corporation in Santa Monica, California is the associate contractor responsible for the SCF system software.

- SSC
  - Sensor System Contractor. Perkin-Elmer, Danbury, Connecticut is the associate contractor responsible for the panoramic search camera systems.

- AC
  - Aerospace Corporation in El Segundo, California is the technical support organization for the Air Force System Program Office. Aerospace Corporation is responsible for the HEXAGON program technical interface coordination.

Operational Support Data Base

Operational Data Base

An ambitious software program is required to facilitate the accurate and timely achievement of defined operational requirements. The heart of the system software is a large data base. The data base includes information relative to camera parameters, files of operational requirements, orbit ephemeris data, priority thresholds, weather criteria, command sequences, telemetry processing, etc. The data base is updated for each mission with new requirements being added and old requirements deleted or modified.

Configuration Control

Configuration control of the data base is maintained through an organization called the Data Base Subgroup (DBSG). The Data Base Subgroup is responsible for working the details of the various data base maintenance requirements. The subgroup is staffed by representatives from each associate contractor with coordination controlled by the Air Force Systems Program Office. (Refer to Fig. 3-5 for the position of the DBSG within the OIWG organizational structure).

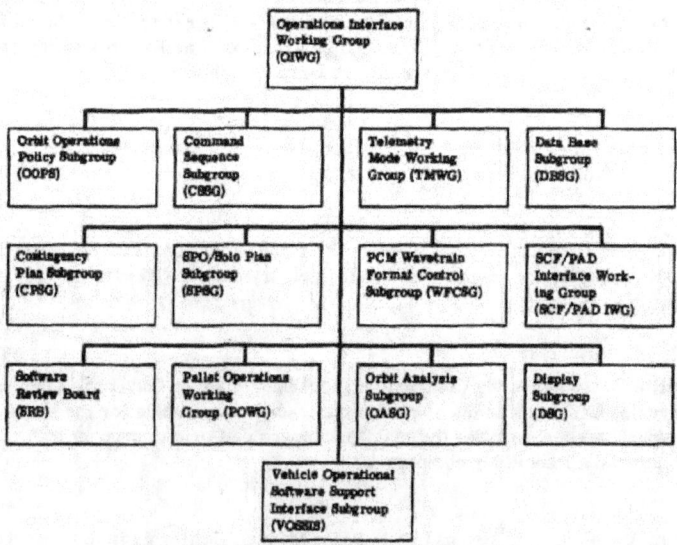

Fig. 3-5 — OIWG organizational structure

Operational Support Activities—Associate Contractors

Operational Support Function Overview

Operational Support Staff

The operational support staff includes resident members of each associate contractor. The support efforts are coordinated by an integrating agency. Each associate contractor is familiar with the operation of his hardware, the software to implement the operational requirements, and the required steps to take should a problem arise.

Here, as with the hardware system integration effort, a team of associate contractor support is essential to provide an efficient operational support capability.

Command Loads

Telemetry Analysis

If the telemetry indicates a system problem, the support staff recommends to the integrating agency what corrective action should be taken.

The approach taken in the anomaly analysis is threefold: first, to properly define the problem; second, to determine the effect on the vehicle and on all other payloads; and third, to establish a corrective action management plan. Anomaly management is a highly coordinated effort by the associated contractor/integrating agency mission support team. (For a typical problem management flow chart, see Fig. 3-6.)

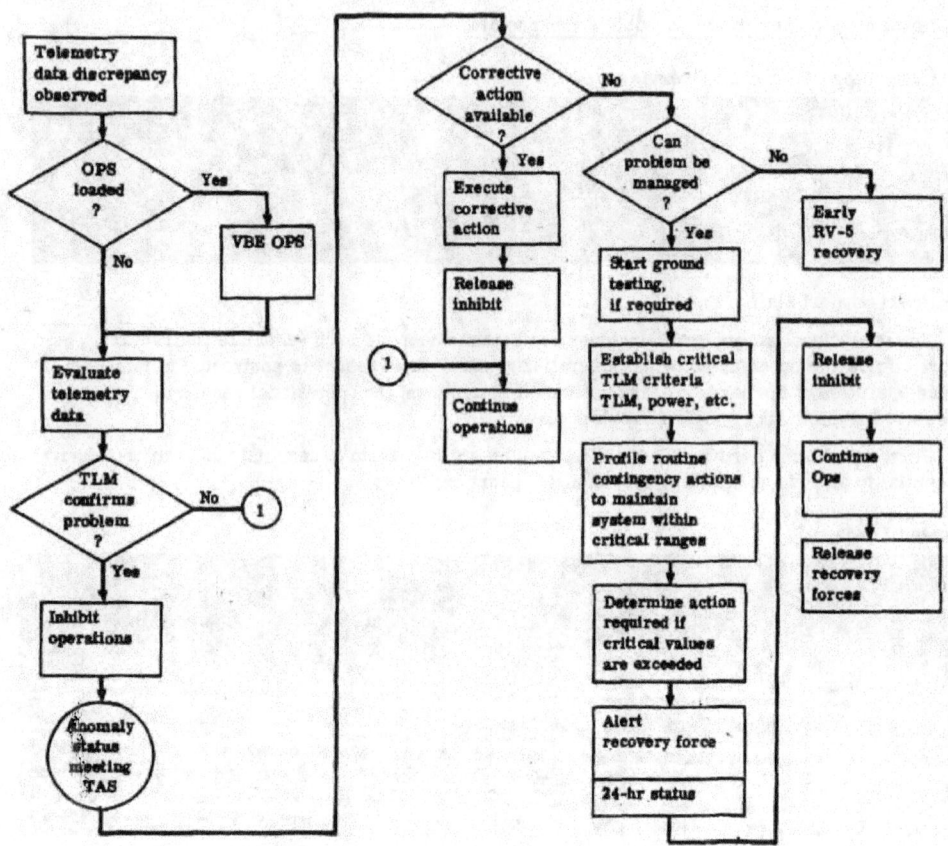

Fig. 3-6 — Problem management flow chart

In determining the implementation of the anomaly management plan, the integrating agency must consider the effects on the vehicle and the effects on other payloads. Because of the magnitude of this type program and the serious impact one failed subsystem can have on the total mission, it is important for each associate contractor to have highly knowledgable operational support representatives.

OPERATIONS DATA FLOW

With the delivery of the mission objective file (MOB) from the intelligence community 7 days prior to launch, the operational data flow begins and continues throughout the flight and into the post-flight evaluation phase.

Throughout a HEXAGON mission, data flows between the intelligence community in Washington and ▨▨▨▨▨ at the Satellite Test Center in Sunnyvale, California.

MCS HISTORY

Prior to 1977, the Satellite Operations Center (SOC), a staff agency at the Pentagon, provided the interface between the user and the west coast. Automation within the intelligence community has centralized requirements and permitted direct interface between the user and the west coast. Updates to the MOB occur on an almost daily basis and are transmitted to the STC for incorporation into the data base.

Data also flows between the STC and Global Weather Center (GWC). Prior to on-orbit activities, climatology is used to conduct mission simulations and mission rehearsals. During on-orbit activities both predicted and verified weather from weather satellites (DMSP) and other sources are used. Predicted weather requests are generated by the TUNITY software program TBAT based on the vehicle ephemeris and the MOB which when merged forms a target acquisition file. The software is run testing the World Aeronautical Chart (WAC) cells to determine which are candidates for mapping photography. Those candidates selected define a geographical area for which a predicted weather request is produced and sent to GWC. These messages request the probability of cloud-freeness over the candidacy areas.

GWC responds to the pre-weather request with forecasted cloud-freeness probabilities at 90 minutes prior to the acquisition of the HEXAGON vehicle at a remote tracking station (RTS). The pre-weather data is merged in TUNITY with all the other operational requirements and if conditions are satisfied, camera operations are generated.

After one or more camera operations have been selected, a command message is generated which produces the time tagged commands needed to execute various hardware systems. Following the successful loading of the command message into the satellite vehicle, a request for verified weather is sent to GWC. Response to this request is the actual cloud-freeness during, or as close to, the time of the photography as possible.

During the time that the command message is being loaded into the SV, the vehicle down link is returning data from previously executed command messages which had been stored on the vehicle tape recorder. Included in the down-link data are the frame reference times (FRT) of the camera operations. The FRT data is collected on an operation-by-operation basis and is merged with the command history, a daily collection in chronological order of executed commands. This data is then combined with the best-fit ephemeris to produce the mission performance report (MPR).

The MPR specifies what photography has been taken by defining each operation and each frame within the operation. It includes the corners by latitude/longitude for each frame and operation and also includes vehicle time, attitude and velocity, sun elevation and azimuth, right ascension and declination, and the film used.

Once the MPR is completed, the mission performance evaluation (MPE) is executed. The MPE takes the verified weather and incorporates it into the MOB to "deweigh or modify" the "worth or value" of the ground photographed. Depending on the requirements established by the user, the modified or deweighted ground may be satisfied, in which case no further photography will occur, it may be a candidate at some future time, or it may still be active and require repetitive coverage.

In addition to the "countdown" of photographic requirements, the MPE provides statistical reports on the status of the MOB. Mapping category satisfaction ratio (CSR) and number of looks (NL) are standard reports and provisions that exist to define specific factors to completely analyze the MOB.

Both the MPR and MPE are provided for each operational day to the user. With the MPE completed, the next operational day is started. Predicted weather is requested, payload selections are made, command messages are generated, verified, and loaded into the vehicle, vehicle data is down-linked and merged with command history, mission performance is reported, verified weather is combined with the MPR to evaluate the mission performance and modify the MOB. Data flow is depicted in Figs. 3-7 and 3-8.

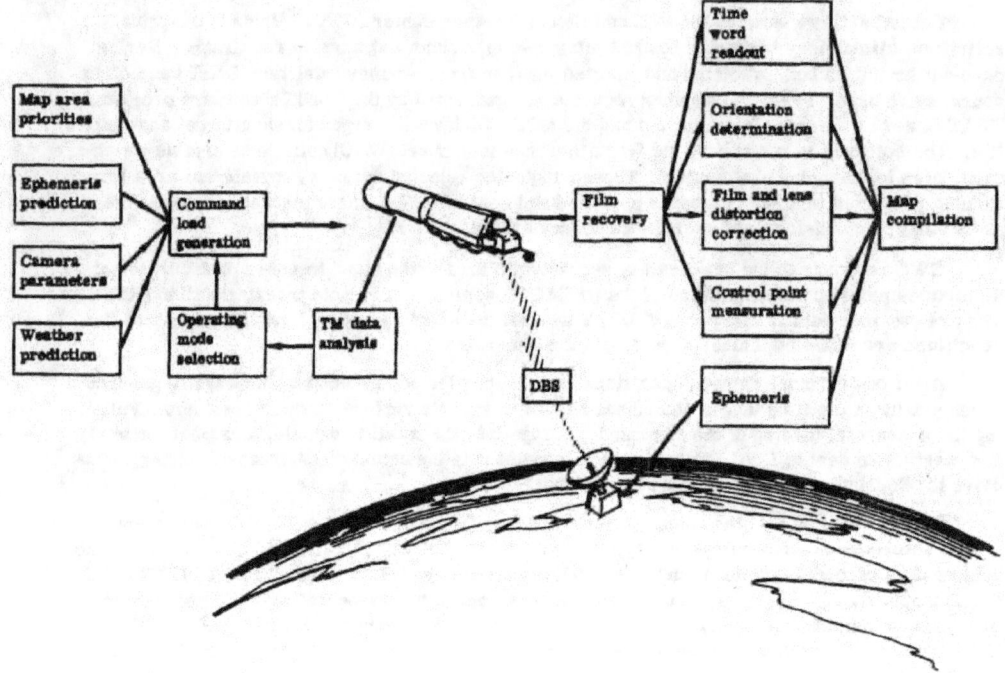

Fig. 3-7 — Mapping Camera operational data flow

And so it goes for the operations analysts pulling shift duty during the long HEXAGON missions. Following the excitement of launch and early orbital health checks, the activities settle into hours upon hours of sheer routine, interrupted occasionally by moments of stark panic. But approaching each recovery, the activity picks up again. In the case of the MCS, as mission completion grew near, the inflight calibration maneuver had to be carefully profiled and coordinated. This was in addition to the usual activities associated with end of mission, such as special film experiments, recovery preparations, "solo" engineering profiles, and data consolidation for the post-flight analysis (PFA) reviews which were conducted at the processing site the week following recovery.

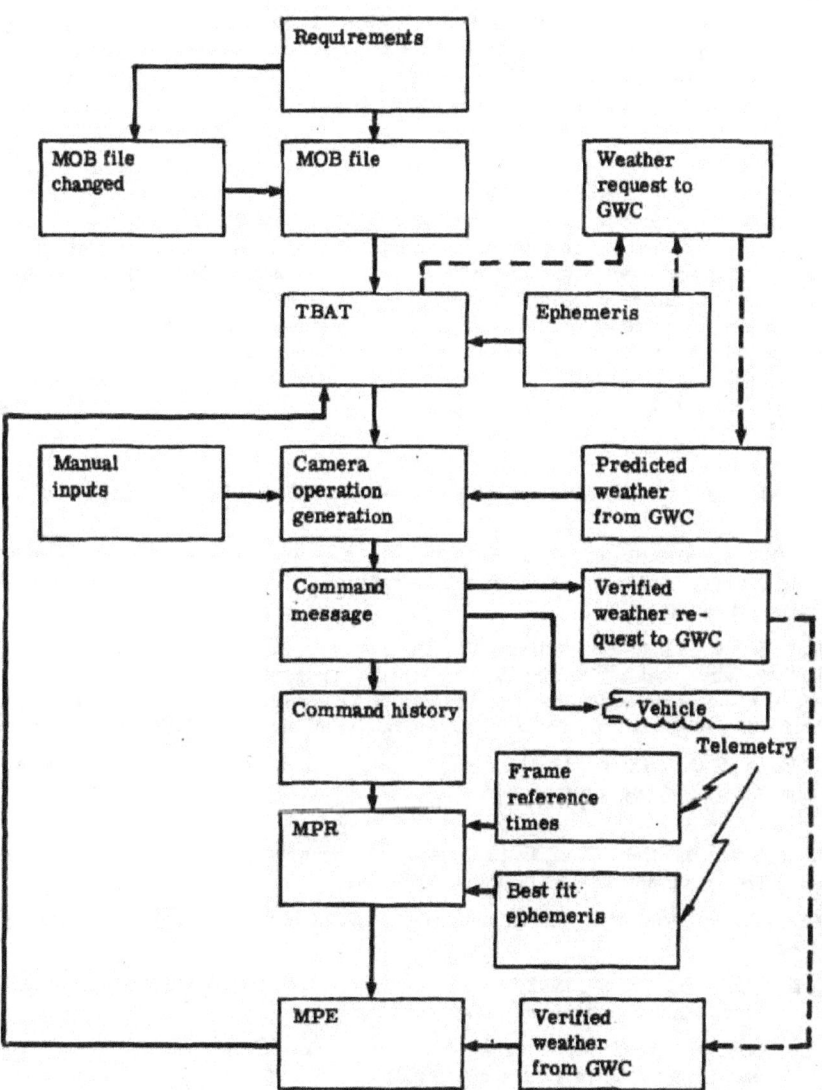

Fig. 3-8 — Mapping Camera operational data flow

DE-ORBIT/RECOVERY SEQUENCE

The performance record of the General Electric Mark V SRV and the Air Force recovery forces was 100 percent successful on the MCS Program. So routine was this operation, there may have been a tendency to take success for granted. The brief description which follows is a reminder of the complexity of this important phase of the operation.

De-Orbit Preparations

After all the exposed film was transferred into the SRV, the cutter-sealer was activated by SV command to seal the recovery capsule and to sever any residual film strands in the tunnels. Approximately 3 revs before de-orbit, power was supplied to the recovery battery heaters to ensure that this component was at its required operating temperature before activation. Up to this time the capsule temperature was controlled by the SV. Finally, the SV was yawed aft and pitched down so as to position RV-5 in the optimum attitude for deboost. The required pitch-down angle was computed for each recovery and was optimized to minimize dispersion. The angle could be adjusted to minimize heating if required due to an unbalanced payload, a condition which fortunately never occurred.

De-Orbit

The RV-5 system was remotely activated via the ARM, TRANSFER, and SEPARATION commands from the SV.

ARM—This signal activated the two recovery batteries. Recovery beacons were wired directly to the batteries and began operating when the battery voltage reached the proper level, i.e., about 25 seconds after ARM.

TRANSFER—This signal fired the de-orbit subsystem thermal batteries, started the interval (backup) timer, and initiated the two time-delayed in-flight disconnect (IFD) squibs. The time-delay squibs ensured that the signal had sufficient time to perform its required functions on the RV side of the interface before IFD separation initiated the RV de-orbit programmer.

SRV SEPARATION—The two MCM pin-pullers holding the RV to the MCM were fired and the separation impulse was provided by four pusher springs located on the MCM side of the interface.

SPIN—3.4 seconds after IFD separation, the RV was spun up to 57 rpm by a cold gas blow-down system. The spin event initiated the retro-fire timer.

RETRO FIRE—A deboost impulse of about 800 to 1,000 fps was provided by the Thiokol TE-236B rocket.

DESPIN—Following deboost, the SRV was de-spun to about 13 rpm. This rate was compatible with reentry dynamics and minimized aft-end heating to the RV.

THRUST CONE SEPARATION—Following the de-spin event, the RV was on a reentry trajectory and, therefore, the thrust cone subassembly was no longer required. It was discarded by electrically and mechanically separating it from the RV. Springs provided the required ΔV between the T/C and the RV. The total elapsed time from transfer to thrust cone separation was less than 30 seconds; this was compatible with the thermal battery activated life.

Using recovery battery power, a backup timer can also eject the thrust cone. This event was provided in case the destruct function was required.

Reentry

The RV separation orientation required to achieve a successful deboost resulted in an initial reentry angle of attack greater than 90°. A nose-forward attitude was achieved by aerodynamic stabilization at a high altitude so that the forebody (heat shield) protected the RV from reentry heating. Before the turn-over was complete, however, the aft end received direct flow impingement and could realize significant heating. Operational limitations were defined to preclude aft-end heating conditions in excess of the design capability.

Redundant vent valves were located on the capsule cover to control the pressure differential across the capsule structure during reentry. Ascent venting occurred through the open film tunnels. The reentry vent valves were self-sealing in the event of water impact.

Inertial switches were located in the recovery capsule to sense axial deceleration. As the deceleration exceeded 3 g, increasing, the switches closed to apply power to the recovery programmer, but maintained a ground, precluding programmer start. As the deceleration dropped below 3 g, decreasing, the ground was lifted, initiating the recovery timer.

Twenty-six seconds later, power was applied to the charge adapter/ejection piston assemblies to eject the thermal cover. The thermal cover was ejected aft through the wake and deployed the drogue chute. This deployment mechanically initiated the pyrotechnic time-delay cutters on the lines securing the main chute bag. In the assembled configuration, the recovery capsule was sandwiched between the forebody and thermal cover by the ejection pistons. Thermal cover ejection released the forebody which fell free with either drogue or main chute deployment, depending on the system weight. This reduced the suspended weight and descent velocity.

Activation of the bag-line cutters released the main canopy which was deployed reefed. This deployment mechanically activated pyrotechnic reefing line cutters which disreefed the canopy 4 seconds later.

Parachute deployment started at about 60,000 feet and the capsule descended on the main canopy to about 15,000 feet where air retrieval operations started. Recovery beacons and parachute coloration were provided to help the aircraft crews acquire the capsule. After retrieval, a connector plug on the capsule cover was removed to deactivate the beacons.

If air retrieval was not successful, the capsule would survive water impact and float. The beacons continued operating for about 10 hours. If water retrieval was not successful, a sink valve functioned, sinking the capsule. A valve on the cover released trapped air during the sink phase.

Backup Timer Events

This timer was included within the recovery subsystem to provide for capsule destruction in the event of an unsuccessful deboost. Should recovery not be initiated within a predetermined time, this timer would initiate the destruct sequence, prematurely separating the forebody and deploying the parachute. Subsequent reentry heating/loads would destroy the recovery capsule and the film load. The timer also provided for thrust cone separation to ensure that this subassembly would not interfere with subsequent recovery subsystem deployment.

Post-Flight

After retrieval, the capsule with its film load was returned to the processing site for film retrieval. Subsequently, the recovered capsule was returned to General Electric's Philadelphia facility where the takeup and cutter-sealer were removed and returned to the associate contractors. The parachute and films taken of the recovery operation were evaluated by GE. The capsule was inspected and recycled into the program flow for re-use. (Water retrieved hardware was never recycled for re-use.)

Typical reentry and recovery sequences for the Mark V SRV are shown in Figs. 3-9 and 3-10, respectively.

Fig. 3-9 — Flight profile

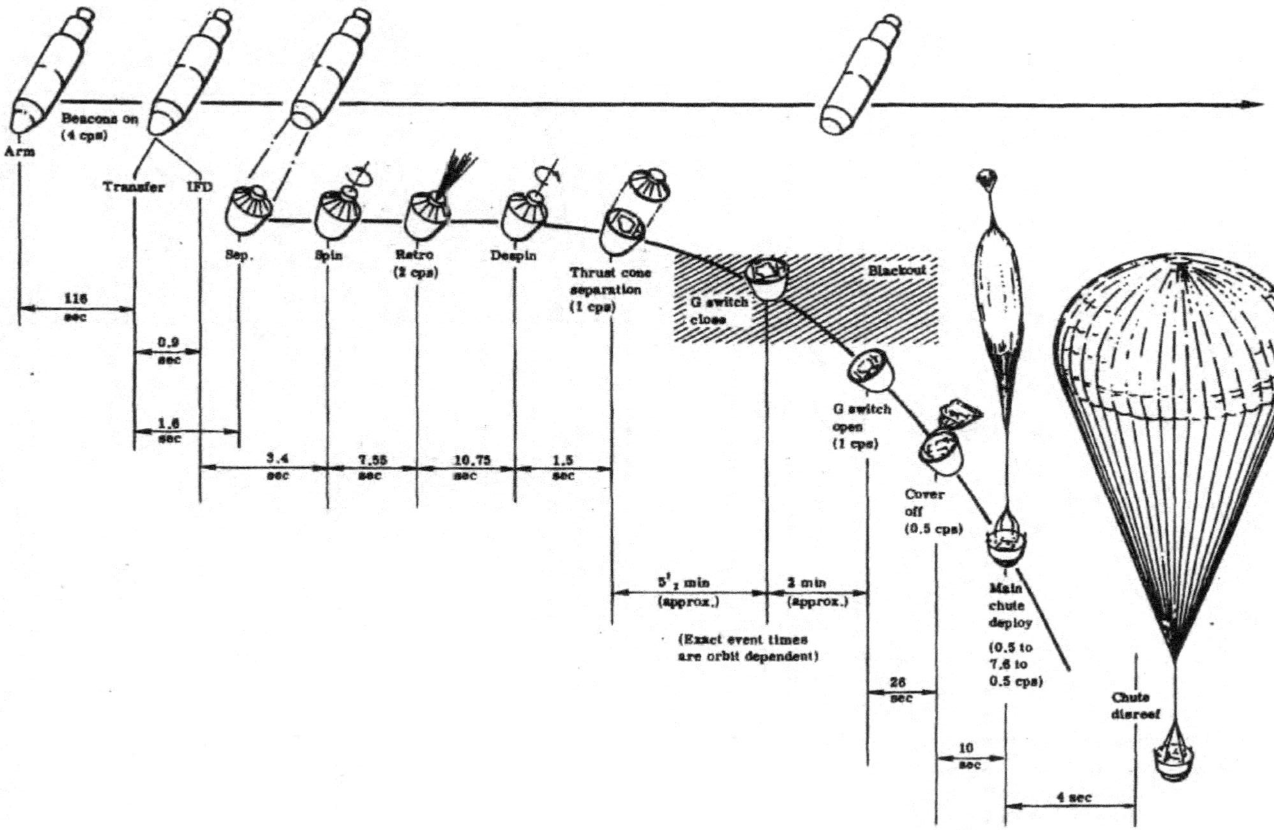

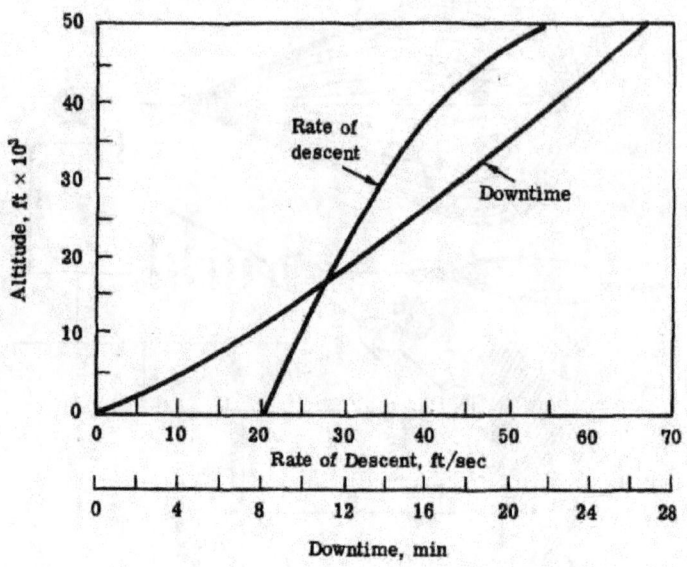

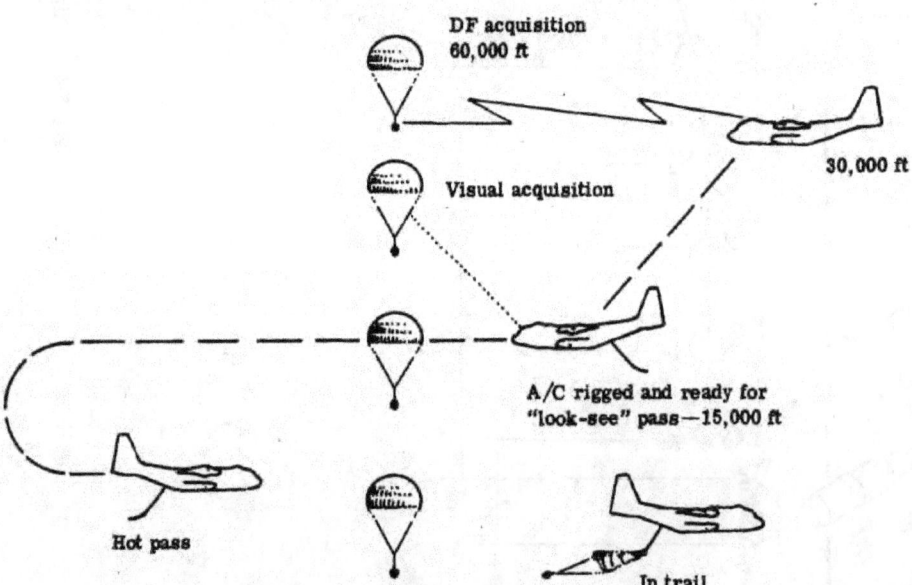

Fig. 3-10 — Typical air recovery sequence

MCS HISTORY

SECTION 4
PROCESSING/DUPLICATION

PROCESSING SITE PRODUCTION WORK FLOW (Fig. 4-1)

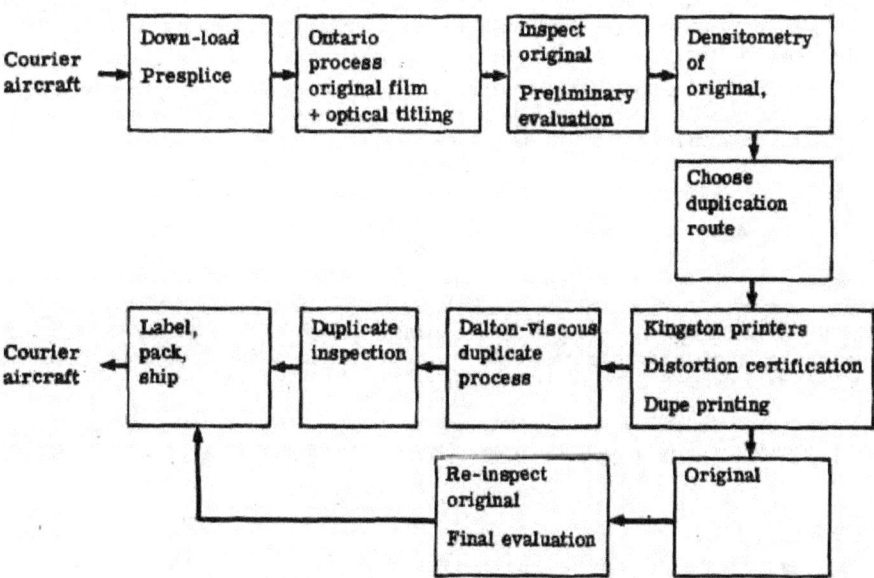

Fig. 4-1 — Processing site production work flow

Receipt of Mission Film

The film payload, in the Satellite Reentry Vehicle (SRV), arrived at a local airport on a USAF courier aircraft and was transported via truck to the processing site. Upon receipt, the SRV was removed from its shipping container, weighed, and placed in a special holding frame which aligned the SRV with the Presplice Complex (see Fig. 4-2).

Downloading/Presplice

The downloading/presplice operations consisted of despooling the SRV payload to permit inspection for defects that might create problems during processing and to configure the film roll into "as required" or "heads out" order for the processor. This operation took place in the dark. Non-contacting infrared detectors and viewers were utilized to assist the operators. The film payload was transported across the Presplice Complex under controlled tension onto a takeup spool dolly which was later mated with the feed end of a processor (see Fig. 4-3). After the entire roll had been despooled, a lighttight cover was placed over the takeup spool dolly and it was moved to a processor room.

Original Negative Processing

The transport dolly that was loaded during presplice was positioned at the feed end of an Ontario Processor, room lights were turned off, and the cover was removed. The head of the roll was spliced to leader which was already threaded through the machine. Previous "start-ups" had established that the machine was in sensitometric control, producing product free from physical defects such as scratches.

During the processing operation, critical machine parameters such as temperatures, solution pressure, and transport speed were monitored by a computerized system to assure stable photographic control throughout the processing run.

Optical Titling

Titles, consisting of mission and frame numbers, date of photography, security classification, and other information, were exposed on the film optically during the processing operation. Development proceeded to the point where camera-exposed framemarks could be detected using sensors sensitive in the infrared to provide frame location information. Immediately following framemark detection, the optical titling system exposed man-readable characters along the edge of the film. The developer action continued to process these characters to a legible and reproducible density. Fig. 4-4 displays an example of typical optical titling.

Titling was controlled by a computer that had previously received a film format and titling information "roadmap" that was prepared from information received from the mission operations center.

Preliminary Evaluation

Following the post flight analysis (PFA) mission review and planning (Section V), each camera roll was fed into the processor as a continuous length of film, but was removed in segments to allow early introduction into the reproduction cycle. These machine cuts were quickly given a preliminary evaluation to check for possible processor-induced anomalies that would dictate immediate corrective action. The accuracy of the optical titling was then checked, and if corrections or additional manual titling were required, it was noted for subsequent action.

Densitometry/Duplication Route Determination

After preliminary evaluation, each original negative part was measured on a macrodensitometer to obtain density range information. Minimum, maximum, and average density computations were utilized to determine initial printing conditions that would yield optimum print density and contrast for each part. The original was then broken down into printer parts, usually less than 300 feet, to facilitate handling and most efficient dupe stock usage during the printing cycle.

Fig. 4-2 — Satellite Recovery Vehicle (SRV)

Fig. 4-3 — SRV interfaced with presplice complex

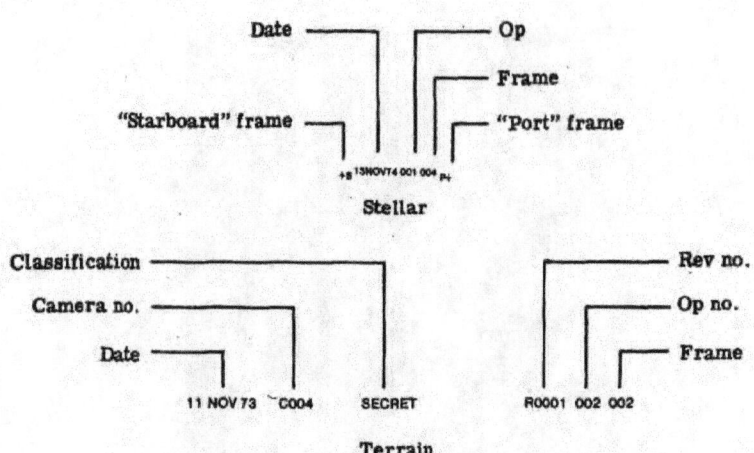

Fig. 4-4 — Examples of optical titling exposed during development

Printing

After breakdown, each roll of original negative was run through a film cleaning device and taken to the printing area. Previously established printing conditions were used to produce desired exposures on the duplication film material. The special low-distortion Kingston Printers (Fig. 4-5) were certified periodically both sensitometrically and for distortion to assure they were meeting established criteria.

Duplicate Film Processing

The exposed film rolls from the printing area were processed on Viscous Dalton Processors. The processing sequence was the same as on the Ontario Processor and all the critical parameters were monitored.

Duplicate Copies Inspection

The processed rolls of duplicate copies from the Dalton were inspected for both sensitometric and physical quality before shipment to customers.

Final Original Negative Inspection

When all reproduction requirements had been satisfied, the original negative received a final quality inspection to document its condition as it was shipped from the processing site.

ACQUISITION MATERIAL

Records from both stellar and terrain cameras had improved significantly during the life of the program. A major contributor to these evolutionary improvements was the testing and eventual use of higher quality acquisition films that were made available to the program (see Tables 4-1 and 4-2).

Stellar Camera

The first Mapping Camera System on mission 1205-5 (April 1973) utilized 3401* film. This relatively coarse-grained product was selected because it is photographically fast (very sensitive to light) and would meet the requirement of recording faint star images. The high speed of this product resulted in a film record plagued with excessive fog due to solar radiation as well as unwanted density caused by static and corona discharges. These high fog levels, coupled with the inherent opacity of the product, made it impossible to successfully optically title 3401 stellar originals as part of the process cycle primarily because of insufficient density discrimination of the framemark against the background.

Table 4-1 — MCS Acquisition Products by Mission

	Stellar	Terrain	
		Mapping	Calibration
1205-5	3401	3400	3401, 2403
1206-5	3401	3400 SO-131	3401, 2403
1207-5	3401	3400 3414	3401
1208-5	3401 3400	3400 3414	3401
1209-5	3401 3400	3414 QX-801	3401
1210-5	3401 3400	3414	3401
1211-5	3400	3414 3401 3411	
1212-5	3400	3414 3411	3401
1213-5	3400 SO-344	3414 1414	3401
1214-5	QX-817	1414 SO-315 SO-208	3401
1215-5	QX-817	SO-315	3401
1216-5	QX-817	SO-315	3401

*Table 4-3 lists the complete film designations.

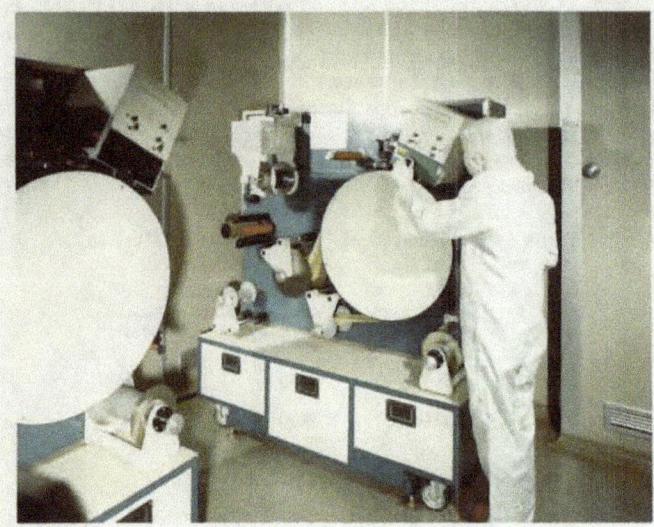

Fig. 4-5 — Kingston continuous printers

Fig. 4-6 — Distortion analysis system

On missions 1208-5, 1209-5, and 1210-5, testing was conducted with short lengths (tag ends) of 3400 film to determine if a slower, finer grain product could adequately record star images while reducing the potential to fog. A special viscous process for 3400 film was developed to provide an optimum speed/grain relationship for this application. The Post Flight Analyses (PFA) team stated in the 1209-5 PFA report and again in the 1210-5 PFA report that testing evaluation had shown that an adequate number of star images could be recorded on stellar cameras using 3400 film with "forced" processing. It further recommended that 3400 replace 3401 as the primary load for the stellar cameras starting with mission 1211-5. Each stellar camera on 1211-5 recorded sufficient 6th magnitude and fainter star images. The reduction in fog from unwanted energy allowed optical titling of the majority (84%) of the stellar record. The 3400 type emulsion remained the primary product for the stellar cameras throughout the life of the MCS. The use of a 1.5-mil base version of 3400 (SO-344 and QX-817 starting with 1213-5) permitted flight load increases of up to 86%.

Terrain Camera

Calibration Tests. On 1205-5, both 3401 and 2403 were utilized for calibration test purposes. The speed and granularity of 2403 (see Table 4-2) is much greater than 3401. The excessive grain size of 2403 affected the analog and automated stereo compilation, but was primarily found unacceptable by the PFA team because of the fact that the grain size approximated the star image size. On 1207-5, it was determined that an increase of one stop exposure combined with the increased transmission gained by replacing the Wratten 21 with a Wratten 12 equivalent filter would permit the use of 3401 as the sole film for calibration purposes. A liquid process was developed to enhance the photographic speed of 3401 using a VERSAMAT roller transport processor. (Starting with 1215-5, the 3401 calibration film was processed on the Ontario Processor using a viscous process which produced lower fog and better physical quality.) The replacement of 2403 with 3401 starting with 1208-5 resulted in significant improvement in resolving power and granularity in the terrain calibration imagery.

Terrain Imagery. The terrain record had the greatest potential for improvement as a result of product substitution because of its requirement for high resolution and acuity. The initial (1205-5) terrain prime load film was 3400. Testing of a higher quality 3414 film began with 30-foot tags on missions 1207-5 and 1208-5. The PFA team indicated that the results of these tests, considering resolution, granularity, smear, and exposure, supported the use of 3414 as the primary film load for 1209-5. (A small quantity of product QX-801 having an intermediate speed between 3400 and 3414 was tested on 1209-5. Films 3401 and its replacement 3411 were also tested on 1211-5 and 1212-5 for similar purposes. The PFA team analyses of these products indicated that they did not provide a level of information content comparable to 3414, therefore, no further testing was performed.) Mission 1213-5 introduced test quantities of 1.5-mil ultra-thin-base (UTB) films. The 1.5-mil 3414 (designated 1414) was tested on the terrain camera (while the 1.5-mil 3400 (designated SO-344) was also tested on the stellar camera. The PFA found that samples of UTB films on the stellar and terrain indicated that film distortion was within acceptable limits, and the consensus was that UTB film was acceptable as the primary loads for 1214, 1215, and 1216.

Mission 1214-5 flew with primary loads of UTB films on the stellar (QX-817*) and terrain (1414). The terrain camera load contained small (tag) quantities of 1.2-mil [ultra-ultra-thin-base (UUTB)] film SO-208 (1.2-mil 1414) and a new product, SO-315. SO-315 represented the most

---

*QX-817 was another designation for 1.5-mil 3400 film.

Table 4-2 — MCS Films

Film Number	Process Developer	Film Speed	Resolving Power		RMS Granularity Value, 2μ	
			1.7:1	1,000:1		
MCS Acquisition Films						
2403	V641	654	25 ± 6	80 ± 19	553	Initial terrain calibration product
3401	V641	260	35 ± 8	95 ± 23	310	Initial stellar product Final terrain calibration product
3411	17DN	212	56 ± 5	160 ± 40	380	Replaced 3401
SO-131	E-31-3	7.5*	52 ± 11	160 ± 46	—	Color infrared film
3400/SO-344/QX-817	20DN	67	75 ± 18	159 ± 38	220	Final stellar product Initial terrain product
QX-801	29DN	18	193 ± 46	469 ± 113	160	Intermediate speed product tested for terrain on 1209-5
3414/SO-208	19DN	15	281 ± 67	723 ± 174	105	Improved terrain product
SO-315	46DN	5.7	441 ± 106	924 ± 222	80	Final terrain product
MCS Duplication Films						
2420	18DN-V		85 ± 20	218 ± 52	196	Initial product MCS
SO-360	EA-5		63 ± 15	125 ± 30	—	Color duplicating film
SO-467/SO-284	18DN-V					2420 variants
2421	18DN-V		100 ± 20	222 ± 65	165	Replaced 2420, SO-467, SO-284
SO-192	18DN-V		459 ± 110	722 ± 173	82	High resolution
QX-822/SO-187	18DN-V		—	—	—	SO-192 on 7-mil base
2422	18DN-V		225 ± 60	551 ± 100	52	Direct reversal film for dupe negs
SO-355/SO-277	18DN-V		135 ± 32	286 ± 69	138	Low contrast film (SO-277 on 7-mil base)

*Effective aerial film speed.

light-sensitive product of a family of high definition, low noise, monodispersed cubic emulsion products recently made available by the Eastman Kodak Company. The PFA team evaluated the UUTB tests and reported that image and metric analyses performed on both films indicated that either film (SO-208 or SO-315) would be acceptable for use, but that SO-315 provided the higher quality image. A recommendation was made to use SO-315 for the primary load on the terrain camera for missions 1215 and 1216, and this was done. The substitution of 1.5-mil-thick 1414 on 1214-5 for 2.5-mil-thick 3414 permitted a film load increase from 3,282 feet to 4,843 feet (+47%) and the use of the 1.2-mil-thick SO-315 on 1215-5 permitted a flight load of 6,244 feet (+90%).

The mission 1206-5 terrain camera utilized 100 feet of false color infrared film (SO-131). The PFA reported that this film produced exceptionally good photography with an estimated 50 to 75-foot ground resolution. The unique spectral characteristics of this film provided scene information not available on the higher resolution 3400 film. The successful use of SO-131 on 1206-5 demonstrated the additional versatility of the system.

DISTORTION IN FILM HANDLING SYSTEMS

Prior to introduction of the mapping cameras on 1205-5, the processing site conducted studies[9] to evaluate the effects and the degree of distortion produced with in-house film handling systems. It was determined that while some dimensional distortion occurs during various phases of film handling, such as processing and duplication, limited distortion can be tolerated if it is uniform, repeatable, and measurable. The processing site studies employed the Moiré technique for measuring such distortion. This technique, whose fundamental relationships are discussed in References 1, 2, and 3, uses a halftone cancellation effect to produce a Moiré pattern. A film image is contact-printed from a master halftone glass plate. After processing, the film is registered to another glass halftone plate producing a Moiré interference pattern. When certified for use as a print master, this plate image can be duplicated for use in analyzing printing equipment and/or other film handling operations. Duplicates produced from such a master are overlayed on yet another glass halftone plate. The resultant Moiré pattern is used to evaluate any non-uniform system distortion in both distance and rotational domains. Fig. 4-6 displays a photograph of the vacuum printing and registration table and the semi-automatic Moiré pattern readout equipment.

These distortion measuring techniques were utilized to initially certify the acquisition film processors [the Yardleighs in 1972 and their replacements, the Ontarios (see Fig. 4-7), in 1977]. Processing site studies indicated that, while the processors and printers apply non-uniform tensions to the original negative during handling, the internal forces do not exceed the product elastic limit, and dimensional fidelity of the original is not impaired in these operations. The same Moiré system is employed to certify and monitor the Kingston* printers throughout MCS duplication operations.

---

*The Kingston printers (Fig. 4-5) were designed specifically for continuous printing with minimum distortion. Essentially, they differ from conventional continuous printers by incorporating devices such as: a large 28-inch-diameter printing drum, independent castered pressure roller, precision spools, and a constant-tension film transport system.

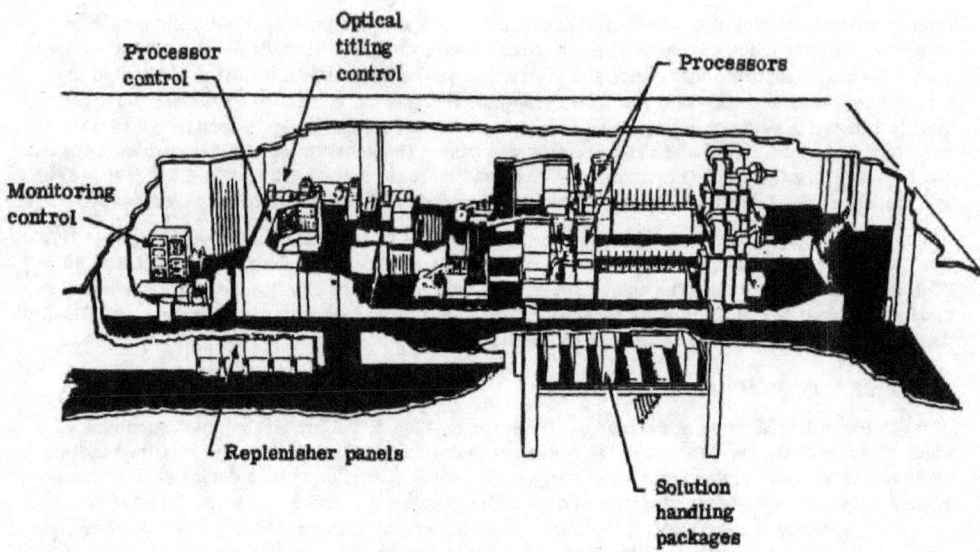

Fig. 4-7 — Ontario processing complex

DUPLICATION

Films

The evolutionary improvement in the selection of duplication films essentially tracked the improvements in MCS acquisition films. The 2420 was the stellar duplicating film until mission 1212-5 when the replacement of 3401 with 3400 acquisition film (which first occurred on 1211-5) permitted the use of the improved SO-284 (type 2421) film.

Early terrain missions were duplicated on SO-467 (modified 2420) and SO-355. The introduction of 3414 on 1207-5 permitted the use of SO-192, a significantly higher quality duplicating film. Starting with 1213-5, additional duplicates on ESTAR thick-base (7.0-mil) film from the terrain camera were produced primarily for customer handling purposes.

System Sensitometry

Duplication criteria for stellar and terrain photography were established in cooperation with customer representatives. The criteria evolved during the reproduction of early RV records.

Stellar. Stellar photography, with its emphasis on detection of 6th magnitude and fainter star images, called for a relatively high contrast duplication system. Density levels of between 0.75 and 1.35 were found to provide good differentiation from the star images while maintaining good grid uniformity from center to corners. A duplicate negative on 2422 was produced utilizing criteria with similar objectives as those produced on the positive.

Terrain. Earlier missions (1205-5 to 1208-5) utilized a normal contrast 2420 duplication system. With the introduction of acquisition film 3414 and duplication film SO-192 on 1207-5, duplication sensitometry became more varied. Mission 1208-5 introduced the Actinic Butterfly Contrast Control (ABCC) printing method. This method provided improved system sensitometry and an in-line production capability for contrast changes on a film part* basis. "Multiple prints" (i.e., a "lighter" or "darker" print in addition to a "normal" print) were produced whenever the acquisition film encompassed a wide range of object densities. In addition, some areas (e.g., snow, sand, or hazy scenes) were reproduced at very high contrast levels in order to provide duplicates with enhanced interpretability. On 1212-5, the duplication system evolved to one with a lower overall system contrast providing a generally superior duplicate for MC&G readout purposes.

An ESTAR Thick-Base (7.0-mil) duplicate positive was produced at customer request to facilitate handling techniques when the positives were cut and used as individual pieces. Starting with 1213-5, SO-277 was utilized as the 7.0-mil material (see Table 4-3) until 1215-5 when QX-822 (later renamed SO-187), a 7.0-mil-base version of SO-192, became available.

Duplicate negatives, used to generate duplicate positives at customer facilities for terrain photography, utilized direct reversal film 2422.

PHYSICAL HANDLING

It is important to provide special handling of the MCS film products in the processing laboratory in order to prevent non-linear distortion, such as local deformation due to handling by personnel. In addition to the special design and procedures for handling film through the processor and Kingston printers previously mentioned, the utmost care must be exercised in several other laboratory operations. These include defilming, evaluation and inspection, densitometry, and manual titling. For example, special procedures call for manual, rather than pigment transfer titling (with a jig to minimize human contact) of stellar imagery not optically titled.

Static discharging onto the film during laboratory handling is minimized by the incorporation of "jet ionizers," a corona discharge apparatus with an air flow stream, located on the defilming/presplice equipment and at the head of the acquisition processors. Corona discharge type "bars" are located on the Kingston printers and web cleaning equipment used to periodically clean the acquisition and dupe film throughout printing operations.

---

*Film parts are sized into lengths up to 300 feet to provide a convenient size for handling in the printing operation.

Table 4-3 — Kodak Aerial Films—MCS Usage

Film Number	Kodak Film Designator	Nominal Base Thickness, mils	MCS System Use
Camera Acquisition Films			
2403	TRI-X AEROGRAPHIC Film 2403 (ESTAR Base)	4.0	Early terrain calibration, 1205-5, 1206-5
SO-131	HIGH DEFINITION AEROCHROME INFRARED FILM (ESTAR Thin Base)	2.5	Test quantity on terrain, 1206-5
3401	PLUS-X Aerial (ESTAR Thin Base)	2.5	Early stellar cameras, 1205-5 through 1210-5, terrain calibration
3411	PLUS-X Aerecon (ESTAR Thin Base)	2.5	Tested on 1211-5 and 1212-5
3400	PANATOMIC-X Aerial (ESTAR Thin Base)	2.5	Stellar, 1208-5 through 1213-5
SO-344	PANATOMIC-X Aerial (ESTAR Ultra-Thin Base)	1.5	Stellar tag, 1213
QX-817	PANATOMIC-X Aerial (ESTAR Ultra-Thin Base)	1.5	Stellar primary load, 1214-5 through 1216-5
QX-801	Special AERO film	2.5	Terrain tag, 1209-5
3414	High Definition Aerial (ESTAR Thin Base)	2.5	Terrain tag, 1207-5; primary 1209-5 through 1213-5
1414	High Definition Aerial (ESTAR Ultra-Thin Base)	1.5	Terrain tag, 1213-5; primary 1214-5
SO-208	High Definition Aerial (ESTAR Ultra-Thin Base)	1.2	1214-5 tag (called UUTB in PFA reports)
SO-315	High Definition Aerial (ESTAR Ultra-Thin Base)	1.2	Terrain tag, 1214-5; terrain primary load, 1215-5, 1216-5
Duplicating Films			
2420	AEROGRAPHIC duplicating film	4.0	Terrain 1205-5 and 1206-5; stellar use
SO-360	EKTACHROME AEROGRAPHIC duplicating film (ESTAR Base)	4.0	Color duplicating film
SO-467	AEROGRAPHIC duplicating film	4.0	Improved 2420
SO-284	AEROGRAPHIC duplicating film	4.0	Improved 2420; predecessor to 2421
2421	AEROGRAPHIC duplicating film	4.0	Replaced 2420/SO-284
SO-192/ QX-822/ SO-187	HIGH RESOLUTION Duplicating Film	4.0/7.0	First Terrain Use on 1207-5 (SO-192); QX-822 Introduced on 1215-5, SO-187 on 1216-5
2422	AEROGRAPHIC DIRECT Duplicating Film	4.0	Duplicate negatives
SO-355/ SO-277	LOW CONTRAST FINE GRAIN Duplicating Film	4.0/7.0	Terrain

MCS HISTORY

SECTION 5

POST FLIGHT ANALYSES (PFA)

CONCEPT AND PURPOSE

As the words "post flight" suggest, analyses of photography recovered from orbiting satellites are conducted to determine the level of system performance, initially from an engineering point of view. For the Mapping Camera program, these analyses were conducted in three locations, first at the film processing site, then in greater depth at the Defense Mapping Agency centers and the Itek Optical System Division's facilities in Lexington, Massachusetts and Sunnyvale, California. The engineering evaluation for each mission was scheduled to be finalized within 60 days following mission completion, at which time a PFA report was issued.

The prime objective of the PFA team was to produce an expedient, credible engineering evaluation of each system, this evaluation then providing the basis for actions as indicated to sustain or improve levels of performance.

PFA BACKGROUND

Of historical interest, the HEXAGON program was not the first to use a PFA team for performance evaluation. The origin of the team evaluation concept was coincident with the early satellite reconnaissance missions, and since that time, post flight analysis has been delegated a significant role in the management of all satellite reconnaissance programs.

With President Eisenhower's decree in May 1960 that there would be no more U-2 flights over Russia, satellite reconnaissance offered the most promising method of probing the alleged "missile gap" between the United States and the Soviet Union; hence satellite systems, under development and soon to be tried, would be vying for longevity in the intelligence collection systems inventory. Crucial decisions would be required to identify the system or combination of systems offering the highest potential.

The most attractive approach to evaluating system performance and comparing one system with another was the formal team concept. Without this, biased opinions and strong recommendations would be coming in from many individual sources, clouding issues and leaving decisions vulnerable.

Exposed photographic film had been recovered from orbiting satellites (CORONA) as early as 19 August 1960, and images had been received from SAMOS via radio transmission—and of course these results were evaluated by government and contractor personnel. But the first occasion in which the formal team concept was employed to evaluate two systems competing for survival was in 1963. These systems were GAMBIT (KH-7) and LANYARD (KH-6), described in Section I. Although by design they would serve different purposes, GAMBIT being a 77-inch spotting (strip) camera and LANYARD a 66-inch panoramic camera, it was nevertheless understood that only one of these systems would be funded into future programming.

Under the direction of Brig. General Robert E. Greer, SAFSP, a team of Air Force officers was appointed to evaluate film recovered from the first GAMBIT mission in July 1963. Then in August 1963, essentially the same team was directed to evaluate the first LANYARD mission photography.

Although each system had experienced problems on orbit, the team's analyses showed that portions of the photography from both systems had achieved design goals.

As stated previously, LANYARD had been developed and flown in the CORONA program as a backup in the event the spotting camera encountered insurmountable problems. Since the KH-7 had met design goals on the first flight, the systems chosen for future programming were the KH-7 as the high resolution system and the 24-inch CORONA panoramic system to satisfy search and surveillance requirements. The emphasis would be on product improvements in these two systems for the near future vice bringing along additional systems, hence, LANYARD was closed out shortly thereafter. Thus, the PFA team, then referred to as the Performance Evaluation Team (PET) had provided to the decision makers a credible analysis of the orbital system performance of these two systems, this analysis being of the utmost importance to the decision that was to be made.

The performance evaluation team (PET) continued to function in the GAMBIT program, totally under Air Force management, with small deviations from the first mission evaluation; however, in the CORONA program, under joint Air Force/CIA management, an official PET was not formalized until sometime later, in mid-1964.

Preceding the formalized PET, the typical approach to CORONA mission evaluation was to get out a flash report via secure teletype from the processing site by a "Tiger Team" comprised of personnel from the National Photographic Interpretation Center (NPIC). A rather informal engineering appraisal by government and contractor personnel would then take place at NPIC during the following week or so, this producing verbal reports and finally, at a later date (several months) a written mission report would be published. This method of evaluating and reporting was soon considered to be inadequate, for the following reasons. By mid-1964 the CORONA "J" dual-bucket series were being launched at a frequency averaging about 1.5 vehicles each month, with frequencies sometimes as low as 12 to 15 days. The two buckets, recovered 4 or 5 days apart, on the average, produced approximately 32,000 feet of panoramic camera film, films from the Stellar Index (SI) Cameras, and vehicle attitude images from the four Horizon Optics (HO) Cameras, all to be evaluated relative to the next flight. The flash reports, albeit timely, were not system oriented, nor were they of sufficient engineering detail to provide the credibility required for decision making—and the final reports, published several months later, had value only as historical documents.

To cope with the problem, Colonel Paul Heran, Director of Program A at SAFSP, directed his payload division manager, then Major M. G. Burnett to formalize and implement a plan for evaluating and reporting that would be commensurate with the accelerated launch schedule. To accomplish this, two significant changes were made, one in the evaluation personnel, the other in the method of reporting. First, personnel were selected from participating governmental agencies and contractors who were intimately knowledgeable with the hardware and film requirements, these personnel to serve as permanent members of the PET. In connection with this, the practice of individuals "floating through" NPIC for a quick look and passing to the community verbal interpretations was discouraged by directors in both organizations, Air Force and the CIA. Secondly, the format for a secure teletype report was formulated which would provide to the community a timely and credible team evaluation of each mission. The proposed message format and details of the plan were presented to Mr. John Crowley, CIA Director of Program B, and two members of the NRO Staff, Colonel Henry Howard, USAF, and Captain Robert Koch, USN. Approval was granted on a "trial balloon" basis, and the plan was initiated on the very next mission evaluation. The mission teletype reports, known as PIER's (Performance Interim Evaluation Reports), proved to be so appropriate and reliable that in a short time the PIER messages were considered throughout the community as the only credible source of information pertaining to system performance.

MAPPING CAMERA PFA ORGANIZATIONS AND PROCEDURES

In the transition from CORONA to the HEXAGON Mapping Camera performance evaluations, there was no loss in theme. Fortunately, some of the key government and contractor personnel, Lt. Colonel William Johnson of SAFSP and Harold Alpaugh, Stephen Herman, and M. G. Burnett of Itek, delegated to organize and serve on the Mapping Camera PFA Team, had served as regular PET members on the CORONA program. Due to differences between the MC and the CORONA contracts, mission frequency and mission objectives, it was necessary only to expand the PFA from the CORONA concept and structure it to accommodate MC program technical and contractual requirements.

As previously stated, there were two phases to the engineering PFA, the initial or "first-look" analyses at the processing site, followed by in depth analyses conducted at the DMA and the Itek facilities. The organizational structures and data flow are shown in Fig. 5-1.

Data gathering for the PFA meetings at the processing site and subsequent post flight analyses began well in advance of mission completions. Any anomalous condition observed during the test cycle and/or the orbital mission which might require post mission special investigation/explanation would be methodically documented and available at the PFA meeting.

Also, and of significant import to mission product identification and exploitation, certain data collected daily throughout the mission was recorded in a Mission Performance Report (MPR). The purpose of the MPR function was to perform the computational analysis required to determine the geographic areas and targeting information during camera operations, and to prepare the data base tables. The specific purposes were to:

1. Identify the World Area Grid (WAG) cells whose centers lay in the photographed area, along with the obliquity sector in which the center laid, and the sub-vehicle point at or near the time of photography. These data were needed for correlation with verified weather data for use in weather countdown in the mission performance evaluation function.

2. Identify which reporting targets lay in each camera frame, calculate their film coordinates, and determine whether they were completely contained in the frame, partially in the frame, etc.

3. Provide the data required to interpret and score the photographs by frame and operation.

4. Provide printout and transmission tapes, as required, of camera operation and ephemeris related data, and reporting target data for use by the user community.

The MPR generation utilized as the starting points for performing its task: (1) a best fit ephemeris to define the vehicle position as a function of time; (2) the history of executed commands; and (3) frame reference times from telemetry for frames exposed by the camera. The MPR was generated on a rev-by-rev basis. Utilizing the ephemeris data, the executed command history data,

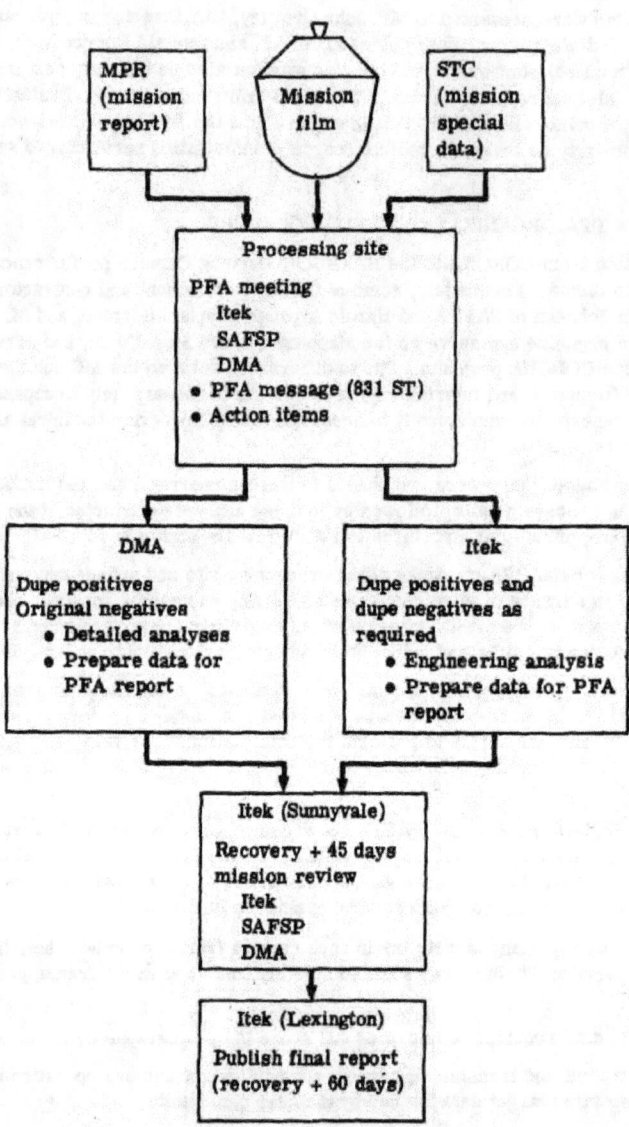

Fig. 5-1 — PFA organizations and data flow

and the telemetry-derived operation data, the camera operations which occurred on the rev were identified. The geographical bounds of the camera operations were then established and output as a portion of the Mission Performance Report. The frame reference times and the ephemeris data were utilized to establish, for each frame in a camera operation, the location on the film of each reporting target. This reporting target data was then output as part of the MPR. Camera operations on the rev were processed in the order of occurrence. Each succeeding rev was processed until all the revs in the rev span of interest had been processed.

Normally, the MPR subsystem was run on a daily basis after the best fit ephemeris had been generated for the time period covered by the rev span of interest.

The data in each Mission Performance Report included:

For each Mission Performance Report:

    Date of report
    Mission number
    Vehicle number
    Vehicle ephemeris identification
    Span of mapping camera operation numbers covered
    Camera identification number
    Lens number
    Calibrated focal length
    Camera filter
    Film type
    Time correlation parameters
    Initial condition parameters
    Physical constants
    Security classification header
    Times of orbit adjust
    Times of reentry vehicle firings

For each rev:

    Rev number
    Date
    Ascending node time (system time)
    Ascending node longitude
    At 20 ephemeris points:
        System time
        Vehicle inertial position coordinates
        Vehicle inertial velocity coordinates
        Vehicle inertial acceleration coordinates

For each mapping camera operation:

    Rev number
    Latitude and longitude of four corners of total area photographed by camera (includes
        mono coverage at beginning and end of operation)
    Time of first and last frames
    Number of frames in operation
    Overlap mode

For each mapping camera frame:

    Rev number
    Operation number
    Frame number (within operation)
    Nadir longitude and latitude
    Time of frame center (GMT)
    Time of frame center (vehicle time)
    Sun angle
    Vehicle altitude
    Inertial velocity
    Ground track velocity
    Ground track azimuth
    Commanded FMC rate
    Vehicle attitude and attitude rate
    Right ascension and declination of center ray for each stellar camera

The MPR was sent on a daily basis from ▓▓▓▓ (Satellite Test Center, Sunnyvale, California) to NPIC (Washington), DMA (St. Louis), DMA (Washington), and the DIA (Pentagon). It was used in the pre-planning efforts of the user community in preparation for the film processing functions, camera performance evaluation, information exploitation, and mapping functions. The daily report was used to generate "road maps" to facilitate film titling, processing, and evaluation.

The nucleus of the CORONA performance evaluation team regulars had seen to it that the PFA was well planned in advance of the first mission. This, together with stability of key personnel throughout the program, enabled the PFA meetings to function with minimal perturbations. Actually, the PFA activities were quite routine. As soon as the recovery date for the Mapping Camera RV was set, the PFA chairman, SAFSP, would send a message to all agencies participating in the "first look" PFA announcing the dates for the meeting, usually of 4 days duration. Since it was considered mandatory that the camera contractor inspect the RV prior to despooling, travel arrangements would be coordinated with the Hickam ferrying aircraft delivery schedule to enable the timely arrival of an Itek PFA member at the processing site for this requirement. Following this, the Itek member, usually Steve Herman, would get together with a processing site representative to compare daily mission records from the STC operation with the MPR. The "road map" compiled from this review would be fed into the optical titling program.

Other members of the team would plan to arrive prior to film processing. An overview of the mission would be presented by Itek from the data compiled during the mission, pointing out any anomalies which might be expected and if indicated, recommendations would be made for special film handling/processing. Action items from previous missions would be reviewed, and agreement reached as to processing chemistry and equipment to be used for the terrain and stellar operational films and test "tag on" films. Resident processing site personnel then proceeded with film processing and duplication.

Following the steps of preparation just described, there were actually four major functions to be accomplished at the PFA. First of course was the processing and duplication. Then upon the availability of the original negatives and duplicate records, team members would start the evaluation. The DMA breakout or "Tiger Team," comprised of a team leader and four members, would examine every frame of the original negative records. Concurrently, the SPO chairman and camera contractor team would be reviewing the duplicate records as they became available, looking for anomalies, examples of duplicate coverage for comparison with previous missions, examples of exceptionally good photography, and representative samples suitable for rating via the Visual Edge Match (VEM) technique. This technique, developed by Itek in 1970, uses a standard photo-interpreter viewing table with a variable zoom binocular microscope. The photographic image is magnified to the point where the image becomes divorced from instant recognition (the image breaks down due to grain, image resolution, image contrast, and other image degrading factors) as judged by the person viewing the photography. This magnification is then converted to an estimated film resolution number based on a calibrated conversion factor that had previously been determined for the photointerpreter. The conversion factor curve was initially determined by an analysis of the magnifications required to read 2:1 contrast resolution targets from laboratory and operationally generated aerial photographs. This conversion factor was further verified by cross checks where the resolution was withheld, forcing the interpreter to judge the resolution from the images. This evaluation approach closely matches that of the photointerpreter who is also hampered by image scale, contrast, shadow lengths, spectral response, and all other factors that effect the final intelligence that may be extracted from the aerial photographs. Photographs of VEM equipment and a VEM matrix array are shown in Figs. 5-2 and 5-3.

During the course of the PFA meeting, two individuals, usually ▊▊▊▊▊ from DMA and Harold Alpaugh from Itek, selected and evaluated a representative sampling of mission terrain photography via the VEM technique. In the MCS program, VEM results were used for two purposes: (1) to show the resolution pattern over the entire 9 × 18-inch format; (2) for comparison, mission-to-mission.

As the meeting progressed, sections of the PFA message (Rebound 831ST) would be written in preparation for the final review and message sign-off on the last day. Among those items reported in the PFA message were general mission statistics, resolution from VEM analysis, exposure analysis and processing, preliminary cloudcover assessment, quality and adequacy of stellar imagery, results of special engineering photography, results of in-flight calibration, conclusions, and action items.

The "extended" PFA, consisting of more detailed analyses, required the use of more sophisticated mensuration devices. This part of the PFA would be performed at DMA. Three reports would result from these analyses and evaluations, i.e., the final cloud cover assessment, the PFA Summary, and the In-Flight Camera Calibration Report. Although the second of these reports is the only one referred to as "PFA", all were reports of post-flight evaluation activities.

The preliminary cloud cover assessment which was accomplished at the Processing Facility was done by visually estimating the percentage range of cloud cover on each frame. Weighted averages were used to determine the percentage of frames which had less than 50 percent cloud cover and the percentage of frames with less than 10 percent cloud cover. For a quick assessment, this works satisfactorily. However, users are concerned with where the cloud cover occurs. Consequently, the first analysis that occurred when the film was received at DMA was the digitizing of the cloud cover of each frame. This digitized data was then used to provide statistics and graphics which were invaluable aids to the users and planners. These results were usually available within 3 days after receiving the film.

The PFA Summary was the DMA portion of the Final PFA Report produced by the Mapping Camera contractor. This summary included:

1. Film distortion analyses and comparisons with previous missions.
2. Fiducial stability analyses which included distance checks between fiducials and fiducial calibration.
3. Densitometer analyses including both macro- and microdensitometry.
4. Various dimensional measurements such as format size, interframe distances for both terrain and stellar camera, and intraframe distance within a stellar pair.
5. Measurements to determine if the film tracked satisfactorily.
6. An evaluation of the stellar film to determine the stellar cameras' ability to image 6th magnitude stars.
7. Timing studies to check cycle times and to determine differences between terrain and stellar exposure midpoints.

When the contractors and DMA had completed their analyses, the PFA chairman (from SAFSP) would convene a review meeting during which all of the various inputs to the Final PFA Report would be reviewed and edited. The status of all action items would also be reviewed, and a unified set of conclusions and recommendations would be agreed upon. The final report would then be published.

x x x x x x

IN-FLIGHT CAMERA CALIBRATION PROCEDURES

This important aspect of the program consisted of two major phases: (1) reduction of data acquired before flight, referred to as "pre-flight calibration" (see Section 2, page 2-46); and (2) reduction of data acquired during flight, referred to as "in-flight calibration." Both phases, beginning with the pre-flight calibration procedures, are described here in the post-flight analysis section since the final solution to the calibration of each MCS was reached only after reduction and analysis of all calibration data had been completed.

The purpose of calibrating a camera is to define certain stable elements which the photogrammetrist categorizes as lens distortion, interior orientation, and relative orientation. The starfield exposures obtained at Cloudcroft were well suited for this stable element definition since stars provide essentially point source imagery and their positions on the celestial sphere are known to a high degree of accuracy. The configuration of the Terrain and Stellar Cameras permitted simultaneous exposures by all three cameras. This was necessary for the determination of the relative orientation of the Terrain Camera to the Stellar Cameras. Split-vertical simultaneous exposures, with the Stellar Cameras only, provided constraining information for relative orientation. Vertical exposures, with the Terrain Camera only, provided constraining information for the Terrain Camera lens distortions, focal length, and principal point offset, i.e., the coordinates of the point in the focal plane intercepted by the principal axis of the lens.

Fig. 5-2 — Visual Edge Match equipment (VEM)

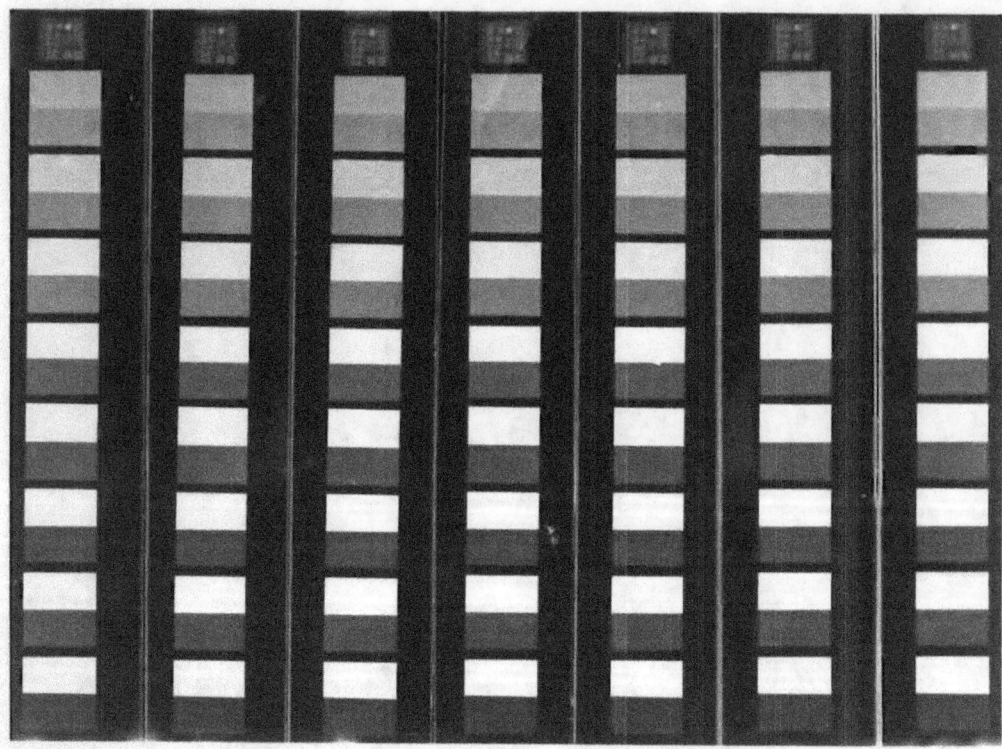

Fig. 5-3 — VEM matrix array

The theory behind camera calibration using starfield exposures is comparatively simple. However, putting theory into practice often becomes a monumental task. Corrections for the effects of the windows on the vacuum tank had to be obtained from Itek. The calibrated coordinates for the reseau intersections also had to be obtained from Itek. Meteorological data obtained at Cloudcroft had to be checked for consistency. Star observation data had to be reduced for atmospheric refraction modeling. Data had to be obtained to correct refraction for an air-to-vacuum interface at the windows of the vacuum tank.

Since the fiducials of the Terrain Camera are stationary relative to the lens, and the reseau is not, the camera and film coordinate systems had to be defined relative to the fiducials. Consequently, a set of calibrated coordinates had to be derived for them. This had to be done before any star image measurements could be completely pre-processed. Fortunately, this calibration is independent of any other except the reseau calibration which is necessary to remove the effects of film distortion.

Ten frames would be selected for the fiducial calibration. The fiducials and the four reseau intersections surrounding each one were measured manually on a Mann comparator. The frame would then be rotated 180° and measured again. Each frame would be measured in this manner by three different operators. Film distortion effects were removed with a six-parameter transformation which corrected for translation, rotation, a scale factor in the x direction, a scale factor in the y direction, and for the non-perpendicularity of the film coordinate axes. The resulting coordinates (now relative to the reseau coordinate system) were matched with the corresponding set from the opposite rotation, and averaged to remove the effects of operator bias. This was done for each operator, and then, corresponding results for the three operators were averaged. At this point, the "best frame" was selected on the basis of the lowest transformation variance, and all other frames were transformed to it by means of a three-parameter (translation and rotation) transformation. This resulted in having all coordinates in a consistent system, and thus all corresponding coordinates could be averaged. The final step was to adjust the coordinates by translation only, such that the average of the x coordinates would be zero and the average of the y coordinates would be zero.

Other activities proceeded simultaneously with the fiducial calibration. Frames were selected for the other calibrations to be performed. Then, the first step was to identify known control stars. Given the time of exposure, location of the camera, and the approximate azimuth and elevation angles of the principal axis, an approximation of the starfield seen by the camera could be made. An overlay would be made showing the location of a number of stars as they appear on the frame. This overlay would be oriented to a star chart of approximately the same scale as the photograph. A small number of stars, usually 10 to 15, would be identified and their Boss General Catalog numbers recorded. These control stars would be used later for a preliminary orientation which would permit the automatic identification of all other star images measured.

Approximately 150 star images on each terrain frame and 60 for each stellar frame would be measured. The four reseau intersections surrounding each star image would also be measured for the purpose of removing the effects of film distortion.

The time measuring was completed, all the ancillary data necessary for processing was compiled, and the pre-processing begun. Pre-processing consisted of correcting film measurements for all known effects which were independent of the elements to be defined by the calibration. Those effects were film distortion, vacuum tank window corrections, air-to-vacuum refraction correction, and the transformation of the image measurements to the fiducial's coordinate system defined by the previously mentioned fiducial calibration.

The Simultaneous Multi-Camera Analytical Calibration (SMAC) computer program written by Duane Brown Associates, Inc. was the primary adjustment program used for calibration. The corrected data was input into the single camera portion of the program. The single camera portion would compute preliminary orientations to starfields using the identified control stars, match the remaining star images to the Smithsonian Astrophysical Observatory (SAO) star catalog, update the star positions, correct for atmospheric refraction, and proceed with a preliminary calibration. Preliminary runs such as this were used to clean up the data by eliminating bad measurements, mis-identified stars and blunders such as a mis-punched computer card. A final single SMAC reduction was made, and the resulting output was used in the multi-camera portion resulting in the calibration of the relative orientation angles between the Stellar Cameras and the Terrain Camera.

The entire calibration effort, except for the fiducial calibration, would be performed at least one more time by a different analyst. If the results were significantly different, the calibration would be done again. If the results were not significantly different, the most consistent set (least variation) was published.

The In-Flight Camera Calibration report consisted of three basic aspects of Post Flight Analysis, i.e., dynamic calibration using Bar XC photography, C-Mode calibration, and calibration verification.

The principles involved in dynamic camera calibration were basically the same as for preflight calibration or C-mode calibration, i.e., a "truth set", in this case geodetic control, is imaged on the photography and from this "truth set" and the tracking available over the Bar XC, the calibration parameters could be determined. Typically, more than 1 operation would be taken over the Bar XC range, and at least 2 would be chosen for calibration. The control points would be identified and marked for measuring. Tracking data would be obtained and a short arc ephemeris computed. Control stars would be marked on the stellar film for the purpose of automatic star identification on the measuring device. All these data were compiled and fed into the computer and the calibration parameters computed. The focal lengths and principal point offsets would be constrained in this solution due to high correlations with position and attitude errors.

In the C-mode operation, the vehicle orientation would be such that all three cameras would photograph stars. Thus, the known right ascensions and declinations of the stars became the "truth set" for calibration. In theory, this method of obtaining calibration data was the best. However, differences between C-mode calibration and pre-flight calibration were met with skepticism. The procedures for C-mode calibration were very similar to those used for preflight calibration. Control stars were identified on each frame for both the Terrain and Stellar Cameras. The control stars were used for preliminary orientation to the stellar field which is used to automatically pre-position the comparator on other stars used in the data reduction.

MCS HISTORY

SECTION 6
EXPLOITATION

EXPLOITATION SYSTEMS DEVELOPED AND/OR UTILIZED

Development of systems and procedures for exploitation of satellite imagery in MC&G production activities was well underway prior to introduction of the KH-9 and its Mapping Camera System.

The U.S. Army Corps of Engineers had principal responsibility for Army MC&G operational programs. The Army Map Service (AMS), being the base plant for production of maps and charts, was an early user of satellite and other SAO materials. The Engineer Topographic Laboratories (ETL) had responsibility for research, development, test, and engineering (RDTE) of surveying, mapping, charting, and geodetic systems and equipment both for use in AMS and for use by Army Tactical Units. Beginning in 1962, ETL undertook development of a Universal Photogrammetric Data Reduction and Mapping System (UPDRAMS) for use by AMS.

The UPDRAMS entailed a radical departure from previous and then-current philosophy: mapping data reduction systems for use with imaging systems had been developed in the past to accommodate the characteristics of specific cameras, with emphasis on the classic 6-inch focal length/$9\frac{1}{2} \times 9\frac{1}{2}$-inch-format aerial cameras in use since before World War II. UPDRAMS stressed use of computer-aided systems for extraction of feature and terrain information from imagery, thereby resulting in equipment which was focal-length independent.

Use of analytical (mathematical) techniques was stressed. Then-current equipment was analog, i.e., pairs of photographs comprising stereo-pairs were manually oriented in stereoplotters; a cartographer then manually plotted, on a paper manuscript, man-made features such as roads and buildings, natural features such as streams and vegetation, and elevation data in the form of contours. In UPDRAMS, measurements made on single photographs using precision measuring systems (comparators) permitted computation of such parameters as camera altitude and orientation (attitude); the computed values could then be used to control high speed, automated compilation systems.

The Army had long been attempting to win approval in DoD for a satellite mapping camera having the long focal length necessary to give reasonably large scale imagery from satellite altitudes. Analyses had indicated that an 18-inch focal length was optimum, but a 12-inch focal length was adequate. Of equal concern was the imaging format. If the focal length was increased, but the format remained $9\frac{1}{2} \times 9\frac{1}{2}$ inches as used in conventional mapping cameras, the resulting ground area covered by a photograph would be very small. Army analyses indicated a 9 × 18-inch format was optimum. Whereas the UPDRAMS was focal-length independent, the imagery format had to be set; the Army thereupon designed the UPDRAMS around a 9 × 18-inch format.

TOP SECRET/RUFF/GAMBIT/HEXAGON

BIF-059W-23422/82
Handle Via
BYEMAN/TALENT KEYHOLE
CONTROL SYSTEMS JOINTLY

The Air Force Aeronautical Chart and Information Center (ACIC) had principal responsibility for Air Force MC&G operational programs. ACIC, like the AMS, was an early user of satellite and other TK imagery in production activities. In particular, ACIC pioneered the application of panoramic photography in MC&G production activities, including the development of mathematical techniques for partial or full compensation of the geometric distortions inherent in that type of imagery. The Rome Air Development Center (RADC) at Griffiss Air Force Base, N.Y., conducted research and development of systems and equipment for use by ACIC. The two organizations collaborated in implementation of a base plant production system which stressed use of computational techniques combined with a computer-assisted analytical stereoplotter adapted from a Canadian invention.

Scientists and engineers at the National Research Center in Canada produced a computer-assisted analytical stereoplotter during the late 1950's. This equipment, although manually-operated, had the capability to produce cartographic products from virtually any photography so long as the camera which produced the imagery could be mathematically described. The automated, computer-assisted MC&G production systems revolutionized cartographic production activities. The MCS alone has made possible the determination of ground feature locations (coordinate values) at a small fraction of previous production costs, while also allowing production agencies to meet most accuracy requirements anywhere in the world.

The Replacement of Photographic Imagery Equipment (RPIE) (see Fig. 6-1) was initially designed and built with the intention that input from the MCS would support major production efforts in producing orthophotos. However, due to changes in programs, priorities, and products, and because of the RPIE's many capabilities, sources other than the MCS were utilized when the RPIE was placed in production. The RPIE's utilization of MCS input was primarily in support of the Large Scale Mapping Program. Other minor and special products have been supported by the MCS on the RPIE.

A Universal Automated Map Compilation Equipment (UNAMACE) (see Fig. 6-2) was developed, the first being delivered in 1963. ETL had been experimenting with automated, electronically-aided systems since the late 1950's wherein a pair of photographs were fused into single stereo imagery. In the UNAMACE, this stereo image is not created in a literally physical sense; rather, an on-line computer receives data as two photographs comprising a stereo pair are automatically scanned by electronic sensors. Incremental imagery displacements necessary to remove all distortions and correctly position image features are computed. These computed corrections are then used in production of an orthophoto, in which every image point is in its correct position relative to every other image point. On an orthophoto, horizontal distance between any two image points can be accurately measured; in conventional aerial photography, this cannot be done.

Even though the UNAMACE was not initially intended for use with input from the MCS, developmental programming efforts resulted in an MCS compilation capability in 1976. MCS input to the UNAMACE supported the Large Scale Mapping Program from approximately 1976 until 1980. Other special projects and MC&G products have been supported by the MCS on the UNAMACE. Seven of the UNAMACE systems were built altogether, the last being delivered in 1969.

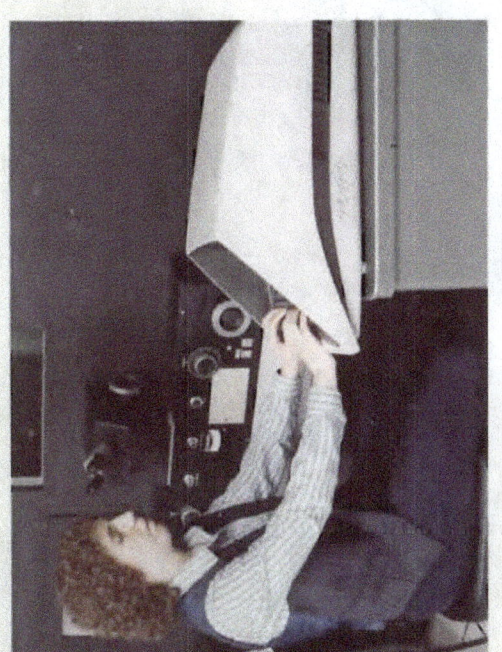
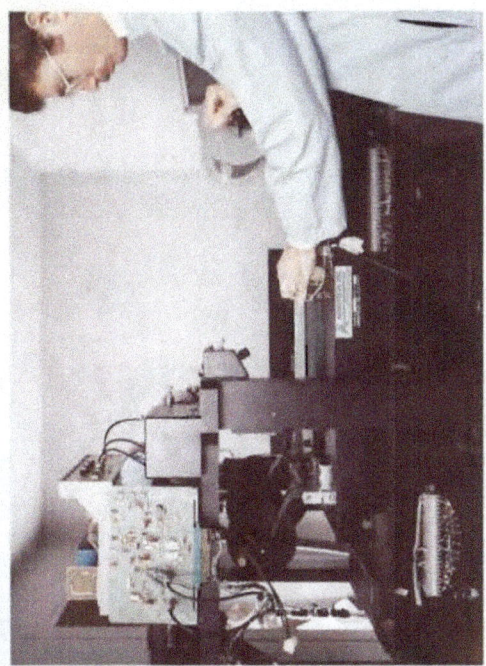

Fig. 6-1 — Replacement of Photographic Imagery Equipment (RPIE)

MCS HISTORY

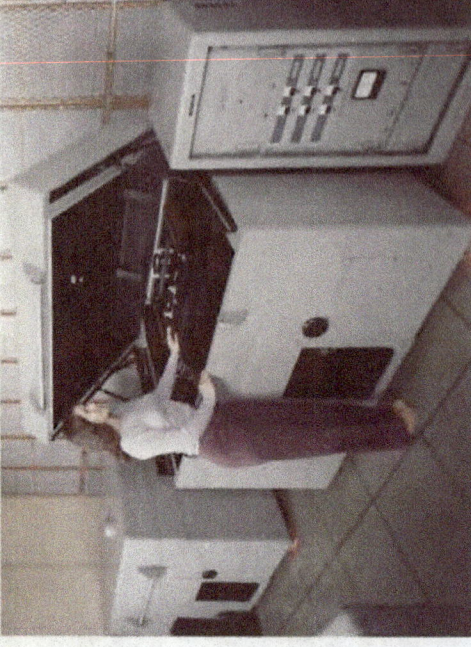

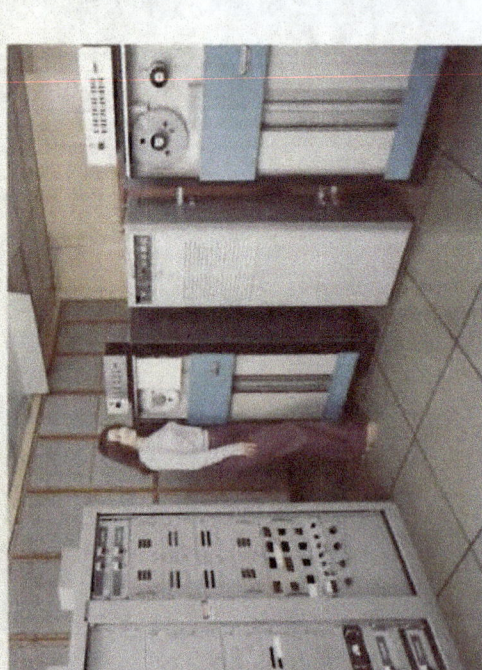

Fig. 6-2 — Universal Automatic Map Compilation Equipment (UNIMACE)

BIF-059W-23422/82
Handle Via
BYEMAN/TALENT KEYHOLE
CONTROL SYSTEMS JOINTLY

TOP SECRET/RUFF/GAMBIT/HEXAGON

The Automatic Reseau Measuring Equipment (ARME) was specifically designed to exploit the MCS characteristics (see Fig. 6-3). It automatically drives to and measures pre-assigned x, y coordinates. It consists of the following:

- Numerically controlled comparator with 1-micrometer resolution, 10 × 18-inch field (H. Dell Foster)
- A 729-line video acquisition system with 0.8-micrometer resolution, 5% contrast sensitivity, and a data matrix of 255 × 255-resolution elements (DBA Systems, Inc.)
- 729-line video point locator system displaying table position versus a contact print of the photograph being measured (DBA Systems, Inc.)
- Functional human engineered control console designed to minimize the effort for operator computer training (DBA Systems, Inc.)
- Central processor unit with a 32K core memory, two magnetic tape units, 491K disc memory, card reader/punch, high-speed printer/plotter, CRT and TTY terminals (Varian Data Machines)

The video image acquisition system replaces visual/manual centering for reseaus, grid intersections, stars, fiducials, and marked points via closed-loop control of the comparator table position.

The ARME was purchased from DBA Systems, Inc. of Melbourne, Florida to support DMA's large scale mapping point positioning data base and other special projects.

The AS-11 stereoplotter (Fig. 6-4), consisting of a viewer/comparator, computer, and coordinatograph, is used at DMA for medium scale map/chart compilations and for digital data compilation of equivalent scale occurring using MCS photography.

The photographic laboratory equipment used to support the MCS (some having been procured specifically for this purpose) is listed below:

- Large format copy cameras
  - These cameras produced enlargements and reductions in support of special products.
- Large format vacuum frames
  - This piece of equipment used a step-and-repeat process in its contact 1:1 printing support of point position data bases.
- Strip vacuum frame (see Fig. 6-5)
  - This equipment was purchased specifically to support the MCS system. It is a step and repeat 1:1 contact printer which supported the point position data bases.
- Enlargers
  - These cameras produced photo enlargements in support of the Large Scale Mapping Program and other products related to the MCS system.
- Electro-Optical Rectifier
  - This instrument produced rectifications and enlargements and supported special products.

- 2.5× Latady
    - The Latady is a 2.5× reduction continuous roll printer purchased specifically to support products from the MCS at large scale.
- 3× Latady
    - The 3× Latady supported the PPOM's, image scale, and other products related to the MCS. This instrument is a step-and-repeat printer enlarger.
- Mark II Printer
    - The Mark II is an auto-dodging contact step-and-repeat printer which supported the Special Products, PPDB's, and other products related to the MCS.
- Reseau Printer
    - This printer is a step-and-repeat strip printer which superimposes reseau ticks on the film. This printer supported the Large Scale Mapping Program.
- SP1070 Printer
    - This instrument is a continuous strip printer with dodging capabilities. This printer supported Special Products, PPDB's, Large Scale Mapping, and other products related to the MCS.
- Kingston Printer
    - The Kingston is a continuous non-dodging strip printer with high geometric accuracy. This printer supported the PPDB program.
- Miller Printer
    - The Miller printer is a 9 × 18-inch contact step-and-repeat printer which supported Special Products, PPDB's, and Large Scale products.
- 10×-20×-40× Enlarger (Kodak)
    - This 10×-20×-40× enlarger supported Special MCS Products.
- Microdensitometer
    - This instrument had digitizing and playback to analog capability and supported Special MCS Products.
- A&B Rectifier
    - This rectifier reduced 9-inch-format film to 70-mm format for rectification. This rectifier supported the Large Scale Mapping Program and other MCS Products.
- E-4 Rectifier
    - This rectifier enlarged and rectified chips and supported Special MCS Products.

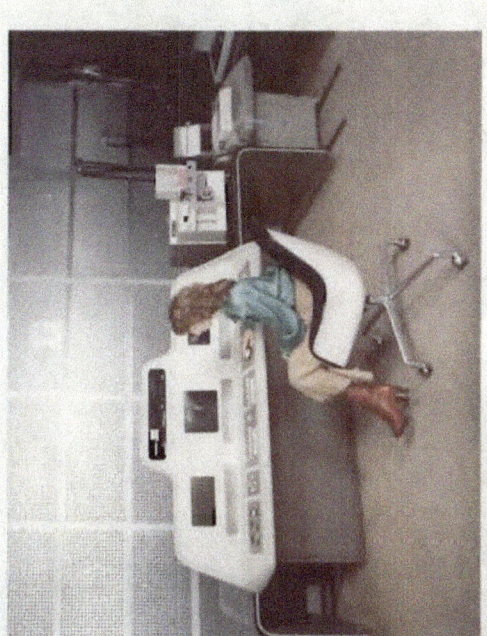

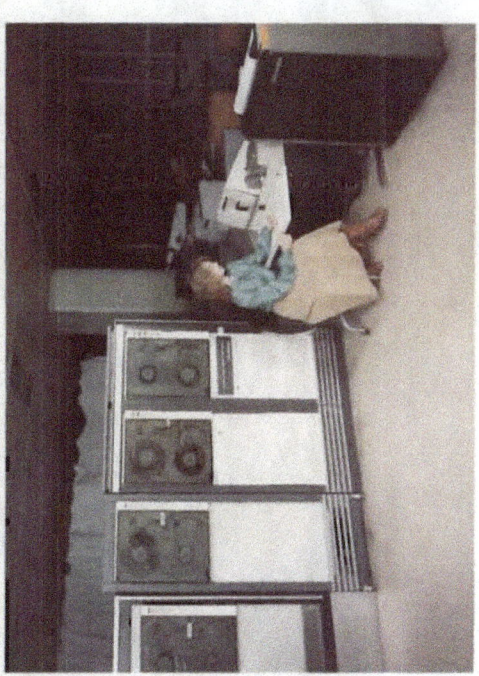

Fig. 6-3 — Automatic Reseau Measuring Equipment (ARME)

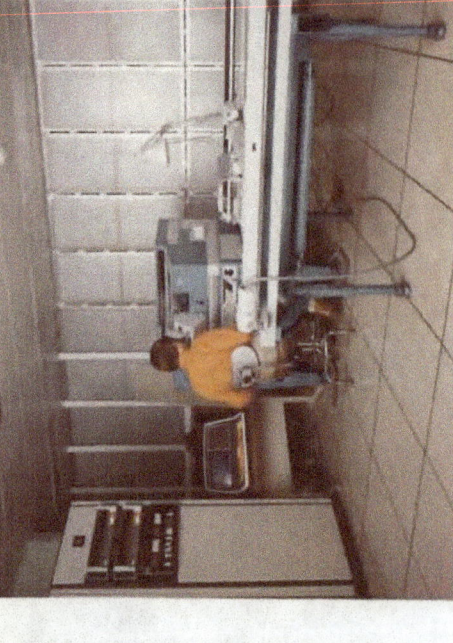
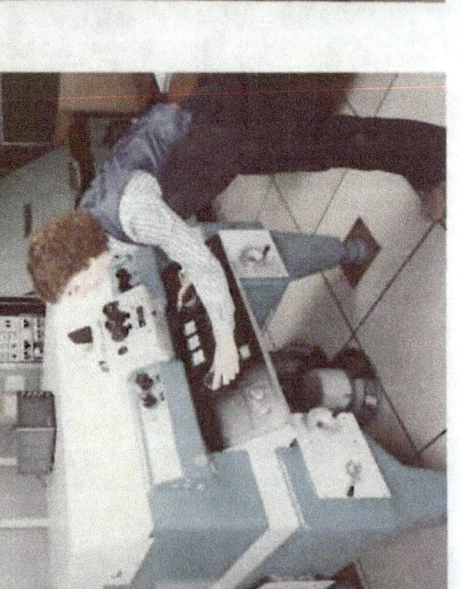

Fig. 6-4 — Analytical Stereoplotter Systems (AS-11)

Fig. 6-5 — Strip vacuum frame

EXPLOITATION OF IMAGERY

KH-9 MCS imagery currently is used by the Defense Mapping Agency (DMA) in a large number of high-priority programs, both for production of conventional maps and charts and, more recently, for products in digital form for direct support of weapon systems. Several key programs will be discussed; these provide a good definition of the breadth of application of the MCS.

Analytical Triangulation

MCS photography is used as a source for analytical triangulation processes performed at DMA. Stereo photogrammetric techniques utilizing previously performed camera calibrations of the MCS are executed to develop ground control for subsequent use in DMA production programs. This control is adjusted on a continental basis to minimize or eliminate differences in ephemeris data between the various MCS missions (this adjustment procedure is discussed in more detail under Continental Control Network Data Base).

Triangulation techniques are based on precise measurements of photographic imagery, performing appropriate physical corrections and transformations to relate image positions to the camera system, and applying the math model relating image coordinates, camera position, camera orientation, and ground position.

Mensuration is accomplished on both stereoscopic and monoscopic comparators (measuring engines). Where the highest order of precision is required [as in the case of the CCN (discussed in a later section), primary control for mapping and charting operations, and target positioning], three-stage stereocomparators are used. The resulting precision for such control is better than 5 micrometers. For supplemental control, monocomparators often are used with measurements made on marked control points. The marking is accomplished with marking devices which remove emulsion at the image providing a record of the measured point in addition to flagging the image for the comparator to set on. Marked images are control points, pass points, and targets of interest. The resulting precision for such control is approximately 30 micrometers. Other points that require measurement are reseaus, fiducials, and star images. All comparator measurements are corrected for film shrinkage, transformed to the camera system, and further corrected for refraction, aberration, and lens distortion.

The corrected image coordinates, along with ephemeris, orientation, camera calibration information, and available control data are reduced by employing least-squares adjustment techniques to determine updated positional and orientation values and ground coordinates of pass points and targets. These output data are then used as input data to the photogrammetric compilation function. In the case of compiling from MCS photographs, these data are applied directly. In the case of compiling from other than the MCS photographs, a transfer of control data from the MCS photographs to the photographs to be used must take place prior to the data leaving the triangulation function.

Continental Control Network (CCN) Data Base

A CCN has been developed for the Eurasian Continent. CCN's for Africa and South America are in planning. The objective of the effort is to improve the accuracy of ground control used for DMA production programs. This improvement is achieved by minimizing the uncertainties induced in ground control coordinates by uncertainties in ephemeris data. Products using the CCN control include: digital terrain elevation data (DTED), digital feature analysis data (DFAD), vertical obstruction data (VOD), target and point positioning data, maps and charts, elevation matrices for systems using Terrain Contour Matching (TERCOM), etc. Points positioned using the CCN have an absolute accuracy of 23 meters horizontally (CE 90%) and 17 meters vertically (LE 90%).

Those points, derived from orbital data without using CCN, have an accuracy of 41 meters horizontally and 23 meters vertically at 90% probabilities. Adjustments of models to the CCN control for use in compiling digital data or point positioning data bases can be made to a relative accuracy of 4 meters horizontally and 3 meters vertically (90%).

The CCN is accomplished with a planetary block adjustment system which involves the construction of a photogrammetric data base composed of a sampling of two to four stereo models from sidelapping operations of KH-9 MCS photography. The common surface area between models from two operations is called a relative geometry point (RGP) area. Each RGP area is processed in a triangulation program and the derived ground coordinates of common image points and control points become the basic input to the adjustment program. A weighted least-squares adjustment minimizes the bias between common points derived from sidelapping operations. Once the translations are determined for an operation, then all exposure station positions in that operation are updated.

The results of CCN diagnostic testing show that, using MCS imagery, ground control and point positions on WGS-72 can be derived to an accuracy that is commensurate with the error budget established for the basic data components of the CCN, as cited above.

Photogrammetric Compilation

The primary photogrammetric compilation instruments at the Defense Mapping Agency are the Universal Automatic Map Compilation Equipment (UNAMACE) and the Analytical Stereoplotter Systems (AS-11).

The input to the UNAMACE consists of camera orientation, ground identified control from the CCN, ground center point, and film positives of the stereo model. Implementing the inputs from triangulation, along with the MCS photographs, provides the operator with a stereo model of the ground coverage suitable for up to 1:250,000-scale mapping.

The basic UNAMACE consisted of an operator console, four tables, one controller, and one magnetic tape unit. Of the four tables, two were input tables (for film positives), one was for output orthophotos, and one was for the output altitude chart. With the advent of the MCS, the UNAMACE was reconfigured to include the console, computer, upgraded magnetic tape units, and a three-table configuration. Two tables are used for input (film positives) and the third table is used for orthophoto production. The elevation information is placed on magnetic tape. This reconfiguration of the six UNAMACE systems left six surplus tables (one per system). With two of the six surplus tables, a special Off-Line OrthoPhoto System (OLOPS) was designed. The OLOPS-I has the capability of producing an orthophoto with 70 to 80 lines per millimeter resolution and at a much faster rate than the UNAMACE, since computations associated with compilation are not being done simultaneously with orthophoto production. The OLOPS-I is a two-table configuration; console, computer, and magnetic tape units. One table uses an MCS film positive and the second table produces the orthophoto. The basic control provided to the OLOPS-I is UNAMACE-generated elevation data stored on magnetic tape. An updated OLOPS-II will have resolution improved to 97 lines per millimeter.

Elevation data produced from MCS imagery is also used to drive the Replacement of Photographic Imagery Equipment (RPIE) to produce an orthophoto with up to 110 lines per millimeter resolution. The RPIE is a multi-purpose piece of equipment that can produce an orthophoto, rectified photography, or, using MCS-derived control information, replace high resolution imagery into geometrically correct positions. The RPIE consists of a computer, printer with a laser light source, magnetic tape units, console, and an analytical stereoplotter viewer converted to automatic correlation.

AS-11 stereoplotter instruments are used at DMA for medium scale map/chart compilations and for digital data compilation of equivalent scale accuracy using MCS photography. The AS-11A stereoplotter instrument consists of a viewer/comparator, computer, and coordinatograph. The computer analytically solves the relative and absolute orientations of a stereo pair (pan, terrain, or a combination) from measurements read from the comparator and automatically drives the comparator carriages to maintain a parallax-free viewing model. The AS-11A has 9 × 9-inch carriages and maintains a calibration accuracy of 3 to 4 micrometers. The AS-11B1, a later version of the AS-11A, has expanded memory, 9 × 18-inch carriages, and automatic correlation equipment to permit automatic contouring. Additionally, the AS-11B1 has expanded software enabling a real-time correction for reseau calibration. Input to the AS-11 systems comprises MCS stereo imagery and precomputed camera position and attitude data; outputs include chart manuscripts and digitized terrain and feature positional values. Two automated stereoplotters, the AS-11BX and ACE, provide automatic scanning of MCS stereo models at speeds of 6 to 8× the AS-11B1s. Their use is restricted to areas with little or no snow, and where image quality is good; manually operated AS-11 systems referred to earlier are used to fill in areas where image quality did not permit automatic correlation.

SECTION 7

OPERATIONAL CONSIDERATIONS AND STATISTICS

This section is presented under three major headings: (1) Highlights, from aspects of engineering and user accomplishments; (2) Mission Planning, Strategies, and Accomplishments; and (3) Mission Engineering Summaries and Major Changes or Proposed Changes (intermixed as they occurred by date during the operational period). Statistics are presented in tabular and chart form.

HIGHLIGHTS

Since the first MCS mission (1205), 9 March to 20 April 1973, the mission duration increased in increments from 42 days on the first mission to 118 days on missions 1215 and 1216.

To further enhance the data gathering capacity of the MCS, on mission 1214 the terrain camera film load was changed from 2.5-mil (STB) film to 1.5-mil (UTB) film, providing a 50 percent increase in footage. Then, on the last two missions (1215, 1216) the film load was changed to 1.2-mil (UUTB) film providing another 25 to 30 percent increase on the average load. Thus, during the course of the program the terrain film load had been increased by a factor of approximately 1.8* and the mission duration increased by a factor of approximately 3. Further, by changing to a finer grain film on the fifth and subsequent missions, the resolution approximately doubled (see Table 7-1 and Fig. 7-1).

MISSION PLANNING, STRATEGIES, AND ACCOMPLISHMENTS

Initially, the total land area of the world, less the ice caps of Greenland and Antarctica, was approved for once-over trilap MCS coverage. The area was approximately 39.4 million nautical square miles (mnsm).

Missions 1205 through 1207 were programmed against a four-priorities system. Some large areas of the world were deferred from collection to place emphasis on production programs within the Sino-Soviet bloc. Total unique area collected was 6.7 mnsm.

The mission plan for 1208 and 1209 increased the areas of the world deferred from collection of MCS photography. Also, all water areas considered for collection were deleted. Emphasis was placed on program requirements in the Sino-Soviet bloc. The strategies for MCS priority collection were changing. The initial 39.4 mnsm approved for MCS coverage had been reduced by 11.0 mnsm from priority collection. The redundancy of MCS coverage from mission to mission was

---

*The stellar camera film capacity was increased as required to provide adequate stellar frame pairs for the increased terrain loads.

Table 7-1 — Mission Statistics

Mission	1205	1206	1207	1208	1209	1210	1211	1212	1213	1214	1215	1216
Launch date	09 Mar 1973	13 July 1973	10 Nov 1973	10 Apr 1974	29 Oct 1974	08 June 1975	04 Dec 1975	06 July 1976	27 June 1977	16 Mar 1978	16 Mar 1979	18 June 1980
Recovery date	20 Apr 1973	24 Aug 1973	07 Jan 1974	09 Jun 1974	27 Dec 1974	30 July 1975	02 Feb 1976	06 Sept 1976	17 Oct 1977	11 July 1978	12 July 1979	14 Oct 1980
Days of operation	42	43	58	60	59	52	60	62	112	117	118	118
Operational Summary												
Operates	141	153	148	200	167	198	202	279	252	386	488	529
Terrain frames	2,026	2,118	2,145	2,120	2,077	2,090	2,066	2,090	2,109	3,144	3,947	3,840
Terrain footage	3,345	3,393	3,442	3,402	3,333	3,354	3,316	3,354	3,385	6,046	6,335	6,163
Terrain film type	3400	—	—	—	3414	—	—	—	—	1414	SO-315	SO-315
Stellar footage	1,823	1,910	1,930	1,908	1,869	1,881	1,859	1,881	1,898	2,829	3,552	3,437
Stellar film type	3401	—	—	—	—	—	3400	—	—	QX-817	—	—
Terrain VEM, l/mm	70	58	51	60	80	75	83	95	94	91	99	99
Factory specification, l/mm	54	—	—	—	85	—	—	—	—	—	—	—
Percent of trilap	100	83	100	—	—	—	—	99.6	93.6	67.1	86.9	86.3
Percent of bilap*	N/A	—	—	—	—	—	—	—	—	22.0	9.7	9.8
Percent of mono	0	17	0	—	—	—	—	0.4	6.4	10.9	3.4	3.4

* Bilap capability did not exist until mission 1214.

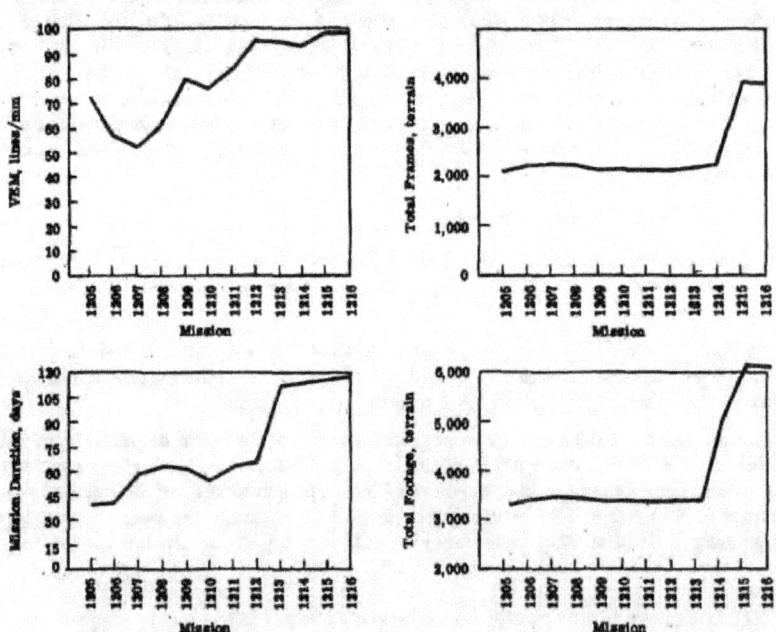

Fig. 7-1 — Mission statistics

increasing. The total redundant coverage (mission to mission) as of Mission 1208 was 0.978 mnsm. This increased to 1.437 mnsm through mission 1209 due to the collection of slivers or small gap areas. The total MCS/MC&G acceptable (trilap 90 percent clear WAG cells) coverage through mission 1209 was 9.7 mnsm.

Starting with mission 1210, new strategies in the DMA MCS planning and collection techniques were used. New data was added which divided WAG cells into 20 separate homogeneous weather areas. These data were provided by CIA from historical climatological records. These areas were used to create mapping categories (MCATS) for collection priorities which were given names and sent to CIA as part of the total collection ("MOB") film. MCATS divided collection priorities by geographic area, thus giving planners at DMA the option to delete or direct MCS operations against a smaller area to increase efficiency. The other strategy in MCS collection planning was the introduction of the Planetary Bloc Concept (PBC), later known as the Continental Control Network (CCN), which resulted in deleting those WAG cells from collection which were three WAG cells or less in cross track to a mission operation and were covered by MC&G pan photography (stereo) suitable for bridging control (cantilevering). This strategy resulted in deferring from collection approximately 30,000 cells (approximately 6.1 mnsm). Mission 1210 netted 1.3 mnsm of MC&G acceptable coverage. The total MCS collection through mission 1210 was 11.1 million of the original 39.4-million-nsm area.

Mission 1211 MCS collection strategies were the same as mission 1210. The Probable Acquisition (PACQ) plan for 1.8 mnsm was selected and MCS coverage (90 percent clear WAG cells covered by trilap) netted 1.4 mnsm coverage. Studies were completed prior to the launch of mission 1211 (Nov 1975), which concluded that metric pan would be adopted in the KH-9 system and the last two MCS buys for missions 1217 and 1218 would be dropped.

Mission 1212 MCATS were developed to maximize priority one and two areas. These were mainly in the Sino-Soviet bloc. All priority three and four areas in Africa were deferred from collection for this mission, which reduced the collection from 344,000 to 28,000 nsm. Beginning with this mission, all clear trilap MCS coverage of WAG cells which equalled 50 percent (12 of 24 sub-cells) or greater would be counted down as successful collection of MC&G priorities. This new criteria also affected earlier mission coverage of WAG cells at the 50 percent satisfaction level. The total collection of MCS coverage through mission 1212 at 50 percent satisfaction was 16.6 mnsm of the 39.4-mnsm requirement.

With mission 1213, the NAVPAC system (satellite-to-satellite tracking) was added to the KH-9 vehicle. Initial planning provided for 300 frames to be used over the Sino-Soviet area to create a network base to improve the CCN positioning. Hydrographic charts which could utilize mono MCS coverage were added to the required programs (separate MCAT). The priority division of 1 through 4 was changed to a low and high priority scale. (Previously, priorities 1 and 2 were high and priorities 3 and 4 were low.) High priority included all programmed production areas. The low priority was for data base collection areas.

<u>Summary of Collection Interests for Mission 1213 MCS</u>

1. CCN/NAVPAC
2. Eurasia land mass—CCN
3. Africa land mass—CCN
4. Production programs
5. Hydro-mono areas.

Mission 1213 achieved approximately 2.2 mnsm of coverage against requirements, bringing the total MCS collection to 18.8 of 39.4 mnsm.

Mission 1214 was the first mission programmed to collect against bilap (55 percent forward overlap) requirements. Mono, bilap, and trilap collection requirements for mission 1214 were approximately 17 mnsm of priority area. The use of bilap coverage (0.82 base/height ratio) provided a one-third increase in coverage of MC&G requirements over trilap. The areas/programs which required the most MCS photography were the CCN gap areas in the Eurasian land mass, (628 frames planned for collection), and the Canadian/Greenland area (481 frames) which was in support of a joint mapping program. The South American continent high priority area was also in support of joint mapping projects (521 frames). The African area of collection (562 frames) provided for production requirements and to develop CCN capability of that continent. Mission 1214 achieved 3.507 mnsm of coverage against an objective of 4.367 mnsm.

With the introduction of ultra-ultra-thin-base film on mission 1215 the coverage potential increased from 3,151 frames to 3,968 frames. Reprogrammed WAG cells previously satisfied or deferred from collection within the Eurasia land mass were submitted for collection. This was due to support required for the Cruise Missile Program and the Point Positioning Data Bases (PPDB) which required 90 percent cloud-free coverage of WAG cells. Previous mission requirements within the Eurasian land mass, which had been satisfied by 50 percent (12 to 24) coverage on cells, were deferred if gaps between MCS operations were three cells or less. An additional diagnostic area was established within the CONUS, the Southwest Geodetic Control Net (SWGCN), to support the metric pan. A total of 2,913 MCS frames were designated against the collection of the Eurasian area. The planned MCS collection was 3.737 mnsm. Mission 1215 achieved 3.271 mnsm (mono, bilap, and trilap) coverage of the requirement areas.

The final mission (1216) MCS collection requirements were generally the same as those submitted for mission 1215 MCS. Priority emphasis was given to completing coverage of the Southwest Geodetic Control Network (SWGCN) which was important to the upcoming metric pan flights, the Continental Control Network (CCN) land mass gaps in Eurasia, Africa, and South America, and coverage of the Cruise Missile Program's test and evaluation in Canada and operational areas in Europe. Other priority requirements for MCS collections included current medium and large scale mapping programs.

The 118-day 1216-5 mission returned 3,840 terrain exposures. There were approximately 2 mnsm of requirements satisfied (19 percent) with mono, bilap, and trilap coverage. The clear/usable MC&G criteria coverage grossed by mission 1216 MCS was approximately 6.1 mnsm; however, the effects of camera operations over slivers and gaps (i.e., areas less than three WAG cells wide) resulted in a relatively small return of unique area coverage. The original MCS coverage requirement was 39.4 mnsm, with each WAG cell seen at least 50 percent clear. The shortfall against this requirement was 12.1 mnsm. For many production purposes, 90 percent of each WAG cell must be seen clear. The shortfall against this requirement was 14.5 mnsm.

MISSION ENGINEERING SUMMARIES AND MAJOR CHANGES

Mission 1205 (SV-5)

The first MCS mission was operational from 9 March to 20 April 1973. Photographic analysis and inflight telemetry data indicated that a high level of success was achieved from both functional and metric aspects.

The terrain photography was judged excellent throughout the mission. Photographic resolution level significantly exceeded the predicted values based on factory acceptance test levels. Visual Edge Matching (VEM) produced an average of 70 lines per mm for the frames measured. Terrain camera quality provided additional value for search and surveillance requirements.

The stellar photography provided adequate star images in both magnitude and quality. The average number of star images in most frames was between 50 and 100 with many frames recording as high as 150 images. Film sensitivity was reduced by fogging from solar radiation which degraded the stellar reseau imagery.

The terrain thermal shutter malfunctioned at the start of the mission, failing to open when it was in the coldest sector of the orbit. Operations were initially limited to latitudes below 50°N. The latitude restriction was periodically updated based on the daily increase of solar elevation.

A second malfunction, an intermittent failure of the tape stop switch, resulted in improper shutdown of the camera. The subsequent camera operations were made in the backup mode of operation. The backup mode circuits utilized a different tape stop switch.* This problem was corrected in subsequent systems by modification to the command sequences to provide a functional backup to the tape stop switch.

Both 3401 and 2403 films were included on the end of the regular terrain film load (3400) for the calibration photography. The in-flight calibration, requiring the vehicle to be pitched so that all three cameras could photograph the stars simultaneously, obviously exposed the terrain thermal shutter to the cold environment of space. Since the failure mode of this particular shutter was temperature oriented, it failed to open, therefore there was no in-flight stellar calibration film exposed for evaluation.

Based upon retrievals from the Mission Performance Report (MPR) data, a total of 728 Category I Missile Target Data's (MTD's) (170 WAG cells) were covered by the mapping camera system. This programmed coverage equated to 30 percent of the photographic coverage required for the Category I inventory. In the actual evaluation, only 481 Category I MTD's (87 WAG cells) were adequately photographed. This equated to only 20 percent of the coverage requirements.

For general information, 2,674 Category II targets, 837 offset aim points, and 676 short range air missiles (SRAM) reference point graphics (RPG's) were covered per programmed coverage. This equated to 19, 24, and 25 percent of their respective inventories. In terms of Mini-Bloc coverage, the equivalent of 3.1 ONG areas were covered, which represents 11 percent of the Mini Bloc requirements.

---

*The tape stop switch is a control function that provides proper shutdown of the MCS following an operate OFF command.

Major action items resulting from the 1205 mission included:

1. Redesign of the terrain thermal shutter mechanical drive system to reduce the coefficient of friction.
2. Modification of the tape stop switch cam to allow more positive actuation of the operate interlock switch.
3. A study to determine if the use of 3414 film could be utilized as the primary terrain film in order to provide sufficient exposure latitude over the range of sun angles encountered in mission photography. (The 1205 mission exposure results indicated a reserve of available illumination existed when using 3400 film.)
4. A study to determine if it would be feasible to increase the stellar exposure time to permit the use of the slower 3400 film in order to decrease the sensitivity to static and corona discharge and solar radiation fogging.

Use of SO-131 (False Color Film)

In April 1973, Robert Kohler of the CIA informally initiated an investigation as to the possible use of infrared sensitive color film (SO-131) in the Mapping Camera to provide special-purpose coverage over selected areas. Having been asked for their evaluation, Eastman Kodak reported that optimum exposure of SO-131 in the MCS was unlikely because of the high T-number and limited shutter times, and provided a table showing the predicted underexposure under several combinations of solar altitudes and shutter speeds.

On 21 May 1973 a message sent from the SOC (Lt. Colonel Clark Lehmann) to SAFSP (Lt. Colonel Albert W. Johnson/Captain James Collins) and the NRO confirmed the requirement to have 100 feet of SO-131 film aboard the terrain camera film supply spool on mission 1206, and stated that a formal collection requirement was forthcoming. Using a combination of the EK data and Itek (Harold Alpaugh/Stephen Herman) generated exposure data, SAFSP then directed that 100 feet of the SO-131 be incorporated for exposure near the end of the mission at predetermined solar elevations and shutter speeds. SAFSP also asked EK to send 200 feet of SO-131 to Itek for test purposes.

Mission 1206 (SV-6)

This mission was operational from 13 July to 24 August 1973. The terrain photography for this mission exceeded the predicted quality levels which were based on acceptance test results. Visual Edge Matching (VEM) produced an average of 57.5 lines per mm.

The stellar photography was comparable to mission 1205 with the majority of frames recording approximately 100 stars and many frames with up to 150 star images.

The quality of the experimental photography with the SO-131 (false color) film was about as predicted, i.e., about 50 to 75 feet GRD. Subjective analysis indicated that correct exposure was achieved but the solar balance was less than optimum.

Two minor anomalies occurred during the mission. Telemetry indicated phase lock dropouts throughout the mission and abnormal operation of the stellar press on two frames; however, no adverse effect on photography was observed. All other telemetry was normal.

Both 3401 and 2403 films were included on the end of the regular terrain film load for the in-flight calibration. The 3401 film produced 20 to 25 star images on each frame. The results of photography on 2403 were considered unacceptable due to film grain size, which approximated star image size, and over-exposure of terrain fiducials and reseau intersects.

Prior to launch, DMAAC submitted 1,980 Category I MTD's (439 WAG cells) for Mission 1206 MCS coverage. Based on programmed MCS coverage taken from MPR listings, a total of 407 Category I MTD's (100 WAG cells) were photographed. Based on visual review, 153 Category I MTD's (58 WAG cells) were adequately photographed by the Mission 1206 MCS. For Mini Bloc requirements, the area equivalent of 3.1 ONC's (operational navigation chart) was effectively covered by Mission 1206.

New Launch Schedule

On 21 August 1973, SAFSP (Captain James Collins) sent a message to DMA (████████) advising of new launch dates. The dates given reflected the plan for launching two HEXAGON vehicles per year:

System	Shipping	Launch Date
001	04 Nov 73	Mar 75
007	05 Feb 74	Sept 75
008	19 Aug 74	Mar 76
009	06 Feb 75	Sept 76
010	17 Aug 75	Mar 77
011	04 Feb 76	Sept 77
012	20 Aug 76	Mar 78

These delivery dates for the Mapping Cameras corresponded to the contractual date they would be shipped from the manufacturer to Cloudcroft, and therefore would serve as a guide for DMA to use in scheduling their Cloudcroft support.

Then, on 12 December 1973, again to aid DMA in their Cloudcroft support planning, Collins sent another message to ████ advising of an advancement in delivery schedules, these new dates being required to comply with revisions to systems integration/test timelines.

System	Delivery Date	Launch Date
007	06 Jan 74	Sept 75
008	23 July 74	Mar 76
009	08 Dec 74	Sept 76
010	21 June 75	Mar 77
011	06 Dec 75	Sept 77
012	24 Jan 76	Mar 78

Mission 1207 (SV-7)

The third ST system was fully operational for 58 days on orbit, from 10 November 1973 to 7 January 1974. This was an extension of mission length by 16 days over the first two MCS missions.

The imagery acquired from all cameras was comparable to past missions. The terrain camera performed at expected levels based on acceptance test results, and the stellar units recorded an adequate distribution of sixth magnitude stars.

This was the first MCS to use 3414 film in the terrain camera. Thirty feet of this film was "tagged on" for special engineering tests. Engineering objectives were accomplished and the results substantiated the optimism for using this film as the primary load for future missions, 1209 being the earliest that could be considered.

Several minor anomalies were noted which did not impact the photographic results.

The in-flight stellar calibration operation was successful.

The following recommendations were made by the post flight analysis (PFA) team:

1. Useful data was obtained from the special test of 3414 film, and additional data to be acquired on the next mission (1208) should support the present belief that using 3414 film with a Wratten 12 (or equivalent) filter and extended exposure time on mission 1209 would improve terrain camera performance. It was therefore recommended that another section of 3414 film be included in the mission 1208 terrain film supply to be exposed as on mission 1207.

2. The advantages of increasing the sensitivity of the 3401 film used in the terrain camera for starfield calibration warranted continued use of this special "forced process."

3. An increase of exposure time by approximately one stop combined with the increased transmission gained through the use of a W-12 (or equivalent) filter (effective mission 1209) should result in an adequate display of stars to, and including, 5th magnitude.*

4. Since the results of the 2403 film (terrain calibrate mode) had been unsatisfactory, 2403 would be eliminated from future flight loads (1209 and subsequent).

5. At film depletion, as the film end leaves the terrain supply spool, film tension is lost with the likelihood that film jamming will occur. This did occur on mission 1207, hence the 3400 film tag-on for stellar exposure evaluation was not recovered. Therefore, it was recommended to continue with tests to provide additional exposure information before committing this film for the stellar camera.

Based on actual review, 665 Category I MTD's were adequately photographed by the MCS. For Mini Bloc requirements, the area equivalent of 5.2 ONC's was effectively covered by Mission 1207 MCS photographs.

---

*On 20 May Jean R. Manent, then MCS Program Manager at Itek, notified SAFSP that Itek would start the necessary work to change "C" time from 2.2 seconds to 3.37 seconds. This decision was based on analysis performed at Itek and information derived from experiments run at the Cloudcroft calibration site.

MCS HISTORY

Project 80 Study

In 1973, there was a study known generally as "Project 80," performed by DMA in cooperation with the mapping, charting, and geodesy working group (MC&G-WG) of the COMIREX. Its objective was to estimate the ability of the panoramic and frame camera imagery from the HEXAGON Program to satisfy current and projected MC&G requirements. A 15 August 1973 message to SAFSP (Colonel Raymond A. Anderson/Colonel Ralph H. Jacobson), from Lt. Colonel Hayden Peake on the NRO staff stated that the Project 80 draft report has been completed but not released. The conclusions and recommendations had been made known, however, and the MC&G-WG chairman had asked each participant to comment formally on the conclusions and recommendations before being allowed to review the study itself and before its submission to the COMIREX.

The Project 80 conclusions and recommendations as quoted were:

A. "DMA has concluded that the combined panoramic and frame camera subsystems satisfy all current and projected MC&G requirements. The potential for meeting all DMA requirements in one camera does not exist in HEXAGON frame camera movements. However, minor improvements in resolution would help in the short-term use of the frame material. The greatest long-term payoff for the 1978 time frame and beyond lies in a single panoramic camera system possessing high resolution, wide angle coverage, adequate metric quality, precise orientation and absolute timing capability."

B. "Based on these conclusions, DMA recommends initial studies be undertaken by NRO as outlined below:

1. A study to determine the major practical improvement in ground resolution possible to the frame camera subsystem, but limited to changes that can be retrofit to the present undelivered series without major cost increases.*

2. A study to consider the following improvements to the HEXAGON panoramic camera from the standpoint of feasibility, cost, and their potential to support total MC&G requirements.

    a. Improved pan camera calibration to reduce recording distortions to less than 10 micrometers (one-sigma).

    b. Improved time readout resolution to 0.10 microsecond.

    c. Continuous attitude readout of each pan camera to a resolution and precision of 1 arc-second accurate to 5 arc-seconds absolute on each axis (one-sigma) throughout limits of scan."

The conclusions, although offering little new, did affirm for the first time in writing the intent of the mapping community to accept only panoramic coverage for their needs, assuming conditions in Paragraph 2, above, could be met. Therefore, it was stressed by Colonel Peake that "before any official response to the reference above, or to the final Project 80 report, it is important that we agree on the current situation and the future as it impacts MC&G actions in relation to HEXAGON." (This subject is addressed again later in the section.)

---

*This had already been done in a sense; the program office, following encouraging results from previous tests, had made the decision to modify the filter and exposure conditions to permit the use of 3414 film starting with SV-9 (mission 1209).

A Look at MCS Block III Buy Possibility. On 1 November 1973, a message was sent from SAFSP (Colonel Anderson and Lt. Colonel John C. Bricker) to the NRO (Mr. ▮▮▮▮▮▮▮▮ and Lt. Colonel Peake) describing the minimum level of effort required to protect the MCS for a third buy, should that requirement materialize. Looking at the possibilities of reduced flight frequencies or delaying the block III procurement as cost effective measures, five options with associated ROM costs were suggested. It was proposed that a block III buy should include some product improvement measures, increasing the supply and takeup capacities to extend mission coverage and mission life through the use of ultra-thin-base (UTB) film, and reworking the prototype unit to make it usable as a viable engineering test bed.

Increase in MCM Film Load. On 18 March 1974, Lt. Colonel William Powell and Captain James Collins, SAFSP, advised Mr. James Ousley, the current MCM program manager in Lockheed, that Itek and General Electric had developed a new packaging scheme for the takeup assembly that would permit a significant increase in the takeup capacity. An increase of 3 pounds in the stellar film, 17 pounds in the terrain film, and 10 pounds of takeup/RV structure was predicted. Lockheed was requested to evaluate the effect of the weight increases on the APSA structure, on the RV separation sequence, and on any other area they felt should be examined.

On 5 April, Itek notified SAFSP and General Electric that they were proceeding to prepare for an increased capacity of approximately 500 additional frames of terrain photography, and they provided the estimated weight increases for the launch condition and recovery condition. After evaluating the data, Lockheed provided SAFSP and General Electric on 3 July with a detailed report which said in effect that the weight increase would have no adverse impact at launch (APSA structure) or at recovery (SRV clearance at separation).

Mission 1208 (SV-8)

The fourth mission was fully operational and anomaly free for 60 days on orbit from 10 April 1974 to 9 June 1974. The terrain imagery was comparable to the best of past missions and an adequate distribution of sixth magnitude stars was acquired on the stellar frames. Terrain resolution by VEM analysis was 60 lines/mm. The in-flight calibration mode was successfully completed.

As on mission 1207, a 30-foot length of 3414 was "tagged on" to the terrain film supply. The results of this test supported the decision to use 3414 as the primary load for mission 1209.

In the mission 1208 post flight analysis report, the PFA team made the following statements and recommendations, and assigned action items as indicated:

1. System 007 for mission 1209 would have a W-12 (or equivalent) filter and the capability to expose frames at 6, 12, and 24 milliseconds adapted to the terrain camera. Exposure analysis of engineering tests using 3414 film on missions 1207 and 1208 provided empirical data for the following exposure recommendations:

Sun Angle, degrees	Exposure Time, milliseconds
0 to 10	24*
11 to 46	12
47 to 90	6

*Under low solar altitudes, exposures would be "manually" controlled.

2. Investigation of the "forced" processing of 3401 film should continue in order to obtain the optimum sensitivity from 3401 film for terrain starfield calibration.

3. Efforts should continue in the development of an optimum process for 3400 film that would produce acceptable results when used in the stellar cameras. DMA should evaluate and report results of the three special process techniques used on the tag end (3400) on mission 1208. One-hundred feet of 3400 film would be tagged on the stellar camera supply for use during the run-out phase of mission 1209. This film would undergo process investigation similar to that conducted on mission 1208.

4. Approximately 30 feet of QX-801 film would be added as a tag-on in the terrain camera supply for mission 1209.

From the actual review of the Mission 1208-5 terrain photographs, it was determined that 180 Category I National Target Base (NTB) points were adequately photographed on this mission. For Mini Bloc requirements, the area equivalent of 2.2 ONC's was effectively covered by Mission 1208 MCS photographs.

Mission 1209 (SV-9)

The fifth MCS flown was fully operational and anomaly free for 59 days on orbit, from 29 October to 27 December 1974. This was the first system to use 3414 film as the primary terrain camera film supply and an equivalent W-12 filter. Preflight exposure analysis done by EK and Mr. Steve Herman (Itek) had provided a two-step recommendation for this combination to be used on mission 1209:

Solar Altitude	Nominal Exposure Time
Less than 46 degrees	12 msec
Equal to or greater than 46 degrees	6 msec

The use of a single time for medium and low solar altitudes would appear to encourage underexposure at the latter condition. However, it was felt that snow cover would be present at very low solar altitudes (below 20 degrees) and that the 12-millisecond exposure time would in effect be applying a snow bias in these areas.

The film/filter combination in practice performed as expected. Resolution performance on 3414 film was 40 percent higher than the three previous missions utilizing 3400 film. The average resolution derived by VEM analysis was 80 lines/mm.

Special engineering tests were conducted using a 20-foot length of QX-801 film (having intermediate speed between 3400 and 3414) tagged onto the terrain film supply. This experimental film was developed to provide a higher speed than 3414 with (hopefully) comparable resolution. The evaluation of this photography indicated QX-801 did not provide a level of information content comparable to 3414 and, since there were no significant offsetting advantages for using QX-801 as the terrain primary film, this was pursued no further on the MCS program.

The in-flight terrain/stellar calibration was normal. This MCS was the first to be modified for the terrain camera extended exposure time in the calibrate mode. Exposure time was increased from 2.2 to 3.4 seconds in order to increase the number and density of terrain camera recorded star images.

Star image evaluation indicated that EK 3400 film could be used as the primary film load for the stellar cameras on the next mission (1210).

Special "solo" testing (in-flight MCS tests conducted after the film was recovered) demonstrated that the MCS (without the film transporting) could remain operable during an extended mission length (138 days).

From the actual review of the MCS photography, it was determined that 206 Category I NTB points were adequately covered. For Mini Bloc, 2,176,000 square nautical miles were effectively photographed.

MCS HISTORY

Mission 1210 (SV-10)

The sixth mission was operational from 8 June to 30 July 1975. This was the first mission where a major MCS anomaly occurred that required significant operational work-around management. When a camera power control relay failed in the closed position, certain camera electronics remained powered as long as the MCS main power was on. This condition caused increased temperatures in the terrain film supply and the takeup systems in the recovery vehicle, resulting also in elevation of the recovery battery temperature. Temperature control was maintained by periodically cycling off the MCS main power and limiting MCS operations.

Successful operational problem management was verified when it was demonstrated by post flight analysis that the performance level achieved was comparable to past MCS missions. The average VEM analysis was 75 lines/mm. Followup analyses and actions for the relay failure were handled through normal failure report procedures.

The in-flight stellar calibration operations were successful.

This system was the first to include six layers of aluminum tape as part of a continuing study of density levels versus radiation exposure of the stellar film. The effect of the tape was not determined because of different stellar base-plus-fog levels from previous missions, plus the possible reduction of film sensitivity from the increased temperatures resulting from the power relay failure.

The PFA team made the following recommendations:

1. A new exposure algorithm should be used on mission 1211 that combined the microdensitometry data from missions 1209 and 1210. The exposure values for 1211 would be as follows:

Sun Angle, degrees	Exposure Time, milliseconds
$23 \geq SA$	24
$23 < SA \geq 66$	12
$66 < SA$	6

2. Evaluation of film tests conducted during flight showed that EK 3400 film should be used as the primary stellar camera film load on subsequent flights.

3. Taping of the first stellar chute section should be done for mission 1211.

4. A tag end of EK 3411 film should be included along with the EK 3401 film for the in-flight calibration of the terrain camera. Since EK 3411 was essentially to be a replacement for EK 3401, a verification test was considered desirable to confirm compatibility.

5. As soon as EK 3410 became available, a tag end should be used in the stellar supply for image evaluation tests. This film was to replace EK 3400 film currently in use.

On missions 1210-5 to 1216-5, all remaining target points (including newer requirements) were slowly covered by the later MCS missions (over 83,000 target points of various types were imaged and geodetically positioned as of March 1981). The shortfall Mini Bloc coverage was merged with newer DTED requirements and systematically collected on the latter seven MCS missions. The overall MCS photo coverage statistics (computed by DMA HTC) are listed in Table 7-2.

Mission 1211 (SV-11)

The seventh mission was fully operational and anomaly free for 60 days on orbit, from 4 December 1975 to 2 February 1976. The terrain camera produced higher image quality and resolution than previous MCS missions. An average resolution derived by VEM analysis was 83 lines/mm.

This was the first MCS to use 3400 film as the primary stellar camera film. Star imagery was good with sufficient 6th magnitude or fainter star images recorded.

Due to a failure in the panoramic camera system, panoramic camera operation for the majority of time was in the mono mode. This placed an unusually high duty cycle on the vehicle attitude control thrusters, therefore the in-flight stellar calibration was deleted to conserve vehicle attitude control operations. The calibration materials 3401 and 3411 that normally would have been used for this purpose were used instead to acquire low solar angle photography.

The PFA team made the following comments and recommendations:

1. A tag end of EK 3411 film should be incorporated in the terrain camera film supply of 3401 film for in-flight calibration evaluation.

2. A tag end of EK 3410 film should be incorporated in the stellar supply to be used during stellar film runout following normal mission operations.

3. Thirty frame pairs of stellar film were evaluated for number and quality of star images. All frames recorded sufficient sixth magnitude and fainter star images. The 17 DN process chemistry with EK 3400 film produced an image quality acceptable for mensuration.

4. The exposure values used for 1211 were recommended for the next mission, 1212.

5. The data derived from the chute taping experiments did not result in firm conclusions regarding the benefits of aluminum tape on the stellar chutes. However, as a precautionary measure, the first stellar chute section (following the supply), the fourth chute section, and the fifth chute section should be covered with six layers of aluminum tape for mission 1212 (the second and third sections are flexible and cannot be taped).

Table 7-2 — Mapping Camera Coverage—By Mission
(Thousands of Square Nautical Miles)

Mission number	1205	1206	1207	1208	1209	1210	1211	1212	1213	1214	1215	1216
Total accessed	5,894	6,282	6,571	6,487	6,773	6,668	6,919	7,363	7,688	13,236	13,782	16,485
Trilap mode	4,457	4,936	5,021	4,418	4,892	4,547	4,652	4,293	4,939	5,933	8,688	9,317
Bilap mode	719	673	825	1,035	941	1,062	1,136	1,536	1,375	4,401	2,554	3,469
Mono mode	718	673	825	1,034	940	1,059	1,131	1,534	1,374	2,902	2,540	3,698
Redundant coverage within each mission	71	154	579	221	188	330	390	307	188	N/A	N/A	N/A
Redundant mission-to-mission coverage	—	166	410	977	1,437	1,450	1,768	770	1,206	222	3,344	N/A

Star Sensor System (S³)

On 16 March 1976, Lt. Colonel Clark Lehmann (SAFSP) sent a message to the DMA (█████████████████████) as follows:

1. The following requirements statement was used in the recent SAFSP Star Sensor Technical Evaluation. Request you coordinate this within DMA and provide a formal requirements statement to the NRO at the earliest possible time.

2. Requirements

    a. The Defense Mapping Agency is responsible for providing precise geodetic positions of predetermined targets for all Department of Defense missile systems. DMA has been tasked to achieve, as a technical objective, point target positioning to within an accuracy of 23 meters horizontal circular error (90 percent reliable) and 17 meters vertical linear error (90 percent reliable) to support the advanced ICBM (MX) system targeting.

    b. A recent decision was made to discontinue the current HEXAGON Mapping Camera Subsystem after SV-16 and modify the HEXAGON system to permit the panoramic cameras to satisfy the aforementioned point positioning requirement. The metric panoramic system requires the following six HEXAGON system capabilities to do that job:

    (1) Attitude rate. The attitude rate of the system must be known on a continuous basis to 1.5 arc-seconds/second at one sigma. This capability exists now.

    (2) Satellite vehicle ephemeris. The position of the vehicle must be known for each photographic exposure to within 30 feet, one sigma, in track, cross track, and radially. This will be possible with NAVPAC effective with SV-13.

    (3) Exposure time. The exposure time of any portion of the pan photograph must be determined to within 0.1 millisecond, one sigma, a tie-in between the NAVPAC clock and the pan camera system on SV-14 will provide the capability.

    (4) Camera calibration. The pan sensor must be calibrated so that photographic distortions can be removed to permit the location of a point on the film format to an accuracy of 10 micrometers, one sigma, in both the in-track and cross-track directions. Calibration will be available with SV-14.

    (5) Absolute attitude. The absolute attitude of each pan sensor line-of-sight must be known to within 5 arc-seconds, one sigma, with respect to the three axes of the local vertical throughout the limits of the scan.

    (6) Relative attitude. The relative attitude of one pan sensor line-of-sight to the other pan sensor line-of-sight for any given set of stereo exposures must be known to within 5 arc-seconds, one sigma, for each axis.

    (7) The star sensors under study for SV-17 and beyond must permit absolute and relative attitude determination accuracies of 5 arc-seconds and 3 arc-seconds, respectively, for the metric pan system.

Ultra-Thin-Base (UTB) Film Implementation

On 3 June 1976, SAFSP (Colonel Lehmann/Lt. Colonel Guy Welch/Major Ronald G. Toman) advised the NRO (Mr. Jimmie Hill/ Lt. Colonel Larry D. Beers) and the DMA (███████████, Mr. ████████████) via message that UTB film feasibility testing on the prototype MCS was in progress and that no problems had surfaced which would preclude using UTB in the MCS.* If the current tests were successful, then a tag-end strip of UTB on mission 1213 and a full load of UTB on 1214, 1215, and 1216 would be feasible. The message further advised that the following additional actions must be taken before UTB could be flown on 1213:

1. Design and build special film path alignment fixtures.
2. Generate new film tracking procedures and verify system tracking capability.
3. Determine alignment sensitivity.
4. Validate interface control document requirements/constraints.
5. Demate the MCS from MCS-13, establish UTB film path alignment, and run a chamber (A-1) thermal vacuum test.
6. Verify RV cut and seal capability with UTB.

The funds required to complete the activities outlined above were identified by amount and date required in order to protect the January 1977 launch of 1213. On 23 June, DMA advised the NRO and SAFSP that the necessary funds would be furnished as requested.

Under the supervision of Grant D. Ross, Itek Program Operations Manager, the special film path alignment fixtures were designed and built by Itek personnel at Sunnyvale, (J. Alfred Shamble) and Lexington (Serge Kunica) and new film tracking procedures were drawn up to verify system tracking capability. Concurrently, Lockheed verified by test the capability of the cut and seal subsystem to perform this most important function with the UTB film.

With the UTB feasibility testing in process, on 29 September 1976, Welch (SAFSP) sent a message to ████████ (DMA) informing DMA of the progress. The sensitivity tests had shown that the UTB mistracking tolerance was less than STB. Additionally, a full-spool tracking test on a flight system at Lockheed had disclosed that a critical roller must be aligned very precisely. From the data presented, it was concluded that the following actions should be taken:

1. Incorporate a wobble roller in the L section.
2. Investigate the use of wobble rollers in the B, E, and G sections of the film path as well, to further desensitize roller alignment.
3. Compare a worst case computer analysis of the mobile structure to identify expected deformations.
4. Incorporate the required modifications and perform another full-spool tracking test on a flight system.

Welch advised that no additional funds would be required at that time to perform these additional tests, and that SAFSP was targeting for a late December date to make a decision as to the MCS's capability to track a full spool of UTB on SV-14.

---

*Investigations and actions which led to the eventual incorporation of UTB in the MCS were initiated as early as September 1971.

Along with the technical feasibility of using UTB film in the MCS, Itek had been looking at the contractual aspects of this endeavor. On 21 January 1977, Itek program contractual and technical management (Dana Jones/R. Manent/M. Burnett) sent a message to SAFSP program contractual and technical management (Harry Potts/G. Welch) addressing both the subjects of UTB utilization and extended operational life for the mission 1213 MCS. The message stated that SV-13 was programmed to contain approximately 100 feet of UTB film to be spliced after the 3401 calibration mode film, but that Itek had not completed the studies or concluded that UTB film could be transported without incident.* Therefore, if UTB were transported by the MCS in SV-13, any anomalies associated with the UTB film would not be considered as proportional critical events in determining the mission performance score. Further, it was also understood that the MCS on mission 1213 would be programmed for operation beyond the time/revolutions (60 days/960 revs) specified in the contract. It was Itek's position that their liability for performance incentive penalties ceased when the mission reached those specified limits. SAFSP concurred with Itek's interpretation in a message dated 31 January 1977.

---

*As recorded elsewhere in this history, UTB was utilized successfully on missions 1213 and 1214, and UUTB on missions 1215 and 1216. But since Itek was under a performance incentive type contract, and the MCS had not been originally designed or qualified for UTB film, they felt they must reach certain contractual agreements in advance of committing a mission to a full load of UTB.

Mission 1212 (SV-12)

The eighth mission was fully operational and anomaly free for 62 days on orbit, from 8 July 1976 to 8 September 1976.

This, the eighth MCS flown, had the longest mission life to that date, and produced the best image quality to that date, with VEM analysis reported at 95 lines/mm with a high of 134 lines/mm. Star imagery was good from both stellar units, with each stellar frame that was evaluated recording sufficient 6th magnitude and fainter star images.

Despite the algorithm tendency for slight overexposure on this mission (0.07 log-E algorithm error), the average measured exposure was virtually at nominal.

The calibrate mode operation was conducted on rev 994. Two separate calibration operates were programmed at 20-degree intervals using 3401 film in the terrain camera and 3400 in the stellar camera.

The PFA team provided the following conclusions and recommendations in the final PFA Report:

1. In an effort to evaluate the effect of adding aluminum tape to the stellar film chutes, density measurements were made to interframe spaces at selected points in the mission and data plotted against inactive time between operates. These plots showed a detectable trend that continues to indicate that the base-plus-fog levels of the stellar film are increasing relative to time in the film path. This practice and evaluation will continue on all remaining systems.

2. The exposure of the 3414 film in the terrain camera was measured to be within 0.01 log-E of nominal, with the algorithm error of 0.07 log-E continuing to be significantly less than the camera error due to the three-step-exposure function. Therefore, it is recommended that exposures for mission 1213 be essentially the same as those used in the past two missions.

3. There will be no further engineering tests using 3410 and 3411 films for the mapping camera. The 3410 and 3411 films are replacements for 3400 and 3401 films, respectively. There are sufficient 3400 and 3401 films in storage to complete MCS program requirements. Continued use of STB 3400 film will require one splice in the operational film.

4. The terrain film load for mission 1213 will contain an add-on of UTB film, 1414. This film is an ultra-thin-base material with the same emulsion characteristics as 3414, and the test strip will be used to provide additional data regarding the use of UTB as a prime terrain camera film.

Mission 1213 (SV-13)

The ninth mission was launched at 11:30:00 (PDT) on 27 June 1977 and recovered via routine air catch on 17 October 1977. The system was operational and anomaly free for 112 days on orbit, the longest mission to that date.

The image quality produced by the terrain camera was comparable to past missions, with an average equivalent resolution derived by VEM analysis of 94 lines/mm. Each stellar camera recorded sufficient 6th magnitude and fainter star images.

Microdensitometer analysis of terrain urban area imagery indicated an average exposure of 0.03 log-E above the nominal. The small error was apparently due to the system shutter granularity since the algorithm error measured only 0.01 log-E. No algorithm change was recommended for the next mission, 1214.

Calibrate mode operation was conducted on rev 1677/1678. Two separate calibration operates were programmed at 20-degree intervals using 3401 film in the terrain camera.

The PFA team reported the following conclusions and recommendations in the PFA report:

1. Analyses performed on ultra-thin-base films (1414, SO-344) for the terrain and stellar cameras indicated film distortions were within acceptable limits.

2. Exposures for the next mission, 1214, would be essentially the same as on the past three missions.

3. The overall concensus of the PFA team was that UTB film would be acceptable as the prime load for the next three missions, 1214, 1215, and 1216.

4. Special film studies would be implemented for mission 1214. PFA analyses would be performed on SO-208 and SO-315 add-ons to the terrain film supply to be operated at the end of the normal mission.

Mission 1214 (SV-14)

The tenth mission was launched at 10:39:37 (PST) on 16 March 1978 and recovered by routine air catch on rev 1902 on 11 July 1978. The system was operational for 117 days, making it the longest mission as of that date.

This was the first MCS to utilize 1414 (1.5-mil) ultra-thin-base film (UTB) as the prime load in the terrain camera and QX-817 film (UTB) in the stellar cameras. Post flight analyses conducted at the processing site, the contractor's facility, and the Defense Mapping Agency showed that mission objectives were met with a high level of success. Tag ends of (1.2-mil) UUTB film (SO-315 and SO-208) were successfully exposed in the terrain camera on this mission.

The image quality produced by the terrain camera on 1414 film was comparable to past MCS missions. An average equivalent resolution derived by Visual Edge Matching (VEM) analysis was 91 lines/mm. Star imagery was good from both stellar cameras. On evaluated frames, each camera recorded an adequate number of star images (6th magnitude or fainter).

Density measurements made on the terrain film at the processing facility indicated exposure levels were within algorithm limits.

The calibrate mode operation was conducted on rev 1659, exposing 17 frames of 3401 film in the terrain camera. Sets 9 through 12 were used for the in-flight calibration.

There were three anomalies during operations:

1. Forward motion compensation (FMC) phase-lock drop-out on one frame each of rev 298 and rev 837. Analysis of telemetry and photography acquired during this anomaly indicated there was no effect on system operation or results.

2. Failure of the terrain thermal door on rev 869. The following summarizes the sequence of events/actions surrounding this problem:

  - Telemetry data from a 6-frame operate on rev 869 indicated abnormal operation of the terrain thermal shutter (also referred to as thermal door).
  - The thermal shutter monitor indicated a delay in reaching the 30-degree position and failure to reach the 93-degree open position on frame 6 of rev 869. During the closing phase, the monitor showed the shutter reached the 30-degree position sooner than normal.
  - The MCS current monitor indicated excessive current during the "thermal shutter closed" time of frames 5 and 6.
  - A one-frame engineering operate was executed on rev 875. The thermal shutter door opened but did not close to the 30-degree position.
  - Real-time analysis indicated a potential for a catastrophic failure of the terrain shutter (closed position).
  - An "Emergency Open" command was executed to open the thermal shutter. MCS operations were completed with the door in the open position.
  - A thermal analysis of the terrain lens was conducted to predict temperature levels and possible effects on lens calibration.
  - "Solo" tests (following MCS RV recovery) were conducted to gain additional data for failure analysis. The terrain thermal door was reset to normal operational configuration on rev 1916.

- The thermal shutter would not respond in a normal manner during solo operations. Temperatures of the terrain window and current levels indicated the door was binding in a partially closed position.

Analysis of thermal data, imagery, and calibration results indicated the following:

- The average stabilized temperature of the terrain lens dropped from 74.0-74.2°F (prior to rev 869) to 69.8-70.2°F (following rev 869).
- There was no noticeable effect on imagery obtained after rev 869.
- There were no differences in metric measurements made before and after rev 869 that could be attributed to the open thermal shutter.

This was the first problem experienced with the terrain thermal shutter during mission operations since the first flight (1205). Because a failure of the shutter in a closed position or partially open/closed position would result in a catastrophic failure, engineering investigation, rework, and test were again substantial, as they were following the 1205 anomaly.

The failure appeared to be of a mechanical bind nature, so emphasis was placed on areas where the mechanism might become misaligned or sluggish due to the temperature environment. Shafts on which gear mechanisms were mounted were modified, spacers added, and a different gear pinning technique incorporated, all of the above directed toward lessening the chance of a temperature-induced bind. Concurrently with the mechanical investigation, tests of lubricants were conducted. It was found in component thermal testing that the lubricant caused sluggish operation of the thermal shutter when exposed to temperatures below 10°F. The Aerospace Corporation recommended that tests be conducted with a substitute lubricant "Braycote," a synthetic lubricant in successful use on other space systems. Thermal testing with this lubricant showed no evidence of sluggishness in the mechanism in the temperature range of −35°F to +125°F. Although there was not conclusive evidence that the lubricant had contributed to the problem on 1214, data from thermal testing of the two lubricants strongly indicated that it would be in order to change lubricants, and this was done for the remaining systems.

Qualification of the refurbished thermal shutter was done at the subassembly level in two steps. First the refurbished prototype shutter with the new lubricant was put through hot and cold cycling, followed with 25,000 cycles in a vacuum chamber. Then, the refurbished shutter for the next flight system (1215) was put through hot and cold cycling followed by 10,000 cycles in a vacuum chamber. There were no malfunctions during any of these tests and the shutter was therefore considered fully qualified for flight.

3. Thermal zone failure. Telemetry data on rev 1305 indicated that one of the MCS heater zones had tripped "off." The decision was made to leave the thermal zone tripped until after the primary mission was complete. All temperatures remained within acceptable limits. During the solo test phase of the mission, the heater zone was reset by command on rev 1895. The average temperatures decreased by the amount they had increased after the heater zone tripped (approximately 1/2°F). The heater zone again tripped on rev 1902. Neither the current monitor nor any temperature monitor provided enough information to identify the faulty heater zone.

The conclusions, recommendations, and action items from the PFA final report were as follows:

1. Analysis completed by all members of the PFA team indicated that metric and photographic performance requirements for mission 1214 were satisfied.

2. Image and metric analyses performed on SO-208 and SO-315 films indicated that either film would be acceptable for use. However, SO-315 provides a higher quality image, enhancing the usability. The PFA team therefore recommended UUTB film (SO-315) for use in the terrain camera for missions 1215 and 1216. The recommendation was a result of analyses performed on available flight/operational data, laboratory test data, and information derived from the completion of action assigned in the mission 1213 PFA report (SO-208/SO-315 evaluation).

3. The exposure values for mission 1215, using SO-315 film, would be derived from the same algorithm as was used on mission 1214. Exposure criteria would be reviewed for mission 1215 to ensure that all questions and considerations had been resolved.

4. Continued analyses would be performed to determine an optimum duplicating film compatible with SO-315.

5. Coordination with operations planning would ensure that the first frame of operation (normally health check) would be programmed over terrain during daylight hours. This would aid in triggering the automatic frame titler during the processing operation.

6. Coordination with operations would be accomplished for planning Bar XC acquisitions on each spliced segment of SO-315 film.

Mission 1215 (SV-15)

The eleventh mission was launched to 10:30:01 (PST) on 16 March 1979 and recovered by routine air catch on 12 July 1979. The system was operational for 118 days on orbit, one day longer than mission 1214. This was the first MCS to utilize SO-315 (1.2-mil-base) film as the prime load in the terrain camera.

The image quality was higher than on past MCS missions. An average equivalent resolution derived by visual edge matching (VEM) analysis was 100 lines/mm. Star imagery was good, with both cameras registering 6th magnitude and fainter stars.

Microdensitometric analysis of terrain camera imagery on SO-315 showed the exposure to be nominal with respect to the aim density of 1.0 above base plus fog.

The calibrate mode operation was conducted on rev 1900 exposing 17 frames on 3401 film in the terrain camera and matching stellar frame pairs. Six sets were used for the in-flight calibration.

Four anomalies occurred during the mission:

1. Terrain transport phase lock drop-out occurred on one frame of rev 65 and three frames of rev 557. Analysis of telemetry and photography acquired during this anomaly indicated there was no effect on system operation or performance.

2. Telemetry data from a 12-frame operate on rev 119 indicated that the terrain thermal shutter did not open on command. The following summarizes the sequence of events/actions surrounding this problem.

- On rev 126 a two-frame engineering op was conducted in the normal and redundant modes. The terrain thermal door remained closed.

- On rev 130 the door was opened using the "emergency open" sequence. Window temperatures were stabilized at 69.7°F by rev 142.

- All subsequent MCS operations were completed with the door open. No noticeable effect on imagery was observed.

- Analyses of flight data, solo test data, and ground simulation test data determined that the most probable cause was a short circuit on the "opening" side of the motor. The engineering aspects of the terrain thermal shutter problem were handled through established failure report procedures.

3. The heater zone tripped off on rev 289. Due to overlapping heater zone design, the effect of one heater zone is minimal; therefore, the decision was made to leave the thermal zone tripped until after the primary mission was complete. All temperatures remained within acceptable limits for the remainder of the mission. Testing during solo operations and flight data analysis failed to identify the faulty zone. There is a probable relation to the opened thermal shutter since a similar occurrence was observed on SV-14 after the thermal shutter was opened by the emergency sequence.

4. On rev 313, the metering length of the first frame of a 6-frame operate was short. This eliminated the interframe space between frames 1 and 2 and caused the data block for one frame to be masked by imagery. Analysis of telemetry and processed film were unable to define the cause of this anomaly. There was no loss of imagery.

MCS HISTORY

Mission 1216 (SV-16)

During the pre-launch briefing at Vandenberg on 16 June 1980 for a scheduled 17 June launch of mission 1216 (SV-16), the final mapping camera mission, Major General John E. Kulpa, SAFSP Director, expressed his appreciation and congratulations to all MCS participants for their roles in this "exceptionally successful" program.

Following a one-day hold to analyze a booster anomaly, SV-16 was launched at 11:30 PDT on 18 June and the RV was recovered by routine air snatch on 14 October 1980. The system was operational for 118 days on orbit, the film load in the terrain camera being expended on operation 529, rev 1901.

Two cycles after the tag end left the supply spool, the camera ceased to operate due to film jamming as a result of the loss of film tension. As on mission 1207, this prevented transporting all of the calibrated photography into the recovery vehicle. Twelve complete frames of terrain calibration film (3401) and four frame pairs of stellar film (QX-817) were recovered.

This was the second MCS to use SO-315 (UUTB) film as the primary load in the terrain camera. Resolution and quality were comparable to the previous mission (1215). An average equivalent resolution of 99 lines/mm was derived by VEM analysis. The exposure of the terrain film as determined visually and by microdensitometer was within algorithm limits. Microdensitometer analysis of 56 urban/industrial acquisitions indicated normal exposure.

Quality and magnitude of star images acquired by the stellar cameras were adequate for mensuration. A visual star count showed an average of 23 images on the port camera photography and 31 star images on the starboard camera photography.

There were two anomalies noted at the post flight analysis meeting. Playback data from MCS operations beginning with rev 123 showed random occurrences of slow thermal shutter open or slow thermal shutter close times. This data was indicative of mechanical binding. To preclude the possibility of shutter hangup in a partially open or closed position, the thermal shutter was commanded "open" on rev 162 and left open for the remainder of the mission. An evaluation performed by the PFA team did not show any image degradation due to the shutter being open.

The other anomaly was light "plus density" marks appearing randomly throughout the mission. The frequency of the marks was noticeably reduced after the first one-third of the mission. The marks were about 1/8-inch wide and perpendicular to the web of the film, starting about 1/2 inch from the non-tilted edge and extending about $2\frac{1}{2}$ inches (variable) into the frame. The cause of the plus density marks was judged to be a film roller pressure mark associated with the start of an operate. These markings appeared to be of little or no consequence to the usability of the photography.

At the conclusion of the post flight analysis message the PFA chairman, Captain David Anderson wrote in the following comments: "Mission 1216-5 is the last of 12 highly successful MCS missions. Since mission 1205, these systems have produced about 29,000 frames of photography covering approximately 103 million square miles of the earth's surface (includes redundant coverage). During this time the resolution and number of frames per system has been doubled from the original specification as a result of minor design changes and photographic film improvements. Those involved in this program deserve to be proud of their contributions to the MC&G community."

CLOSEOUT OF THE OPERATIONAL PHASE

The HEXAGON Mapping Camera Program, initially programmed to involve six flight systems with coverage beginning in February 1973 at the frequency of three missions per year, had been extended to twelve systems decreasing in frequency to one each year, thus extending the active portion of the program to October 1980. The program "stretches" were the results of the ever-increasing mission duration capability of the basic HEXAGON system (31 to 180+ days) and the intermix of community requirements from HEXAGON and other satellite reconnaissance programs.

In 1979, with the completion of the MCS program drawing near, thought was being given to personnel phasedown and other program closeout related activities. In this vein, on 9 July 1979, a message was sent from SAFSP (Colonel Les McChristian/Captain David Anderson) to the DMA and associate contractors relative to the preparation of an MCS program history. The message was addressed to: Itek (R. Manent, D. Jones, W. Williamson, M. Burnett), EK, DMA (███████████ ███████████████████████), and NPIC (Major █████████████████). In the message, SAFSP stated "with the conclusion of the highly successful MCS Program in sight, it seems appropriate that a history be written. Over the past few weeks discussions have taken place between DMA, SAFSP, and Itek personnel on this subject. As a result of these dicussions, an outline and schedule has been assembled. In addition, the agencies best qualified to author each section were identified. We expect this report to be approximately 100 pages and be controlled in both product and Byeman systems. The responsibility for editing and publication will be Barney Burnett's (Itek). It will also be the responsibility of ███████ to assemble those areas that have joint responsibilities."

The outline for the proposed history which had been formulated from the referenced discussions was given in the message with requests that all participants review and provide comments promptly, identifying project officers. The idea here was to work on the history as time permitted throughout the remaining active months of the program and have it essentially complete by the end of the program, rather than the usual practice of starting on the history at some point in time following completion of a program.

In September, both DMA and EK responded with their comments/suggested changes to the outline, and named project officers. The project officer for DMA was to be █████████ and for EK, Joseph Russo.

By November 1979, another idea which had been suggested earlier as a possibility was beginning to receive quite a bit of interest. This was an idea taken from the predecessor CORONA program, i.e., the use of residue subassemblies and components to build up an MCS for a classified museum display. The idea was put on the back burner but kept alive until the middle of 1980 when it became clear that there would be no use for the residue parts for additional systems.* In September 1980, following discussions with ████████ (DMA) and agreements with DMA (who was to be the recipient of the museum piece), SAFSP gave Itek the formal go-ahead to build up the museum piece using the Lockheed APSA qual structure, Itek prototype and spare subassemblies, and a spare General Electric Mark V recovery vehicle. The build-up was accomplished at the Itek Morse Avenue Facility in Sunnyvale, California, being completed in December and delivered and assembled at the DMA Hydrographic Topographic Center in January 1981. (See Fig. 7-2 for a photograph of the museum piece in place at DMA.)

---

*Itek and Lockheed had proposed that by the use of mainly existing parts and existing trained personnel, one additional MCS could be put together for a very reasonable cost for use on SV-17 or SV-18, should a gap develop for one reason or another in the switch from the MCS to the metric panoramic concept (S^3) scheduled for SV-17 and subsequent HEXAGON vehicles.

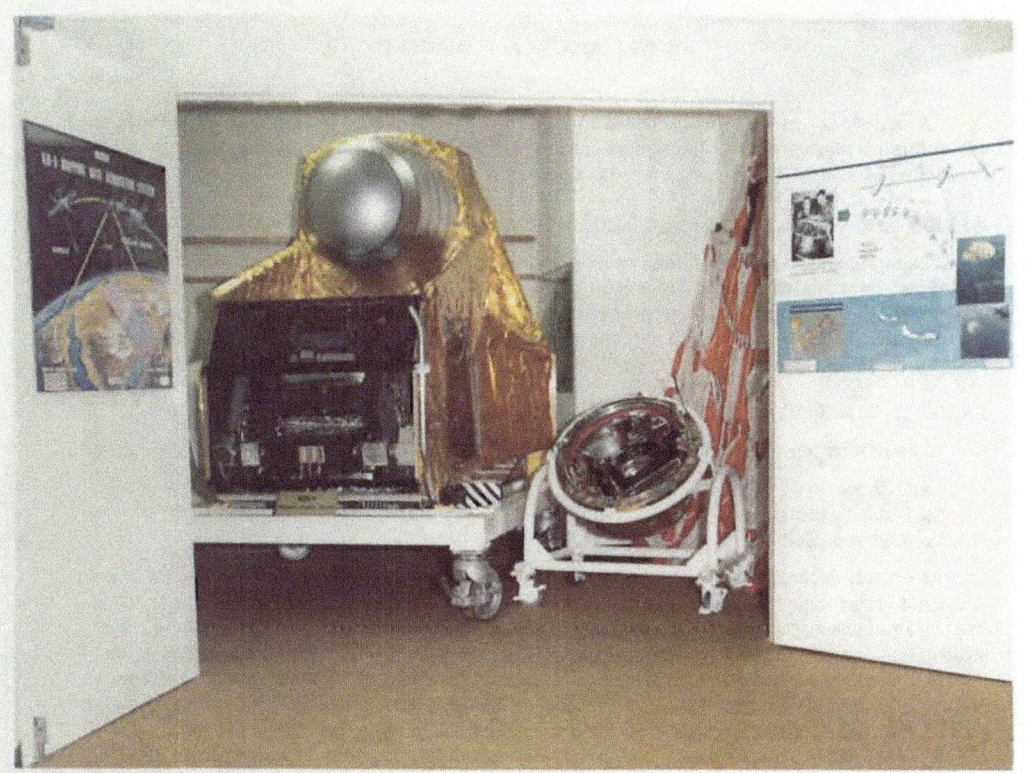

Fig. 7-2 — HEXAGON Program museum display

SECTION 8
KH-9 PRODUCTS AND IMAGERY

A variety of products are produced by DMA directly from MCS photography or through the use of this photography as a control medium for high resolution photography employed in the production process. The following are examples of these products.

MEDIUM AND SMALL SCALE MAPS AND CHARTS

Topographic, aeronautical, and hydrographic map/chart production at 1:200,000 and smaller scales can be compiled directly from MCS photography controlled to the CCN and supplementary control. High resolution photography is used on medium scale products for intensification of features using the features compiled from MCS as control for placement on the map/chart manuscript.

LARGE (1:50,000) SCALE TOPOGRAPHIC LINE MAPS

The production of these maps requires primarily the following intermediate products:

- Orthophoto
- Source package
- Elevation data

Ground control derived from MCS imagery is used to control production of elevation data and an orthophoto from high resolution imagery. This orthophoto process, which involves the RPIE and OLOPS equipment, was referred to earlier. Using the high resolution orthophoto, planimetric information is traced manually on an overlay. From the orthophoto, approximately 85% of the required detail can be obtained directly; the balance is obtained by referring back to the source material from which the orthophoto was produced. The overlays are completed manually, using a zoom transfer scope and rectified photography. Upon completion, the overlay is digitized, edited, and plotted ready for reproduction. The elevation data are plotted and updated, using a zoom transfer scope and original photography. All plots are given a final registration check and made ready for lithographic reproduction.

DIGITAL DATA

DMA produces digital terrain and feature data used separately or together to support advanced weapon systems. These systems include Cruise Missile, Firefinder, Pershing II, SIOP route and mission planning, and Radar Navigation Trainers (e.g., aircraft simulators). Developing the required digital data to support these advanced DoD systems is one of DMA's newer missions. The production of digital terrain and feature data employing MCS photography involves the following:

Digital Terrain Elevation Data (DTED)

DTED is produced to meet various data density and accuracy requirements of DoD users. DTED produced from MCS photography is compiled at 3 arc-second intervals (approximately 300 feet) of horizontal spacing with a vertical accuracy to mean sea level of 30 meters. DTED is used for mission planning (e.g., Cruise Missile), route penetration analysis, and for radar simulations. This data is also used, if the terrain meets the required roughness of approximately 600-foot elevation differences, for producing landfall and enroute TERCOM matrices. The data intervals and direction are extracted and reformatted at various sizes for TERCOM. Mission planners make use of DTED in avoiding offensive weapons systems by flying low, around threats, or behind hills. DTED is also used in conjunction with DFAD (discussed below) for weapon system simulators such as for the EA6B, F-16, B-52, EA-2C, C-130, and the A-6.

Improved production methods have been developed in order to generate the increasing volume of required digital data from the KH-9 system. Two of the major development technologies were in the photographic measurement of the elevation data and in digital data computer flow. New AS-11B1/ACE and UNAMACE automated stereoplotter software and a related Integrated Photogrammetric Instrumentation Network (IPIN) were developed to accomplish these critical milestones and to meet the digital data collection requirements. Approximately 2.5 million nautical square miles of digital data have been generated in priority areas from KH-9 mission photography taken primarily by MCS. In addition, there have been over 1,000 TERCOM matrices produced from MCS and the KH-9 Panoramic Camera System.

Digital Feature Analysis Data (DFAD)

Orthophotos produced from MCS photography are used as a control medium for positioning of features on a manuscript overlay. High resolution source material is used for the identification and classification of features. Height measurements of significant features are also performed with the high resolution material. The manuscripts are digitized with automated line following or scanning equipment and subsequently processed into final product user format with specialized computer software. These data, used in conjunction with DTED, have many uses, including aircraft flight simulators, weapons system terminal guidance reference scenes, shipboard radar navigation, and source data files for automated map/chart compilation.

POINT POSITIONING

Precise positional information is produced using MCS materials for strategic and tactical targeting and navigation purposes.

Target and navigation fix point positioning with the MCS materials can be accomplished using CCN control or by the "direct positioning" method to the accuracies shown above. The direct positioning technique is based upon a constrained analytical photogrammetric solution; that is, constraining the exposure station positions and attitude information and deriving ground positions of photo-identified point(s) using corrected photo-image coordinates. Evaluations are obtained through the propagation of errors associated with exposure station positions, attitude information, photo-measurements, etc. The direct positioning employed with the KH-9 MCS materials is feasible since the required system parameters are determined to a high degree of accuracy.

Before the initial operational capability (IOC) of the KH-9 Mapping Camera System (mission 1205), the worldwide ephemeris accuracy for this system was predicted to be 48.7 meters horizontal CE 90% and 15.2 meters vertical LE 90%. The ephemeris accuracy figures were based on the one-sigma projected capability of 90 feet in-track, 60 feet cross-track, and 30 feet radial. According to the covariance information of the earlier MCS missions (1205-1213), the ephemeris points over Eurasia have approximate accuracies of 60 feet in-track, 40 feet cross-track, and 20 feet radial (horizontal 32.7 meters CE 90% and vertical 10.0 meters LE 90%). When adequate satellite-to-satellite tracking data was added to the reductions (missions 1214-1216), the values reduced to 30 feet in-track, 30 feet cross-track, and 20 feet radial (horizontal 19.6 meters CE 90% and vertical 10.0 meters LE 90%). Combining the correlated ephemeris evaluation with the determined photogrammetric uncertainties gives the direct positioning capabilities of each MCS mission over the Eurasia landmass. The 90% values listed in the table below are presented as overall accuracy estimates. However, in production application, the proper covariance matrices associated with the individual photographic materials are rigorously projected into final product evaluation.

Through the use of the KH-9 Mapping Camera System materials, over 70,000 positional values of various targets were generated and achieved the accuracy requirements of DoD components (refer to Table 8-1).

Table 8-1 — MCS Direct Positioning Accuracy
(Eurasia Landmass)

Mission Number	Horizontal CE 90%		Vertical LE 90%	
	Meters	Feet	Meters	Feet
1205	43	140	20	65
1206	40	130	20	65
1207	38	125	18	60
1208	38	125	18	60
1209	40	130	20	65
1210	37	120	20	65
1211	37	120	20	65
1212	37	120	18	60
1213	37	120	18	60
1214	24	80	20	65
1215	21	70	18	60
1216	TBD	—	TBD	—

Fig. 8-1 — 1:50,000-scale map

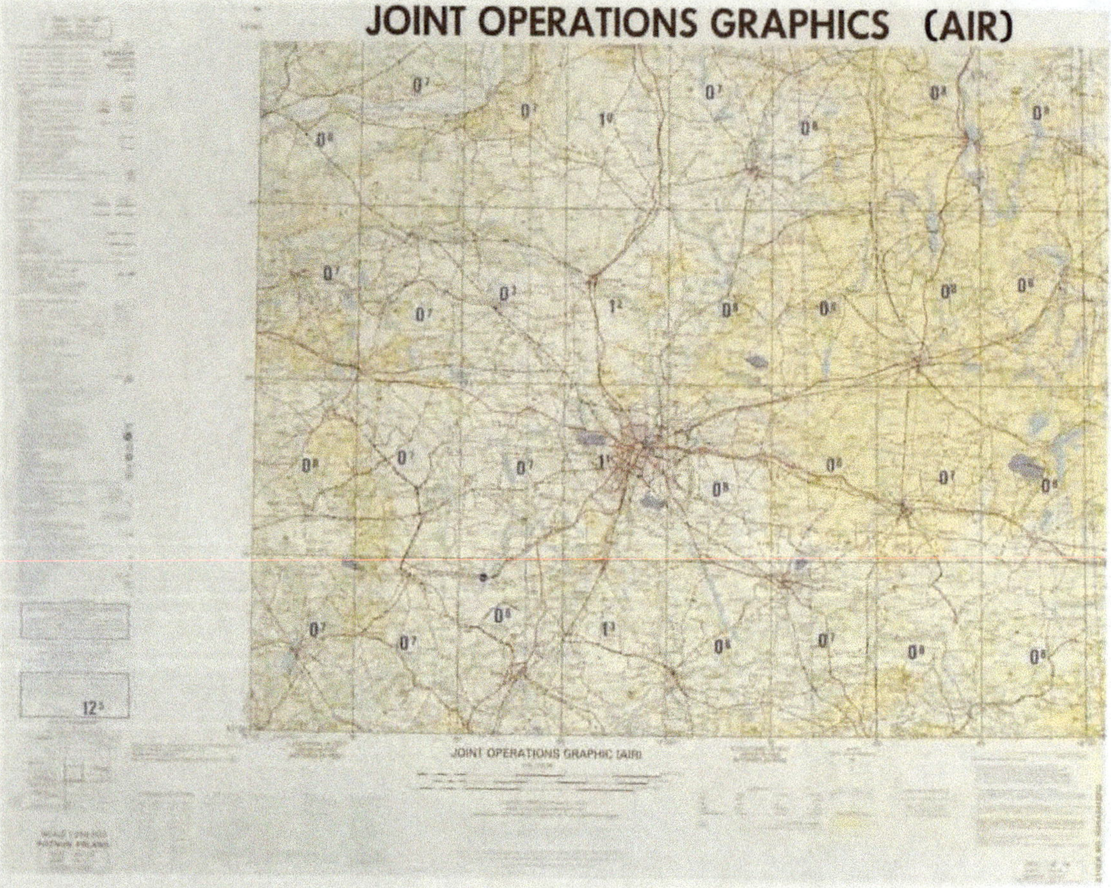

MCS HISTORY

JOINT OPERATIONS GRAPHICS (AIR)

Fig. 8-2 — Joint operations graphics (air) map

TOP SECRET/RUFF/GAMBIT/HEXAGON

BIF-059W-23422/82
Handle Via
BYEMAN/TALENT KEYHOLE
CONTROL SYSTEMS JOINTLY

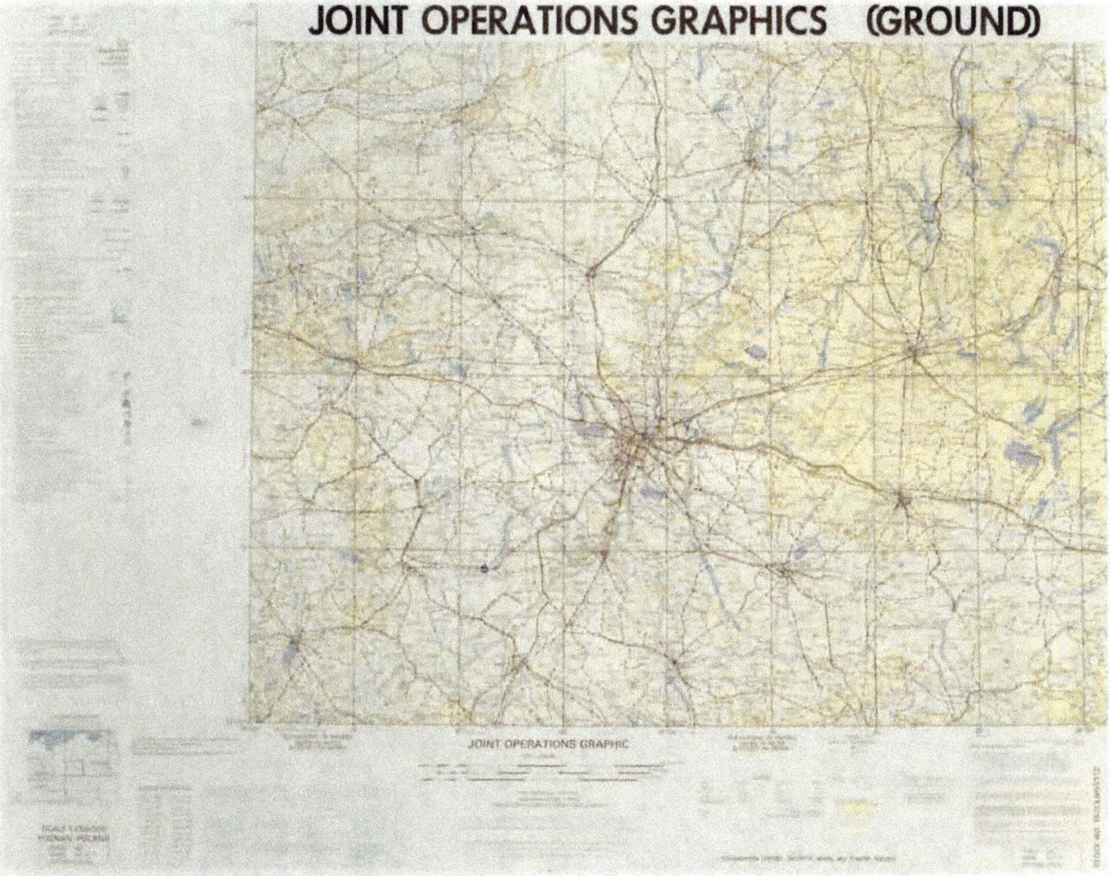

Fig. 8-3 — Joint operations graphics (ground) map

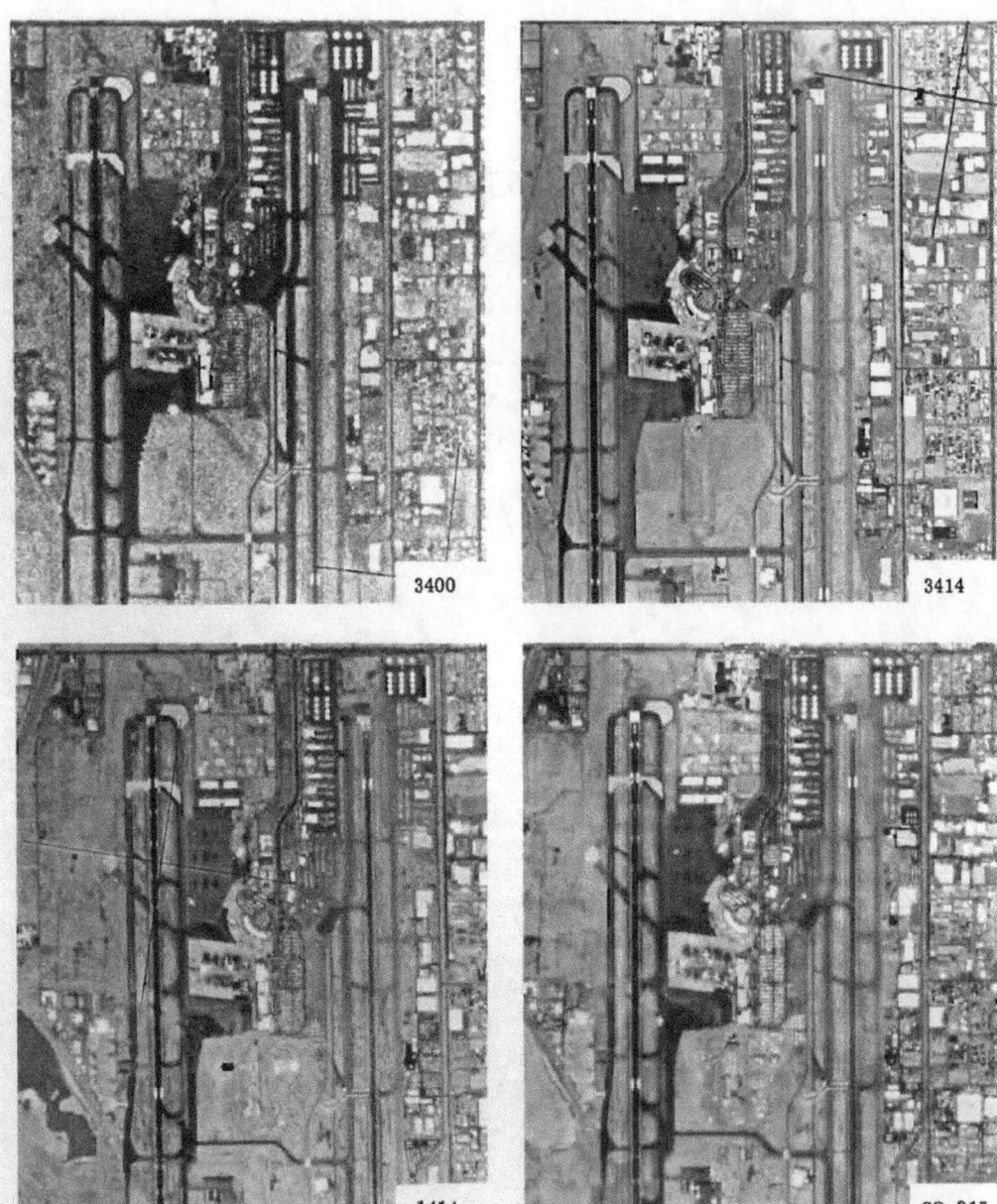

Fig. 8-4 — Comparison, operational films—Sky Harbor, Phoenix, Arizona
20× enlargements

Fig. 8-5 — Tag-on of QX-801 film—Dayton, Ohio; 2.5× enlargement

Fig. 8-6 — Experimental photography—IR color; SO-131; contact print

Fig. 8-7 — Map of San Joaquin Valley, California; graphically depicts area coverage of Fig. 8-6

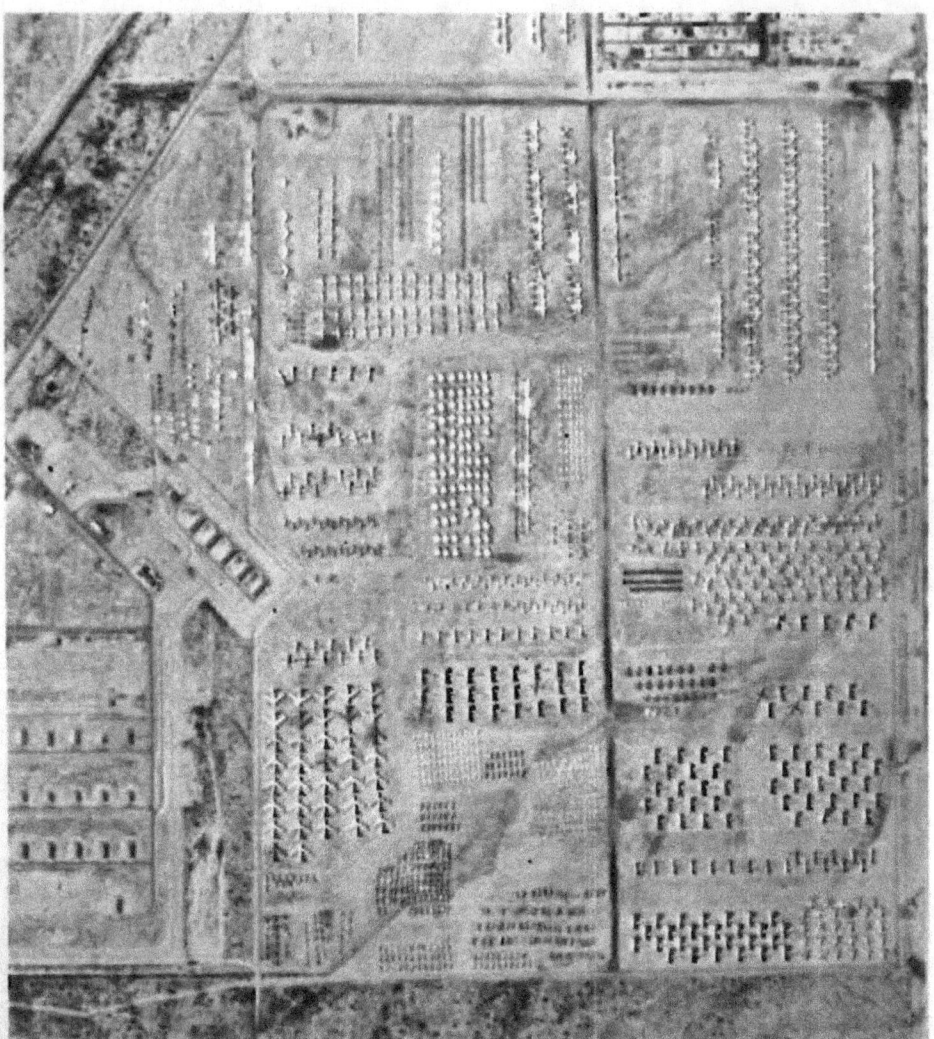

Fig. 8-8 — Terrain photography—DMAFB, Tucson, Arizona; 40× enlargement

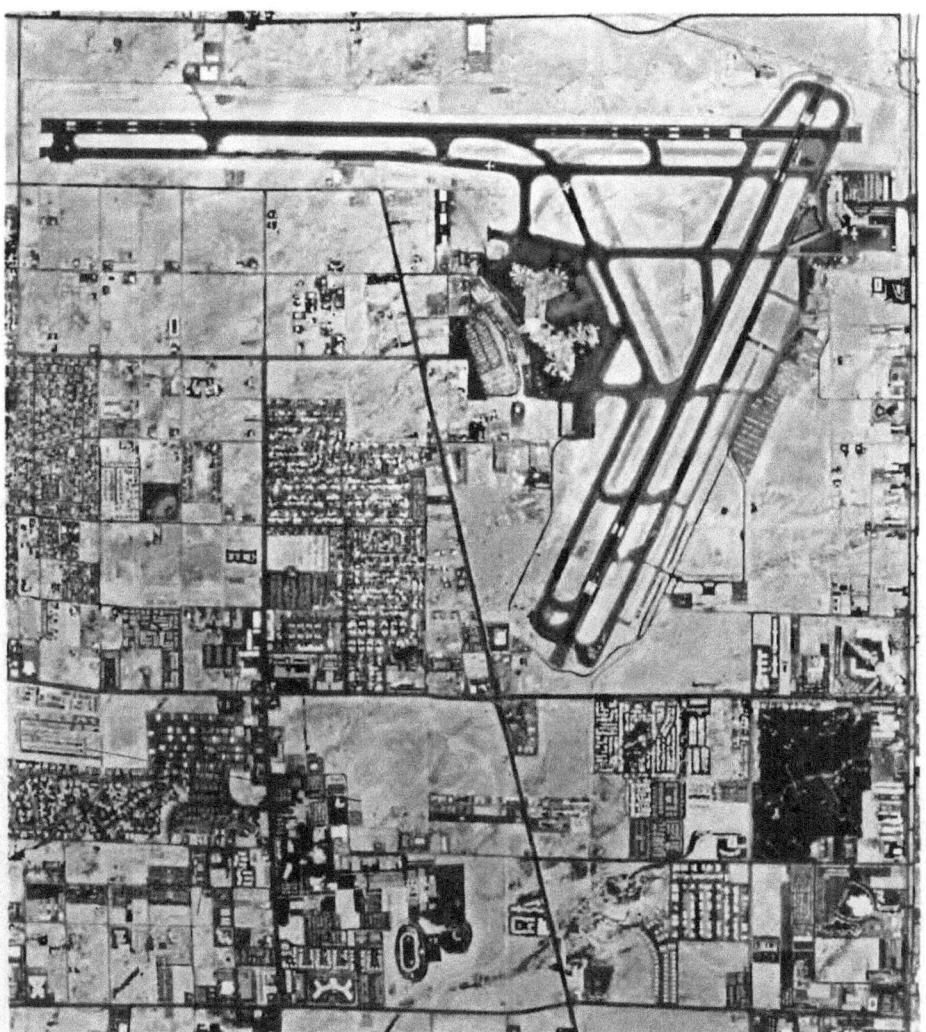

Fig. 8-9 — Terrain photography—Las Vegas, Nevada; 20× enlargement

MCS HISTORY

Fig. 8-10 — Terrain photography—Glen Canyon Dam, Arizona; 10× enlargement

~~TOP SECRET~~/RUFF/GAMBIT/HEXAGON

BIF-059W-23422/82
Handle Via
BYEMAN/TALENT KEYHOLE
CONTROL SYSTEMS JOINTLY

MCS HISTORY

Fig. 8-11 — Terrain photography—Aswan Dams, UAR; 6× enlargement

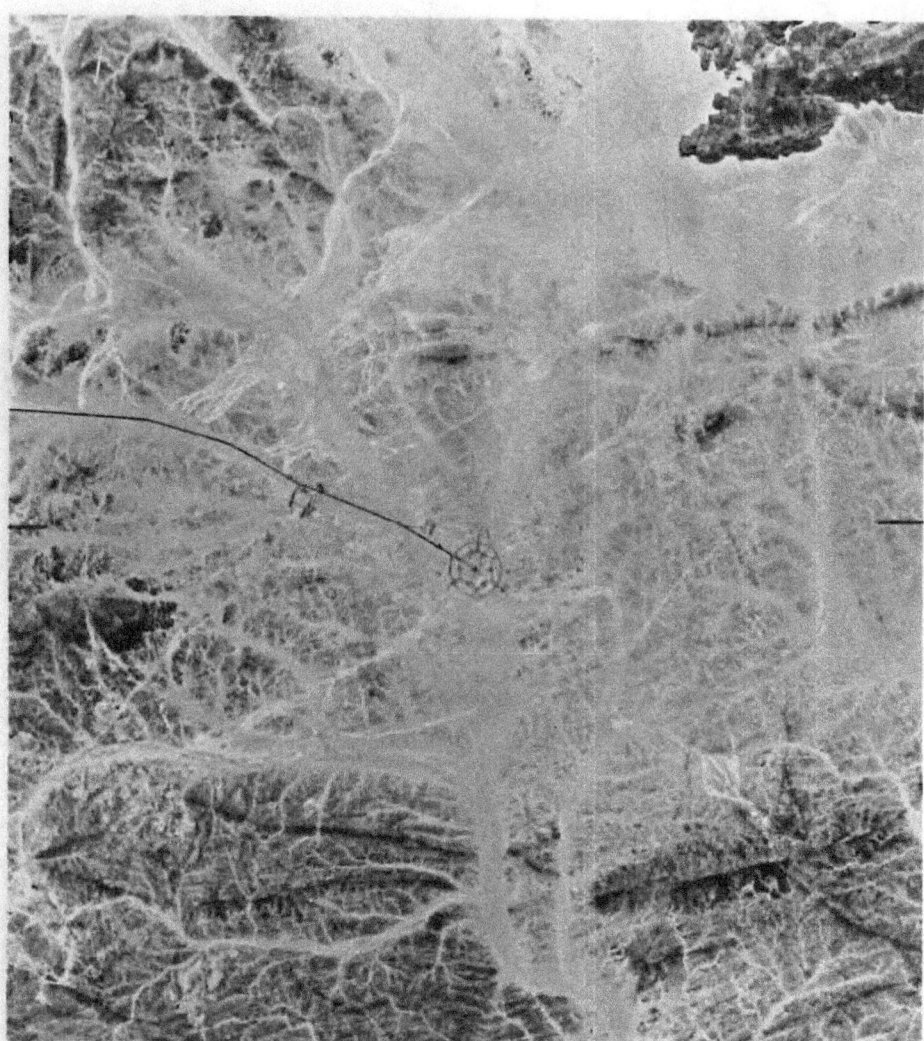

Fig. 8-12 — Terrain photography—Aswan, UAR; 20× enlargement

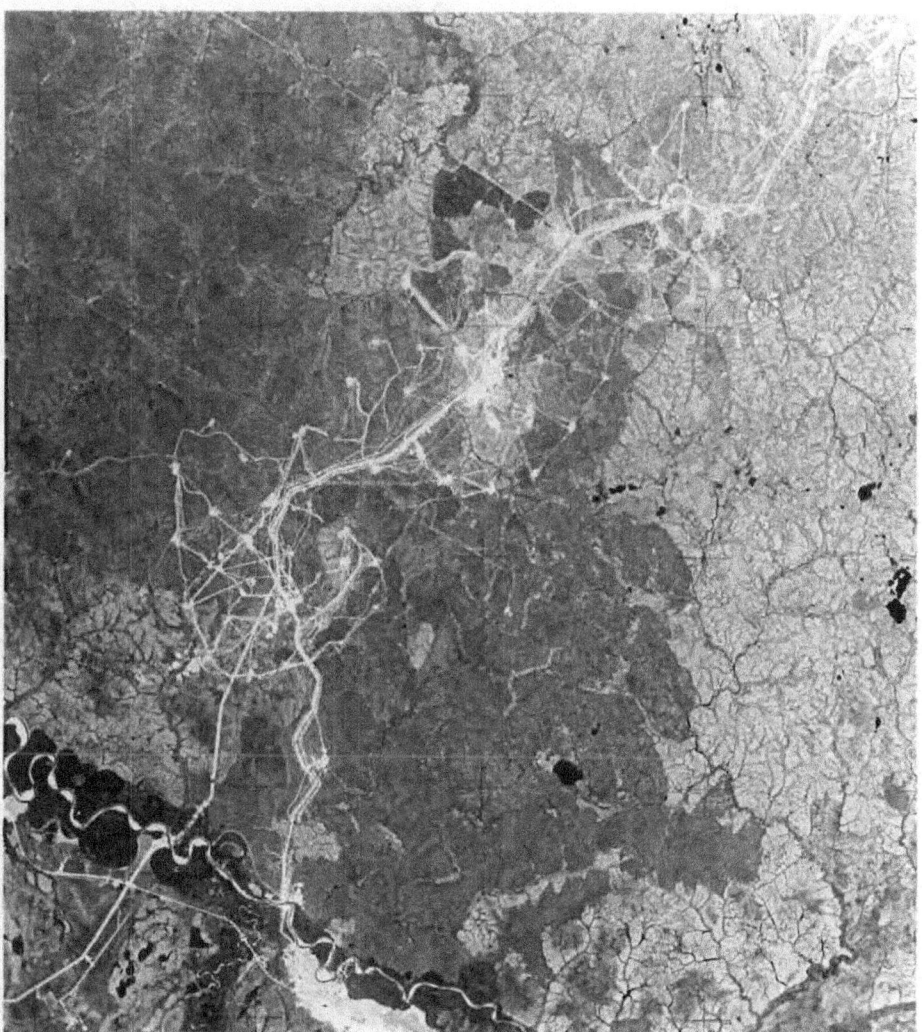

Fig. 8-13 — Terrain photography—Siberia, Asia; 4× enlargement

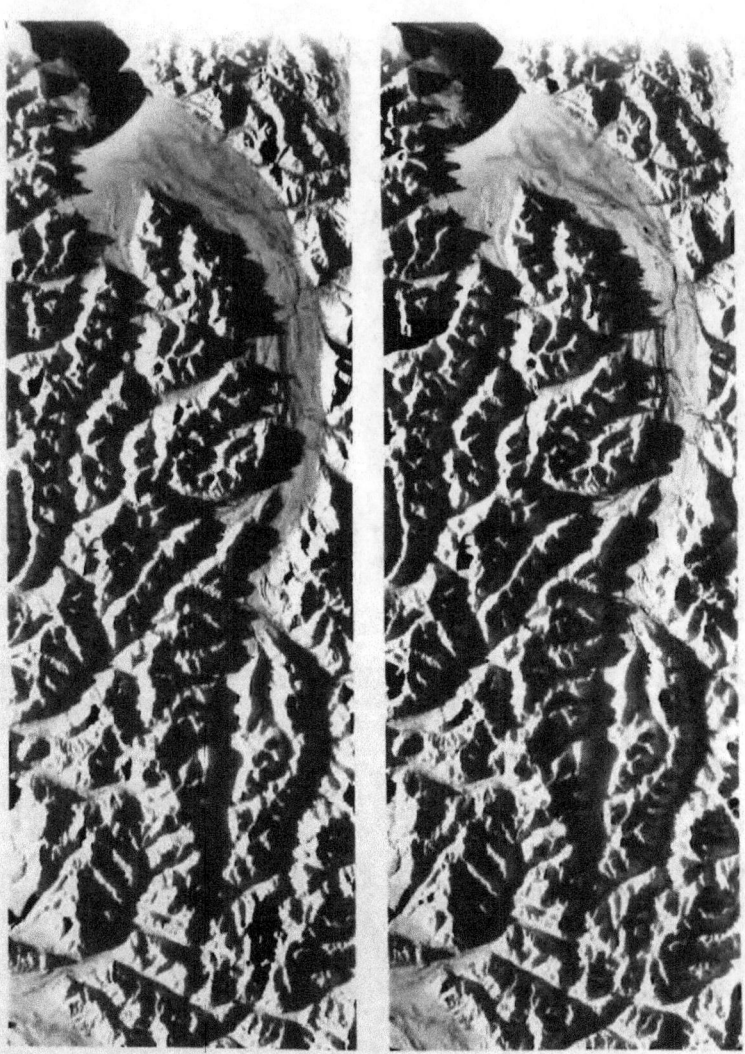

Fig. 8-14 — Stereo pair—low sun angle, USSR; contact print

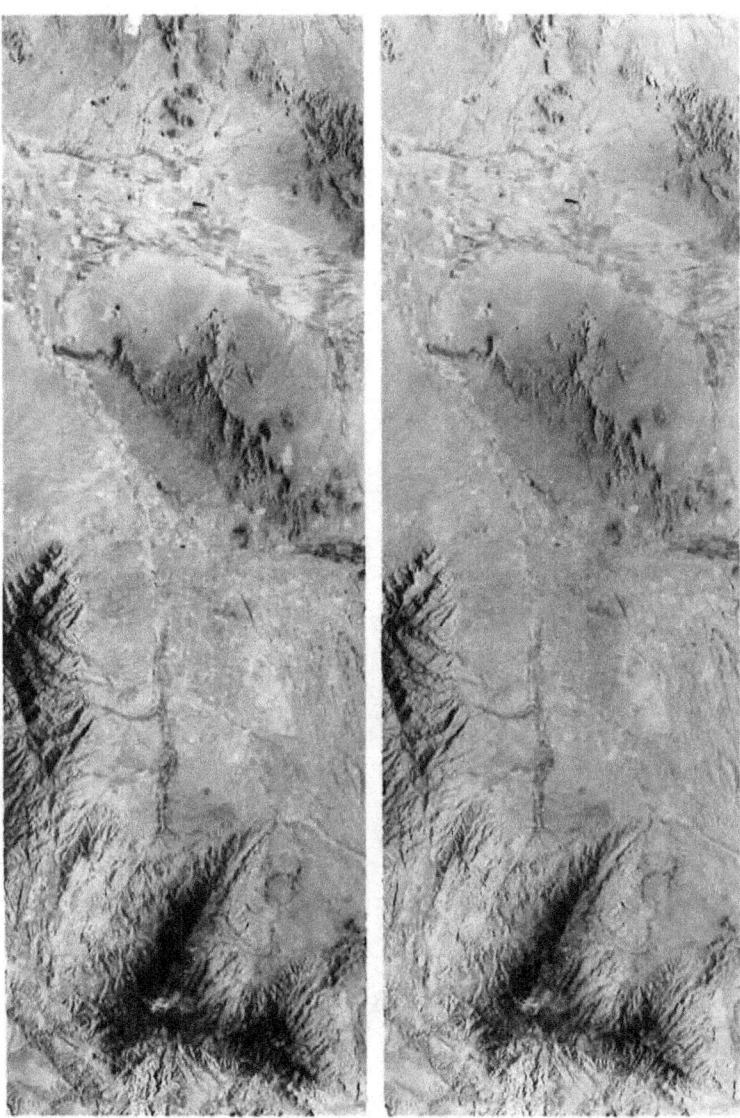

Fig. 8-15 — Stereo pair—Tucson, Arizona; contact print

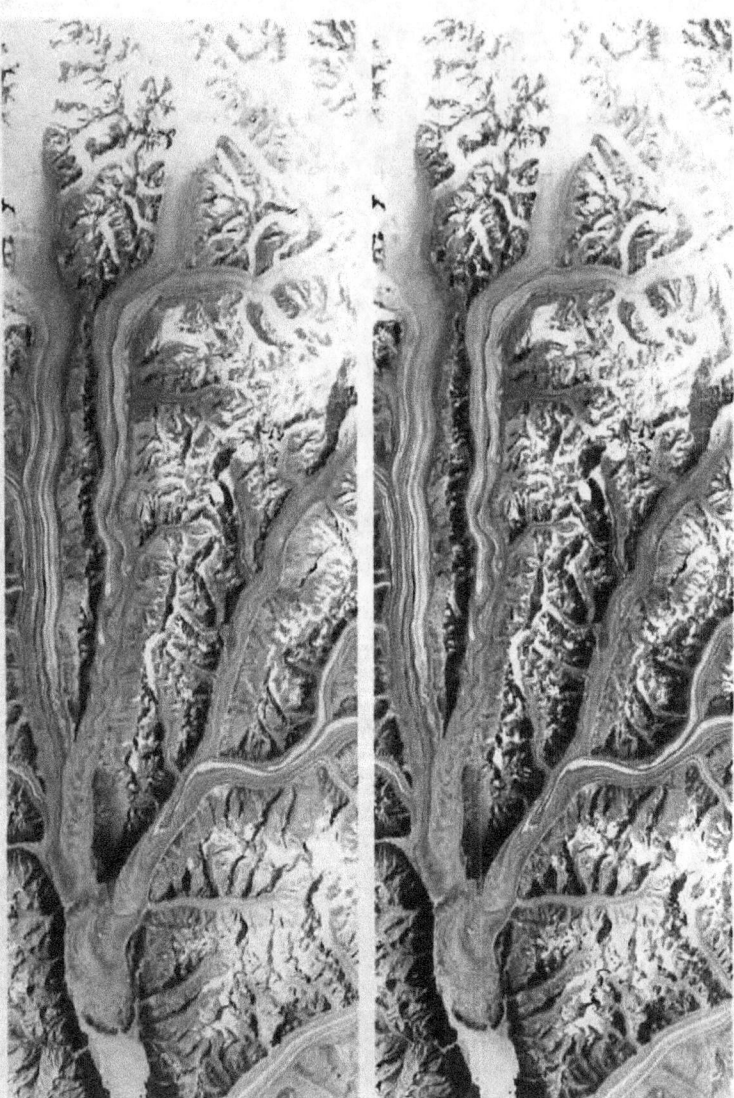

Fig. 8-16 — Stereo pair—Mount Logan, Alaska; contact print

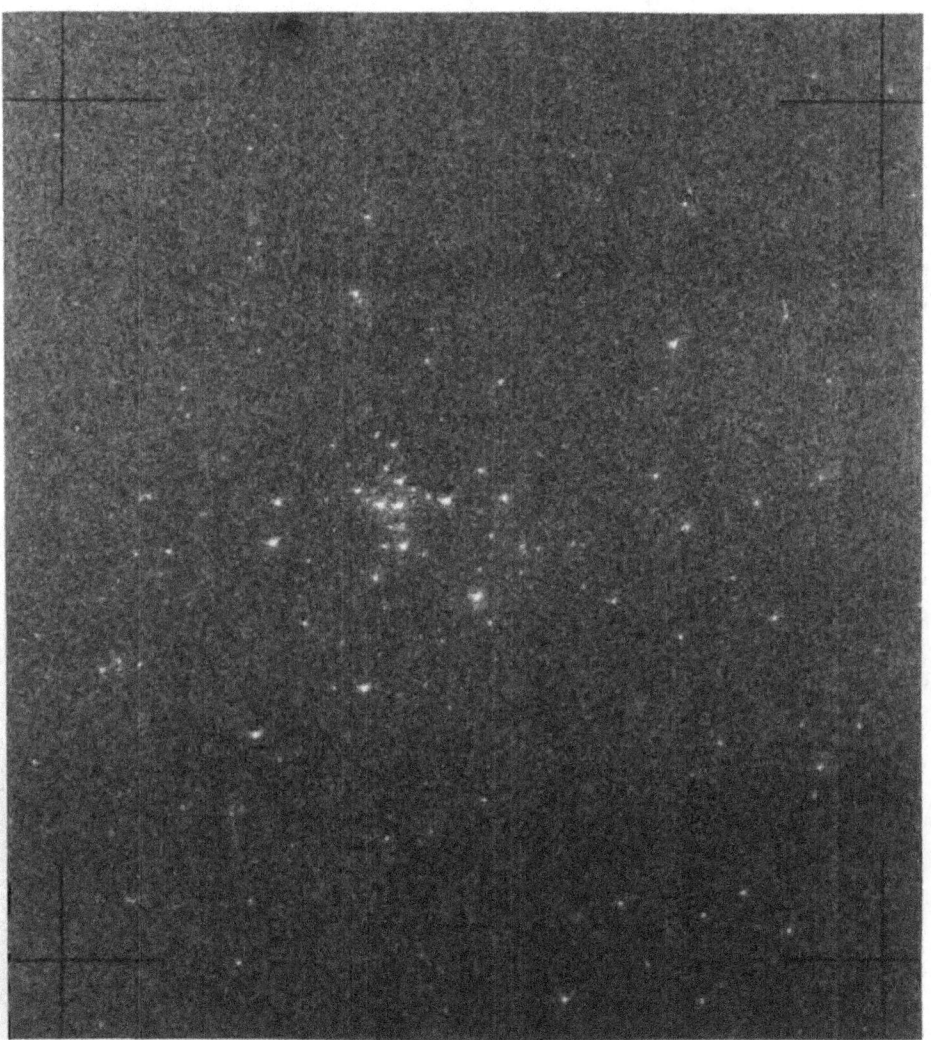

Fig. 8-17 — Stellar photography; 15× enlargement

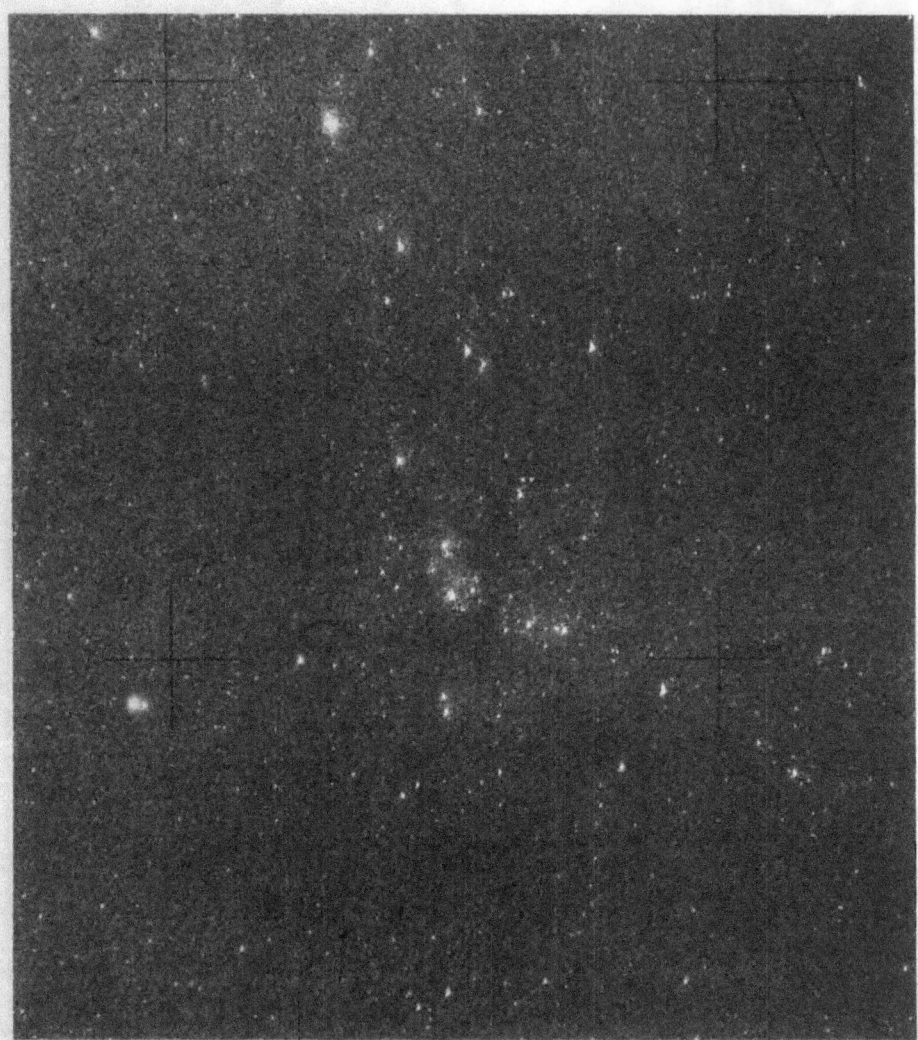

Fig. 8-18 — Stellar photography—calibration mode; 10× enlargement

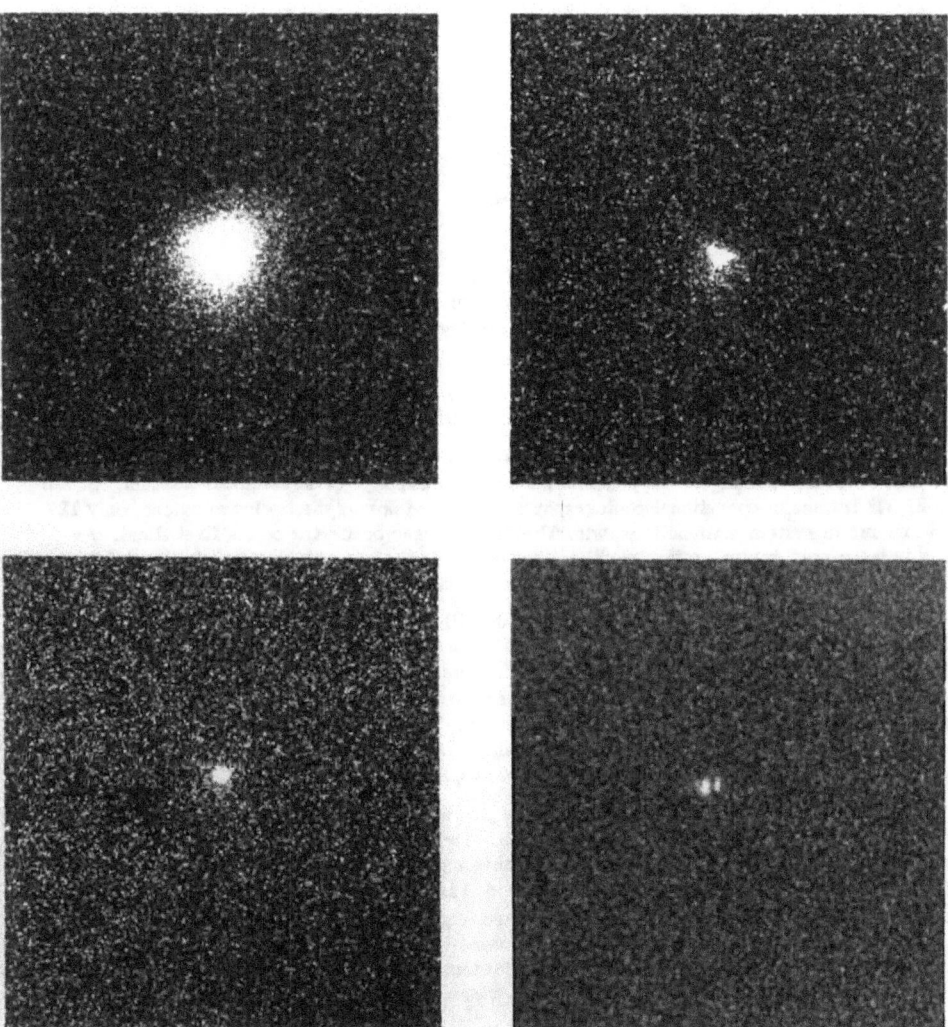

Fig. 8-19 — Stellar photography—40× enlargements

SECTION 9

SUMMARY

The use of earth satellites as a survey platform to acquire large amounts of MC&G data in a relatively short time, without incurring the political implications inherent in conventional survey systems, has been the most significant technological advancement in mapping and geodesy to take place in many years.

The HEXAGON (KH-9) Mapping Camera Program, when evaluated from both operational and productivity points of view, must be rated as exceptionally successful.

From an engineering/operation aspect, there was no occasion during the 7-plus years of operation wherein the MCS was responsible for schedule slippage or mission curtailment. Of the total 29,972 frames of operational photography programmed during the twelve missions, only 127 were lost due to system malfunctions, with 80 of these losses occurring on the first flight. As shown in the report, minor system modifications to enable the use of higher resolution and thinner base films, as they became available, increased performance and productivity dramatically.

From the point of view of accomplishment, the KH-9 MCS enabled production of photogrammetric control for MC&G with approximately a 4-times improvement in the relative accuracy (point-to-point) over the previous system (DISIC/Doppler 3-inch focal length). Although control aspects (CCN) and positioning goals of the MCS were stressed in the original charter, the actual mapping and charting application was the major bonus. The automatic and manual correlation equipment's use of MCS over panoramic camera material to generate required products provided a 10-times saving in effort by the photogrammetric application and process.

The Mapping Camera System has also made possible the development of a Continental Control Network (CCN) across most of the U.S.S.R. This network comprises photoidentifiable ground points for which WGS coordinate values are computed. The unique feature of the Network is that photography from a number of KH-9 missions has been adjusted simultaneously in a single computed network. Using 1980 imagery, DMA plans to extend the network into Africa and South America. This concept demonstrates what is, perhaps, one of the most significant contributions of the MCS: giving the MC&G community a powerful tool for the establishment, through the use of photogrammetric techniques, of a virtually world-wide system of accurately known ground control for subsequent use in a wide variety of programs.

During the course of the program, there was a considerable turnover of director and management personnel in the government and contractor organizations. In spite of this, the highest level of dedication and team spirit was maintained throughout as a continuous thread; this without doubt contributing significantly to program success.

SECTION 10

REFERENCES

1. Perry, Robert, "A History of Satellite Reconnaissance," Vol. IIIB, BYE-17017-74, prepared under the direction of the NRO (TOP SECRET/BYEMAN).
2. "Project HEXAGON Overview," Lockheed Missiles and Space Company, with inputs from associate contractors, under the direction of SAFSP, 25 Jan 1978 (TOP SECRET/BYEMAN/H).
3. KH-9 Search and MC&G Performance Study; Vol. II, "Historical Performance Summary," National Photographic Interpretation Center, Oct 1977 (TOP SECRET/RUFF).
4. "Stellar and Terrain Camera (SI) for the Photographic General Search and Surveillance Satellite System (HEXAGON)," Technical Report, Vol. II, Itek Corp., 15 Oct 1966 (SECRET/BYEMAN/H).
5. Memorandum from Mr. Paul H. Nitze, Deputy Secretary of Defense, Dec 22, 1967, for: Secretary of the Army; Chairman, Joint Chiefs of Staff; Director of Defense Research and Engineering; Assistant Secretary of Defense (Administration); Assistant Secretary of Defense (Comptroller); Assistant Secretary of Defense (System Analysis); Director, Defense Intelligence Agency; Director, National Reconnaissance office. Subject: "Study of Satellite Mapping Camera Requirements" (TOP SECRET).
6. Memorandum from Mr. Sol Horowitz, Assistant Secretary of Defense, 15 February 1968, for the Deputy Secretary of Defense. Subject: "Study of Satellite Mapping Camera Requirements" (TOP SECRET).
7. Memorandum from Lt. Colonel William E. Williamson, USAF to Dr. J. L. McLucas, DNRO, 19 May 1969. Subject: "Justification of the HEXAGON 12-Inch Stellar Index Camera (TOP SECRET/BYE/H).
8. "Engineering Manual for Mapping Camera Subsystem (MC)," (Rev A) Itek Corp., 15 Aug 1974 (SECRET BYEMAN/H)
9. PAR 116S/R2 of Final Report, "Study Distortion in Photo Duplication," Contract ▓▓▓▓▓▓▓, Task 2, 15 July 1968 (SECRET).
    PAR 157S/R1 of Final Report, "Contract Printing Distortion Study," Contract RD 20001, 29 Sept 1970 (SECRET).
    "Physical and Chemical Behavior of Kodak Aerial Films," Kodak Cat. no. 147-7918, pp. 11-16, Apr 1974 (UNCLASSIFIED).
10. Classified message traffic between various branches of the Government and participating contractors during the period 1969 to end of program (SECRET and TOP SECRET/BYEMAN).

SECTION 11
ACKNOWLEDGEMENTS

The "HEXAGON Mapping Camera Program History" is dedicated to the men and women in the Government (military and civilian) and in private industry whose contributions made this program exceptionally successful and productive.

The author extends his "Thanks" to the people listed below for their assistance during the research and preparation of this historical resource document.

Secretary of the Air Force Special Projects

Lt. Colonel Sanford Gallof
Captain David R. Anderson
Captain Peter W. Young
Captain Richard J. Mizgorski

Defense Mapping Agency

National Photographic Interpretation Center

Aerospace Corporation

William L. Griego
David F. Nelson

Itek Optical Systems Division

John T. Wilkinson
Harold R. Alpaugh
George L. Coggan
Publications Department
Security Department

Lockheed Missiles and Space Company

William E. Williamson
Julian N. Dyer

Eastman Kodak

Joseph P. Russo

General Electric

Richard J. Lasher
Edward J. Bonner

www.ingramcontent.com/pod-product-compliance
Lightning Source LLC
Chambersburg PA
CBHW082026300426
44117CB00015B/2366